77 Topics in Current Chemistry

Fortschritte der Chemischen Forschung

Inorganic and Physical Chemistry

Springer-Verlag
Berlin Heidelberg GmbH 1978

This series presents critical reviews of the present position and future trends in modern chemical research. It is addressed to all research and industrial chemists who wish to keep abreast of advances in their subject.

As a rule, contributions are specially commissioned. The editors and publishers will, however, always be pleased to receive suggestions and supplementary information. Papers are accepted for "Topics in Current Chemistry" in English.

ISBN 978-3-662-15468-7   ISBN 978-3-540-35679-0 (eBook)
DOI 10.1007/978-3-540-35679-0

Library of Congress Cataloging in Publication Data. Main entry under title: Inorganic and physical chemistry. (Topics in current chemistry ; 77) Bibliography: p. Includes index. CONTENTS: Bikerman, J. J. Surface energy of solids. – – Wiedemann, H. G. and Bayer, G. Trends and applications of thermogravimetry. – – Huglin, M. B. Determination of molecular weights by light scattering. 1. Chemistry, Physical and theoretical – – Addresses, essays, lectures. 2. Chemistry, Inorganic – – Addresses, essays, lectures. I. Bikerman, Jacob Joseph, 1898 – II. Wiedemann, Hans G. Trends and applications of thermogravimetry. 1978. III. Huglin, M. B. Determination of molecular weights by light scattering. 1978. IV. Series. QD1.F58 vol. 77 [QD475] 540'.8s[541'.3] 78-13471

© by Springer-Verlag Berlin Heidelberg 1978
Originally published by Springer-Verlag Berlin Heidelberg New York in 1978
Softcover reprint of the hardcover 1st edition 1978

2152/3140 – 543210

Contents

Surface Energy of Solids

Jacob J. Bikerman

Dept. of Chemical Engineering, Case Western Reserve University, Cleveland, Ohio 44106, U.S.A.

Table of Contents

List of Symbols

A	Surface area; a constant
A_m	Area of surface or interface per molecule
A, A_s, A_{sl}	Areas of vapor-liquid, vapor-solid, and liquid-solid interfaces
A_1	Combined area of grain boundaries
A_0, A_1	Surface areas of a large and a minute crystal
A_1, A_2, A_3	Areas of different crystal faces
a	The longer half-axis of an ellipse; lattice parameter
B	A constant
b	The shorter half-axis of an ellipse
C, C_1, C_2	Solubilities
$C.E.$	Cuticular energy
c	Depth or half-length of scratch
D	Grain diameter; diffusion coefficient
E	Modulus of elasticity; intensity of electrostatic field
E_0	Internal energy of a bar
E_1	Internal energy of a broken bar
e	Charge of an electron
F	Helmholtz free energy
F_o	Free osmotic energy
F_s	Free surface energy
f	Force; number depending on crystal structure
G	Shear modulus; arbitrary constant; temperature gradient
H	Amount of heat evolved
h	Planck's constant; half-thickness or thickness of a slice or a ribbon; depth of scratch
h_0	Initial depth of scratch
K	Modulus of compressibility; proportionality constant
k	Boltzmann's constant; coordination number of atoms in bulk; compressibility
k_s, k_v	Number of atoms in two-dimensional and three-dimensional unit cells
Δk	Number of broken bonds per atom
L	Length; thickness of a crystal
L_l, L_s	Heat of vaporization or sublimation
l	Length; length of a "roof"; variable length of a filament
l_0	Initial length of a filament
l_1, l_2	Edges of two cubes
M	Molecular weight; atomic mass
MW	Molecular weight
m	Mass of an atom; mass of hydrogen atom; mass
N	Number of atoms in unit cell; Avogadro number
n	Number of grain boundaries in a wire; number of molecules in unit volume
n_+, n_-	Number of cations or anions on unit area
P	Gas pressure
P_c	Capillary pressure
P_o	Vapor pressure above a plane surface
p_1, p_2	Pressure in gas, in liquid
Q	Heat content
Q_0, Q_p, Q_∞	Heat of solution of large and powdered solids, and solids with zero surface
q	Number depending on crystal structure; surface density of electric charge
q_0, q_1	Heat of solution of unit mass
R	Radius of curvature; gas constant
R_1, R_2	The principal radii of curvature
R_∞	Distance at which deformation becomes negligible

R	Radius of wires; distance from an atom or ion; cube root of molecular volume; radius of a void
r_0	Radius at the bottom of a groove; radius at zero time
r_1, r_2	Radius of wire before and after elongation
S. E.	Strain energy
T	Absolute temperature
T_b, T_m	Boiling and melting points
T_l	Melting point of plate of thickness l.
T_{mr}	Melting point of drop of radius r
T_t	Temperature of treatment
t	Time
U	Interatomic potential
U_l, U_s	Total surface energy of a liquid, a solid
u	Height of a ridge point
u_1, u_2	Displacement of neighboring atoms
V	Volume of unit cell; specific volume of a liquid; volume; molecular volume
V_1, V_2	Gas and liquid volumes
W	Load; breaking load
W_0	Equilibrium load on a filament
W_w	Breaking load of wires
$\mathfrak{W}$	Work of extending a rod
$\mathfrak{W}_f$	Fracture energy
$\mathfrak{W}_m$	Work spent on macroscopic deformation
w	Width; distance between two ridges
x	Variable depth of crack
Y	Plate thickness
Z	Valency of an ion; effective number of electrons per ion
α	Angle on a crystal surface; numerical constant; an angle
β	Contact angle at the edge of a solid
γ	Surface tension of a liquid
γ_c	"Critical surface tension"
γ_s	Specific surface energy or surface tension of a solid
γ_s^*	Specific free energy of an interface between UC and U vapor
γ_{sl}	Specific energy or tension of a liquid-solid interface
γ_{sv}	Surface tension of a solid in a foreign vapor
γ_1	Surface tension of a grain boundary
$\gamma_{12}, \gamma_{13}, \gamma_{23}$	Tensions along boundaries between fluids 1 and 2, 1 and 3, and 2 and 3
"γ"	Specific fracture energy
Δ	Thickness of a line on which f acts
δ	Thickness of surface layer; an angle
ϵ_0	Maximum strain
η	Viscosity
θ	Contact angle
θ_0	The characteristic (Debye) temperature
λ	Wave length of disturbances; heat of melting
$\lambda_1; \lambda_s$	Internal heat of vaporization, sublimation
ν	Number of free valence electrons per atom; Poisson's ratio
ξ	Cohesion of a solid
ρ, ρ_3	Density of a solid
ρ_1, ρ_2	Density of a gas, a liquid
σ_0	Average stress
σ_m	Local stress
τ	Thickness of surface layer
φ	Potential energy of a crystal

φ	Work of removing an atom from its neighbor; work function; half the dihedral angle in a liquid medium; electric potential
φ_1	Energy of valency electrons at the Fermi level
ψ	Half the dihedral angle in a gas
ψ^*	Dihedral angle in a foreign vapor
Ω	Volume of a molecule or atom

I. Introduction

Only surface energy is mentioned in the title of this review, but surface tension also is considered in the following text. The dimensions of (specific) surface energy (ergs/cm^2 or joules/m^2 or g/sec^2) and of surface tension (dynes/cm or newtons/m or g/sec^2) are identical. For typical liquids, also the two absolute values are equal; for instance, the surface tension of water γ at room temperature is about 72 dyne/cm, and the (specific) surface energy is 72 erg/cm^2.

For typical solids, not only is the relation between surface tension and surface energy poorly understood, but also the very existence of these quantities questioned. In the following brief historical sketch, an attempt will be made to indicate why scientists felt it necessary or useful to introduce notions of surface tension and surface energy (γ_s) of solids into science.

Apparently, the first reference to γ_s or "superficial cohesion" of a solid is found in an Essay by Thomas Young[1]. The original does not contain illustrations, but Fig. 1 here represents the author's idea. A solid wall (S) is partly wetted by liquid L

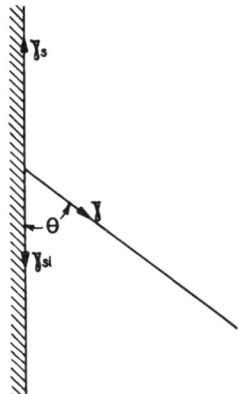

Fig. 1. The hypothetical balance of forces at the 3-phase line. The surface tension γ_s of the solid (shaded) is supposed to be equal to the sum $\gamma_{sl} + \gamma \cos \theta$; γ_{sl} is the tension along the liquid – solid interface, γ is surface tension of the liquid, and θ is the contact angle

while the vapor V is in equilibrium with both S and L. The angle occupied by L is θ. Along the boundary V/L the surface tension γ of the liquid acts; Young imagined an analogous surface tension γ_s to act along the boundary V/S and a tension γ_{sl} to act along the frontier L/S. To achieve equilibrium along the solid surface, it is necessary (if gravitation is neglected) to have the tension γ_s equal (and opposite) to the sum $\gamma_{sl} + \gamma \cos \theta$, or

$$\gamma \cos \theta = \gamma_s - \gamma_{sl} \tag{1}$$

The contact angle θ is always measured in the liquid.

As usual with Young, no attempt to prove his equation, and no attempt to justify the introduction of a tension of solid surfaces are made. A discussion of the above reasoning will be found in Section III.8 of this review, but one difficulty may be mentioned at once. Equation (1) was derived for the forces acting along the solid surface but it disregards the force normal to it. If γ tends to contract the V/L interface and if the vertical component $\gamma \cos \theta$ of this tension is believed to pull the solid atoms (present along the 3-phase line) downward, what effect has then the horizontal component $\gamma \sin \theta$? We would expect it to pull the above atoms out of the plane toward the right. Young[2] became aware of this component many years later but could only suggest that "the perpendicular attraction" is "counteracted by some external force holding the solid in its situation".

Another line of study was started by Quincke[3]: It has been found, experimentally, that the breaking load W_w of wires having different radii r could be expressed by the equation

$$W_w = mr^2 + nr, \tag{2}$$

m and n being empirical constants. Quincke's guess was, that the term nr represented the resistance offered by the surface tension of the wire. Hence,
$nr = 2\pi r \gamma_s$ or $\gamma_s = n/2\pi$. At present it is generally admitted that the breaking stress (i.e., $W_w/\pi r^2$) of thin wires exceeds that of thicker samples because their microscopical structure is different. In particular, the act of drawing elongates many weak spots in the volume of the wire material.

If a thin solid plate contains an elliptic hole of the half-axes a (the longer one) and b, and the ellipse is oriented so that the longer axis is perpendicular to the direction of the tensile force, then the stress at the two ends of this axis is

$$\sigma_m = \left(1 + \frac{2a}{b}\right) \sigma_0 \tag{3}$$

σ_0 being the average stress caused by the external pull. Hence, the local stress concentration (equal to σ_m/σ_0) is greater the greater the ratio a/b. If the plate is extended by the above pull, this ratio decreases: an ellipse perpendicular to the tensile force may become parallel to it. Consequently, σ_m is lowered by drawing, and the average stress σ_0 must be raised to achieve that value of σ_m at which fracture proceeds. Equation (3) is the most probable explanation of Eq. (2) which, on this view, has no relation to any surface effect.

A relation between rupture phenomena and (specific) surface energy "γ" was postulated by Dupré[4] almost simultaneously with the hypothesis of Quincke. Let a cylindrical rod be broken in tension. After rupture, two new gas – solid interfaces of πr^2 each are present, r being the radius of the rupture surface. Consequently, the work of rupture ought to contain a term $2\pi r^2$ "γ". Dupré did not indicate how to separate this term from the main component of the work of rupture, which is the work required to extend the rod to its maximum elongation (or strain). The modern development of Dupré's ideas is reviewed in Section III.3. below.

Soon after Dupré's publication, Gibbs[5] pointed out that, whereas for liquids the work of extending the surface by dA cm^2 is identical with the work of creating a new surface of identical area, the two quantities are different for a solid. In fact, it is difficult to imagine a method for expanding the surface area of a solid without altering the strain energy (or the stored elastic energy) of the bulk. The work required to create a new surface is evidently identical with that considered by Dupré.

Consequently, this kind of surface energy does not imply any superficial tension or tendency to contract. This was clear already to Gibbs and will be stressed again in Chapter V.

II. Theoretical Approach

1. General

Laplace's theory of capillarity (1805), confirmed for liquids in innumerable tests, in principle embraces solids also. It is based on the notion that each material point (or atom, or molecule) exerts an attractive force on all neighboring points but the intensity of this attraction decreases so rapidly when the distance between two points increases, that it is significant only when the above distance is negligible (the attraction is "sensible at insensible distances only"). This notion ought to be as valid for solids as for liquids. However, the visible or measurable consequences of the theory would be expected to be quite different for typical solids and typical liquids. In the latter, the mobility is so great that the equilibrium shape of liquid bodies is to a large extent determined by the capillary pressure existing at present and is not influenced by the pre-history of the specimen. It is possible to pour a Newtonian liquid from a cylindrical beaker into a vessel of a most intricate design, and then pour it back; no inspection of the resulting sample would permit us to detect the past contortion of the liquid.

The behavior of solids is, of course, quite different. Their present shape and their internal stresses and strains greatly depend on their pre-treatment. Thus we can measure the capillary pressure existing in a given liquid sample simply by measuring the shape of the sample in an external force field, such as one due to gravitation; but a determination of the capillary pressure in a solid is an almost impossible task. Chapter III deals with this difficulty.

Laplace could not express any opinion on the absolute values of the attraction assumed by him. At present, our knowledge of interatomic, interionic, and analogous forces is much greater than 170 years ago, and attempts to calculate surface energies and surface tensions are possible.

The usual procedure, however, does not follow Laplace's precedent, but rather uses Dupré's scheme. Consider a bar, of unit area cross section and containing $2n$ atomic layers perpendicular to the axis of the bar; n is a large number. Let the internal energy of the bar be E_0. If the bar is cut in two, so that each half contains n atomic layers, the energy of the system changes to E_1. One-half of the difference between the two energies is considered to be (specific) surface energy of the material, that is "γ" $= 0.5\,(E_1 - E_0)$. The factor 0.5 appears, of course, because two rupture surfaces are formed by one act of separation.

The following Section II.2. is devoted to the attempts on calculating "γ" from fundamental atomic and molecular constants. In Section II.3., semi-empirical expressions for the surface energy of solids are briefly reviewed.

2. Basic Calculations

It is commonly agreed that E_1 is different from E_0 because the resultants of the forces acting on elementary particles (*i.e.* electrons, atoms, ions, etc.) near the rupture surface differ from those in the bulk of the material: the bulk is more symmetrical than the surface layer.

Three tiers of questions arise now.

1. What units must be considered? Is it enough to calculate the forces between molecules or must atoms, ions, nuclei, and electrons all be treated separately?

2. What displacements, as compared with the positions in the bulk, are permitted for each of the centers of force present in the surface layers?

3. Which of the many suggested laws of force (*i.e.*, what dependence of the attractive and repulsive forces on the distance between the particles concerned) should be chosen for the calculation?

From the above paragraph it is clear that very many different models may be subjected to calculation. The numerical results, naturally, vary from model to model (for a given real crystal), and at present it would be too risky to claim correctness for any of them. Consequently, only a few (mainly recent) examples of the approaches to the computation of "γ" are reviewed in this section, chiefly to indicate the unsettled state of affairs at present. Older theories and their numerical results may be consulted, for instance, in Ref.[6].

Inert Gases. The calculation of "γ" should be relatively straightforward for crystals of inert gases, in which only one kind of interaction may be expected. These crystals have a face-centered cubic structure. If each atom is treated as a point source of attractive and repulsive forces, only the forces between the nearest pairs of atoms are considered, the zero point energy is neglected, and no re-arrangement of atoms in the surface region is permitted, then the calculated "γ" still depends on the equation selected to represent the interatomic potential U.

A popular equation, often attributed to Lennard-Jones, is

$$U = C_1 r^{-12} - C_2 r^{-6} \tag{4}$$

r being the distance of an arbitrary point in space from the source of force, and C_1 and C_2 constants which usually are calculated from crystal properties and the second virial coefficients of the gases, assuming, of course, the validity of Eq. (4). With all the above restrictions, "γ" of the {100} faces of Ne, Ar, Kr, and Xe at 0 °K appeared to be 20.5, 45.0, 55.0, and 64.7 erg/cm^2, respectively[7]. When, instead of (4), equation

$$U = be^{-kr} - cr^{-6} \tag{5}$$

was used, in which b, k, and c are constants to be determined from the heat of sublimation of the crystals at $0\,°K$, and the viscosity and the second virial coefficients of the gases, the "γ" for the above four crystal faces became 19.9, 43.6, 55.5, and 64.7 erg/cm^2. If now the external atoms are allowed to rearrange themselves so as to lower the total energy as much as possible, the "γ" decreases by about 0.16 erg/cm^2 for neon and about 0.6 erg/cm^2 for xenon.

A later calculation[8], based on very similar notions, resulted, for instance, in a value of 40 erg/cm^2 (or 0.040 J/m^2) for the {100} face of argon, still apparently at $0\,°K$.

If no re-arrangement of the peripheral atoms is permitted but, in addition to the forces between pairs of atoms, also those between one atom in one, and two atoms in the other fragment (of the initial ideal crystal) are computed, then the "γ" appears[9] to be lowered by 4–15% as compared with the values quoted above from Ref.[7].

The above calculations may be classified as static. The energy is altered by permitting the atoms to vibrate[10]. Let Φ be the potential energy of the crystal, and u_1 and u_2 the displacements of two neighboring atoms in the latter. The derivative $\partial^2/\Phi/\partial u_1 \cdot \partial u_2)$ is computed, that is, the effect of the simultaneous vibrations of other neighbor atoms is disregarded. The dependence of Φ on the mutual distance of two atoms is assumed to be that of Eq. (4). No change in the structure of the surface region is allowed below the melting point. With these (and other) approximations, the vibrational surface energy for the {100} surface of Ar, Ne, Kr, and Xe at $0\,°K$ becomes 2.8, 2.7, 1.8, and 1.6 erg/cm^2, respectively; thus vibrations seem to contribute less than 10% to the value of "γ".

Ionic Crystals. Of the crystals consisting of two different atoms each, those of lithium hydride LiH are perhaps the simplest to deal with. In this system, four kinds of energy are taken into account[11], namely the kinetic energy of every particle, the energy of attraction between the nuclei and the electrons, that of the mutual repulsion between the nuclei, and that of the repulsion between the electrons. In the instance of LiH, there are only two different nuclei and only 4 electrons per molecule. Consequently, in the bulk of the crystal, only 5 variational parameters are important, namely the internuclear spacing, the ordinate of one orbital center (the Li nucleus being the origin of the co-ordinates), the ordinate of the other orbital center, and the two diameters of the orbitals. The analogous equations for the energies in the 3 ionic layers nearest to the rupture surface are more complicated; here 12 variational parameters are needed, namely six for the external, and six for the penultimate layer. There are 6 rather than 5 variables per each layer because the distance of the Li nucleus (serving as the origin of the co-ordinates) from the undisturbed third layer also is permitted to vary.

It is supposed that this displacement can occur only in the direction normal to the rupture surface, not parallel to the latter. When all the above distances are systematically varied until the minimum of the total interaction energy is reached, then the most probable position of all 6 centers of force is found. It appears that the distance between the Li nuclei in the outermost and the H nuclei in the second layer is smaller (by 0.00032 angstrom) than in the bulk of the crystal, and the distance between the H nuclei in the external and the Li nuclei in the penultimate layer is (by 0.0016 angstrom) greater than in the bulk. The energy of these two layers exceeds that of two undisturbed layers in the crystal by 190 mJ/m^2 (or erg/cm^2).

The lattice energy (L.E.) of alkali halide crystals appears to be relatively easy to calculate, because the forces between the regularly spaced ions are chiefly of the Coulomb type. Thus 3 different models of the NaCl crystals yielded values of L.E. between 1.297 and 1.282 x 10^{-11} erg/molecule, that is, in a good agreement with each other. However, no such agreement is observed when the "γ" is computed for the same models[6]. Thus the earliest 9 estimates of the "γ" of the {100} face of NaCl at 0 °K gave[12] values ranging from 77 to 210 erg/cm^2.

The earliest theory (Born, Stern 1919) resulted in the simple equation

$$\text{"}\gamma\text{"} = 0.1173\, z\, e^2/a^3 \text{ erg/cm}^2 \tag{6}$$

for the {100} face of NaCl; the numerical coefficient depends on the geometry of the crystal, z is the valency of each ion (i.e., $z = 1$ for NaCl), e is the charge of the electron, and a is the lattice parameter.

Zadumkin[12] considered 4 kinds of energy, namely the Madelung energy, the Born energy, the dispersion energy, and the zero-point vibrational energy. For the above crystal face, the values of these components at 0 °K appeared to be +367.6, -198.0, +50.5, and -7.3 erg/cm^2, respectively. Thus the expression for "γ" contained two positive and two negative terms. Naturally, the final result was very sensitive to the accuracy of each of them. The displacement of the ions caused by the proximity of the gas phase was disregarded in this computation.

More recently, the temperature dependence of the above vibrational component was studied[13]. Still for the {100} face of NaCl, the contribution of this component to the total surface energy, see Section II.3., was approximately -2 erg/cm^2 and to the free surface energy near $- 16$ erg/cm^2, both at 273 °K. No correction for the polarizability of the ions was employed.

Metals. A review of the published predictions of the "γ" of metals was published, for instance, in Ref.[14]. In that article, the idea of an uniform electron fluid in which metal cations are embedded is rejected; the electrons are concentrated in the spaces of the most intense electric field between the cations and thus form something like a second lattice. This remark seems to be disregarded by the later investigators.

According to two simultaneous hypotheses[15, 16], the principal component of the "γ" of simple metals (from lithium to aluminum) originates in the shift in the zero-point energy of the plasma models of the system when one perfect crystal is broken into two separate crystals. Some simplifications and arbitrary assumptions lead finally to the relation

$$\text{"}\gamma\text{"} = Cr^{-2.5} \tag{7}$$

C is a universal constant (calculable from the model) and r is the density parameter equal to $r = (3/4\,\pi)^{1/3}\, (V/N\nu)^{1/3}$; V is the volume of the unit cell which contains N atoms, and each atom contributes ν "free" valence electrons to the collective motion. Because the actual crystal structure is disregarded, only one value of "γ" is obtained for each metal. For lithium, for instance, "γ" seems to be 490 erg/cm^2 and for sodium: 220 erg/cm^2, both at 0 °K. The values at the temperature of melting are approximately 397 and 186 erg/cm^2.

Similar approaches have been used in Ref.[17] and, apparently Ref.[18] which, however, has not been available in the original. The "γ" of the {100} faces of Na and K was obtained[18] as 210 and 120 erg/cm^2.

A more detailed treatment of the model of Craig and others resulted[19] in values 4 or 5 times as small as those quoted above for the contribution of the plasmon zero-point oscillations to "γ"; it is concluded, consequently, that "γ" must contain also other terms.

The ion lattice model of Ref.[20], applied to metals, leads to the values of 360, 210, and 730 erg/cm^2 for the {111} face of lithium, sodium, and aluminum, respectively, all of which crystallize in the face-centered cubic lattices. See also Ref.[21].

A calculation based on the decrease of the electron density at the surface and on the relaxation of the top lattice plane resulted[22] in values such as 190 and 1234 erg/cm^2 for the {100} faces of sodium and aluminum, respectively. The above relaxation, that is, the ratio of the interplanar distance in bulk to that in the external region was calculated[23], assuming a Morse type interatomic potential. The above ratio appeared to be, e.g., 1.13 for the {100} face of calcium, and 1.016 for the {111} face of lead. The relaxation lowered the energy "γ" by 0.5 to 7% for different metals and crystal faces.

According to Ref.[24], the "γ" of the octahedral face of Al is somewhere between 730 and 760 erg/cm^2. The dispersion force component of the "γ" of metals was calculated[25] for the room temperature and the {100} crystal planes; it appeared to be about 376, 393, 440, and 490 erg/cm^2 for aluminum, copper, silver, and gold, respectively.

3. Semi-empirical Calculations

Several computations of the total surface energy (per unit area) U_s start from the experimental value of the sublimation energy, L_s, of the crystal. When it is remembered that the intensity of interatomic forces frequently is derived from this L_s, the direct use of the experimental data does not appear to be a serious drawback. On the other hand, this approach does not differentiate between different crystal faces and can give only an averaged value for all of these.

The total surface energy per unit area of liquids, U_l, is related to measurable surface tension or the specific free surface energy γ by the thermodynamic equation

$$U_l = \gamma - T \left(\frac{\partial \gamma}{\partial T}\right)_A ; \tag{8}$$

T is the absolute temperature and A is the (constant) extent of the vapor – liquid interface. Equation (8) is valid for reversible processes only. No proof is available for its validity in systems of a crystal and a vapor, but the analogous relation

$$U_s = \gamma_s - T \left(\frac{\partial \gamma_s}{\partial T}\right)_A \tag{9}$$

11

often is relied upon by investigators; γ_s is the specific free surface energy of the solid.

For liquids, Stefan[26] long ago advanced the hypothesis that, when a molecule is transferred from the bulk to the interface with the saturated vapor, its total energy changes just as much as when the molecule moves on from this interface into the vapor. Consequently,

$$U_1 = \frac{ML_1 - RT}{2\,NA_m} \tag{10}$$

M is the molecular weight, N the Avogadro number, A_m is the area of the vapor – liquid interface per molecule, L_1 is the experimental heat of vaporization per gram, and RT is the part of it spent on the increase of volume from liquid to vapor, assuming that the density of the liquid is much greater than that of the saturated vapor. Abbreviate $L_1 - (RT/M)$ as λ_1 and suppose that each molecule is a cube of the edge l. Then the volume of the cube is $l^3 = MV/N$, V being the volume of 1 gram of the liquid, or

$$l^3 = M/N\rho_2 \tag{11}$$

ρ_2 is the density of the liquid. If only one face of the cube is in contact with the vapor phase, then

$$A_m = l^2 = (M/N\rho_2)^{2/3} \tag{12}$$

Consequently

$$U_1 = \left(\frac{M}{N}\right)^{1/3} \frac{\rho_2}{2}^{2/3} \lambda_1 \tag{13}$$

If an analogous equation is written for the solid phase and it is assumed that the ratio γ_s/γ is equal to U_s/U_1, then

$$\gamma_s = \gamma \left(\frac{\rho_3}{\rho_2}\right)^{2/3} \left(\frac{\lambda_s}{\lambda_1}\right) \tag{14}$$

easily results[27, 28]; λ_s is the conterpart of λ_1 for solids, and ρ_3 is the density of the solid phase. The γ_s at the triple point of silver and aluminum, for instance, come out as 1045 and 570 erg/cm^2, respectively.

An early calculation of the γ_s of metals was based[29] on the equation $N\varphi = M\lambda_s + 0.5\,RT$, in which the heat of sublimation per gram-atom was combined with the work φ of removing a surface atom from its nearest neighbor. The work of forming a new surface (per atom) was said to be, for instance, 1.5 φ for the $\{000\bar{1}\}$ face in dense hexagonal lattices. These hypotheses lead to γ_s values for Mg $\{000\bar{1}\}$, Al $\{111\}$, Ag $\{001\}$, and Au $\{001\}$ equal to, respectively, 728, 1674, 1934, and 2539 erg/cm^2. It is worth noting that the results were given as if correct to 1 erg/cm^2.

Equation (13) must be made more complicated whenever the crystal anisotropy is taken into account. The relation

$$U_s = \left(\frac{M}{N}\right)^{1/3} \frac{\rho_3^{2/3}}{2} \lambda_s \cdot \frac{k_s \Delta k}{f k_v^{2/3} k} \tag{15}$$

derived in Ref.[30] is a suitable example. In it, k_s is the number of atoms in the two-dimensional unit cell, k_v this number in the 3-dimensional unit cell, Δk is the number of broken bonds per atom, k the coordination number of the atoms in the bulk, and f a numerical coefficient depending on the crystal structure. From (15), the U_s appeared to be, for instance, 1950 and 1720 erg/cm^2 for the {100} and {110} faces of α-iron at 0 °K.

If the compound dissociates in the vapor, the sum of the heat of sublimation and the heat of dissociation must be used instead of λ_s alone. If also the lattice structure is known, the γ_s can be calculated for metal carbides such as TiC or ZrC[31]. For instance, γ_s of TiC at 298 °C should equal 3.05 J/m^2 for the {100} face.

Several relations for γ_s have been derived by Oshcherin[32]. Thus for fcc lattices

$$\gamma_s = \frac{0.468 \, r \, V}{(d V/dp)_Q} \tag{16}$$

r is the distance between the two nearest ions, and $(d V/V \cdot dp)_Q$ is the adiabatic compressibility of the crystal, so that Q is the amount of heat present in the latter; 0.468 is a coefficient determined by the lattice geometry. Another approximation is

$$\gamma_s = 0.539 \, (k/h)^2 m M \theta_0^2 \tag{17}$$

Here the numerical coefficient is valid for fcc lattices, k and h are the Boltzmann and the Planck constants, m is the mass of a hydrogen atom, M is the molecular weight of the compound, and θ_0 is the characteristic (Debye) temperature of the latter, all in the CGS system. The results for the {100} face of NaCl at a non-specified temperature are 183 according to (16) and 226 erg/cm^2 from (17), taking $\theta_0 = 325$ °K.

What appears to be a strictly empirical rule (the original[33] is not clear on this point) is formulated as γ_s g/sec^2 = 142 cm x K g/cm sec^2, K being the modulus of compressibility of the solid, i.e., Vdp/dV, as in Eq. (16). According to this rule, γ_s of argon at 60 °K is 25 erg/cm^2, of iron at 600 °K: 2240 erg/cm^2, of potassium bromide at 360 °K: 215 erg/cm^2, and so on.

A connection between γ_s and the melting point T_m °K was repeatedly suggested. Thus γ_s is said[32] to be proportional to $T_m/V^{2/3}$, V being the atomic or molecular volume. An almost identical rule has been advocated for metals more recently[34], namely

$$\gamma_s = 4.77 \times 10^{-16} \frac{T_m}{r^2}, \tag{18}$$

r being expressed in centimeters; $r = V^{1/3}$. However, Eq. (18) resulted in improbably small values of γ_s. Consequently, the authors introduced a frankly empirical constant and obtained the relation

$$\gamma_s = 760 + 4.77 \times 10^{-16} \, T_m r^{-2} \text{ erg/cm}^2 \qquad (19)$$

The new equation gives, for the average surface energies of aluminum and copper, for instance, the values of 1100 and 1500 erg/cm^2.

The theory of Benedek[35] also must be regarded as semi-empirical. The authors treat the γ_s of alkali halide crystals as a sum of three terms, namely γ_+, γ_-, and γ_b. The first component represents the energy required to separate the positive ions, and the second: the analogous work for the anions. Both are calculated more or less *ab initio*. On the other hand, the expression for γ_b, *i.e.*, the thermal contribution, has no theoretical foundation. It is

$$\gamma_b = 0.5 \, k(T_b - T_m) \, (n_+ + n_-); \qquad (20)$$

k is the Boltzmann constant, T_b is the boiling temperature of the salt, and n_+ and n_- are the numbers of cations and anions per unit area of the crystal face considered. For the {100} planes, at 0 °K, of LiF, NaCl, and KCl the calculated γ_s are 341, 170, and 158 erg/cm^2.

A close relation between the work function φ of a metal and its γ_s has been suspected by several investigators. Thus[32], the γ_s of fcc metals is supposed to be

$$\gamma_s = \text{const.} \, \frac{m}{h^2} \, (\varphi_e - \varphi)^2 \qquad (21)$$

φ_e is the energy of the valency electrons at the Fermi level, m is the mass of the hydrogen atom, h the Planck constant, and the value of the constant depends on the units selected. A more detailed theory[36] leads to the equation

$$\gamma_s = 1530 \frac{\varphi}{q} \left(\frac{z \rho_3}{M} \right)^{2/3} \qquad (22)$$

if γ_s is measured in erg/cm^2, φ in electron-volts, z is the effective number of electrons per atom, q is a numerical coefficient depending on the geometry of the lattice (its value is, for instance, 0.80 for fcc and 0.93 for bcc lattices), and M is the atomic mass (grams). From (22), γ_s at the T_m of each metal, is 285 for Na, 1070 for Al, 1330 for Au, and 1540 erg/cm^2 for Cu.

The work functions of different crystal faces are different, and their kinship with the surface energies of these faces (for W, Mo, Ta, etc.) was emphasized in Refs.[37] and[38].

Mechanical properties of a material can be used to estimate its γ_s or, more exactly, "γ". When a tensile stress σ_m is applied to a bar, the latter extends along its whole length. However, to find "γ", only the extension of one interplanar distance from its equilibrium value a to the value $a + \Delta a$ (at which rupture occurs) is needed. The corresponding work of rupture is 2 "γ", and

$$2 \text{ "γ" } = \sigma_m, \Delta a \tag{23}$$

if Hooke's law is applicable to the process considered and Δa is treated as a very small length. From the same law,

$$\frac{\Delta a}{a} = \frac{\sigma_m}{E}, \tag{24}$$

E being the modulus of elasticity. Hence, "γ" $= 0.5 \, a \sigma_m^2 / E$. Similar expressions have been obtained by Orowan as early as 1934, and recently in Ref.[8]. However, the values thus deduced for NaCl crystals were different from those predicted from interatomic potentials.

A related relation is

$$\text{"γ"} = \frac{Ea}{\pi^2} \tag{25}$$

Its meaning can be made clear if, in Eq. (23), σ_m is introduced from Eq. (24). Thus "γ" $= 0.5 \, (E/a) \, (\Delta a)^2$ is obtained. Suppose that fracture takes place whenever the ratio $\Delta a / a$ reaches 0.4. In this instance,

$$\text{"γ"} = 0.5 \times 0.16 \, Ea = 0.08 \, Ea \tag{26}$$

i.e., almost identical with Eq. (25). Some values derived from Eq. (25) agreed poorly, for instance, for the {100} of copper, with those calculated *ab initio*.

Many scientists believe that γ_s is only a little greater than the γ of the corresponding melt at a near temperature; see for instance, Eq. (14). In different publications, the ratio γ_s / γ is expected to be near 1.05, or 1.10, or 1.15, and so on. The expectation of such a rule is not obvious: for instance, the viscosity of a solid is not by 5%–15% greater than that of the melt.

Several semi-empirical calculations are based on wetting. They are discussed together with wetting experiments in Section III.8.

The poor agreement between different attempts to calculate γ_s from plausible models or from semi-empirical rules underlines the importance of the experimental determinations of γ_s. The main methods used for this are examined in Chapter III.

III. Experimental Methods

1. The Shrinkage or Zero-creep Method

This method was introduced by Berggren[40]. In its usual form, several thin wires (fibers, filaments) are each loaded with a different load W and suspended vertically in an oven kept at a temperature near 0.90 or 0.95 T_m, T_m being still the melting point of the fiber on the absolute scale. After such a heat treatment for several hours or days, it is found that the filaments weighted with a small W are shorted than before the test, whereas those loaded with larger W's are longer. An interpolation permits us to find that load (W_0) at which the wire does not shrink or extend. It is supposed that in this equilibrium state, the force of gravity W_0 which tends to lengthen the fiber is exactly compensated by the force of the surface tension of the fiber which tends to shorten the latter. The test must be performed almost at the T_m because only at such temperatures the mobility of the molecules, atoms, and so on in the solid is high enough to render the attainment of equilibrium feasible in a reasonable length of time.

Difficulties of two types are encountered in the determination of W_0. Those of purely experimental origin include, for instance,

(a) temperature variations between different wires during the protracted test;

(b) high reactivity of the solid near its T_m so that impurities in the system (such as paint marks used to define the initial length l_0), which would have been innocuous at a low temperature, become troublesome,

(c) high volatility of many solids in the vicinity of their melting point, and

(d) load variations along the filament (*i.e.*, W at the bottom and W+ the weight of the filament at the top).

The other type is related to the interpolation. In some instances, during the test, the fiber length l increases or decreases, at a constant temperature and constant load W, linearly with time t, so that $l = l_0 \pm kt$ and k is a constant. In these systems, k can be plotted as a function of W, and the value of W at which $k = 0$ is taken as that of W_0. Frequently, however, the function $l = f(t)$ is not linear, and the interpolation loses its precision. An extreme instance of l being (for a time) independent of t also was observed[60]. The error of the determination of W_0 is estimated in various publications as $\pm 10\%$ or even $\pm 15\%$.

A higher precision, claimed to be about 1%, is achieved[41] by employing a lever arrangement and only one filament. No sketch of the apparatus used is shown in the original, but Fig. 2 probably gives a correct idea of the instrument. The fiber (*1*) is

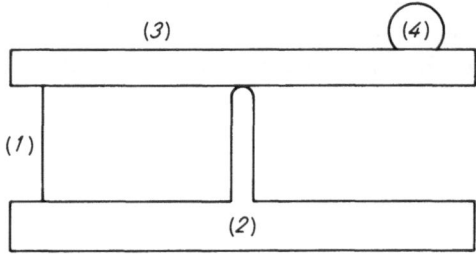

Fig. 2. A method of measuring the contractile force in a hot filament 1. This force is equilibrated by the weight 4 which can be rolled along the balance beam 3. The filament is attached also to base 2

firmly attached to the base (*2*). (*3*) is a balance beam to which the upper end of (*1*) is fixed. When the filament tries to contract, the load (*4*) is rolled to the right so as to keep the beam steady. Thus, l_0 does not vary during the heating, and the force

W_0 necessary for this constancy is given by the mass and the position of (4). The fiber is kept horizontal in a more recent modification[42] of the apparatus of Fig. 2.

The equality, referred to above, between the force of gravity and the force of surface tension leads immediately to the relation

$$W_0 = 2\pi r \gamma_s \tag{27}$$

r being the radius of the filament in equilibrium[43]. The corresponding equation for a thin foil of width w obviously is[44]

$$W_0 = 2w\gamma_s \tag{28}$$

However, in the last 30 years, Berggren's result, namely

$$W_0 = \pi r \gamma_s \tag{29}$$

was generally used. The derivation of Eq. (29) is as follows.

When the load W descends dl cm, thus losing $W \cdot dl$ ergs of potential energy, the surface energy of the fiber increases by $\gamma_s dA$ ergs, A being the area of the fiber surface. Hence, it is said, the change in the (Helmholtz) free energy F of the system is

$$dF = \gamma_s dA - W \cdot dl \tag{30}$$

Since

$$dA = 2\pi r \cdot dl + 2\pi l \frac{dr}{dl} \, dl \tag{31}$$

and $dr/r = -\nu \, dl/l$, if ν is the Poisson ratio, then

$$dA = 2\pi r (1 - \nu) \, dl \tag{32}$$

Berggren and the bulk of his followers assume that $\nu = 0.5$, that is, the deformation of the solid occurs without a change in volume. Hence, $dA = \pi r \cdot dl$ and

$$dF = \pi r \gamma_s dl - W \cdot dl \tag{33}$$

In equilibrium $dF = 0$, so that $W_0 = \pi r \gamma_s$, see Eq. (29). The recent derivation by Heumann[45] is not essentially different.

Both Eq. (29) and its derivation have been severely criticized[46]. The most basic objection to Eq. (30) is that the strain energy of the solid is disregarded there. In the theory of elasticity, the decrease $W \cdot dl$ in the potential energy of the load is numerically equal to the simultaneous increase in the strain energy of the wire. Neglect of this term means that the volume and the shape of a solid can be altered without applying any external force, as long as A remains constant; it is impossible to accept such a conclusion. This comment may also be differently formulated. The cross section of a wire (πr^2) may be represented as the sum of two areas referring to the bulk [$\pi(r - \delta)^2$] and the surface layer ($2\pi r\delta$). Although the thickness of this layer is only a small fraction of the wire radius r (the ratio r/δ is unlikely to descend below 10^6), Eq. (30) assumes that the whole force acts on the annulus $2\pi r\delta$ only, whereas the major part of the cross section remains free of stress.

It may be felt that the notion on which Eq. (30) is based is too improbable to deserve an experimental test. Such a test, however, would not seem to be difficult. If the filament or ribbon is made of a material showing intense stress birefringence, the birefringence pattern caused by the load W would be compared with that calculated from the strain energy. A reasonable agreement would indicate that W is suspended by the whole volume of the solid, not by its surface layer alone. On the other hand, if the specimen shows no birefringence when stressed by the load W, then Eq. (30) would gain in respectability.

As far as the present reviewer is aware, no experiments of this kind have been performed yet. It is easy to think of other possibilities for testing Eq. (30). Thus, specimens of equal volumes but different areas A, and, in another series, specimens of a constant A but different volumes could be compared; according to the theory, the volume is irrelevant for the value of W_0 or for the extension produced by a given W. Would this prediction be confirmed?

The second fundamental criticism refers to the requirement that dF be equal to zero in equilibrium. $dF = 0$ whenever the process visualized is isothermal and reversible. If the process is irreversible, the gain in the free energy of the system, that is, $\gamma_s dA$ is smaller than the work $W \cdot dl$ spent on achieving this gain. It has been shown[46, 47] that Berggren's process always is irreversible and $W \cdot dl > \gamma_s dA$. In brief: when the filament is extended, its radius decreases (unless $\nu = 0$). Let the two radii (before and after elongation) be r_1 and r_2; $r_1 > r_2$. For extension to occur, W must be greater than $2\pi\gamma_s r_1$, and the resulting force is $W - 2\pi\gamma_s r_1$. After the elongation, the actual stretching force is $W - 2\pi\gamma_s r_2$, that is, greater than before. Consequently, the extension proceeds with acceleration, and the work $W \cdot dl$ is spent also on this motion, not only on an increase in surface area.

Hence, if the wire shrinkage would have been a surface phenomenon, as postulated, Eq. (29) still would be incorrect from the viewpoint of thermodynamics. Equations (27) or (28) are the only ones conforming to this science. Moreover, it is assumed in the derivation of Eq. (29) that $\nu = 0.5$. Most solids, especially metals, have ν nearer to 0.3 than to 0.5. If $\nu = 0,3$ is substituted in Eq. (32), then $W_0 = 1.4\,\pi r\,\gamma_s$ follows.

Several additional objections to the method have been raised. It is assumed in Eq. (27) and (28) that surface tension γ_s acts along the vertical direction and thus can fully balance the external force W. But all solid surfaces are rough, and the tension allegedly acting along the slope of a hill on the specimen does not act in the direction opposite to the gravitation. In many metals, a special kind of roughness forms during the heat treatment needed to determine the function $l = f(t)$. The wires recrystallize and form short grains (or "bamboo structures"), see Fig. 3, with grooves ("grain boundary grooves") between each pair of grains. It is not explained, how γ_s operating in the wall of a groove can be just as efficient as the γ_s present along the vertical wall.

As Fig. 3 indicates, the radius of the wire at the bottom of a groove (aa or bb) in many instances is significantly shorter than that between the grooves. If the two radii are r_1 and r_0, obviously, W_0 cannot simultaneously be equal to both $2\pi r_1 \gamma_s$ and $2\pi r_0 \gamma_s$, and no one ever showed that $2\pi r_0 \gamma_s$ should be preferred to $2\pi r_1 \gamma_s$. Several investigators produced samples similar to that pictured in Fig. 3 simply by

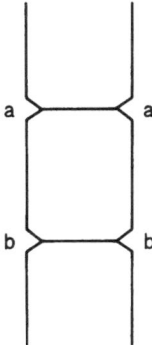

Fig. 3. The "bamboo structure" in a metal wire after protracted heating. The diameter of the wire along the bottoms of the grain boundary grooves, i.e., aa or bb, is smaller than that along the main stretch of the wire

scratching the wires with a razor. This operation rendered the determination of the elongation $l - l_0$ more exact: the distances between two razor marks were carefully measured before and after the heat treatment. The objections raised above in reference to grain boundary grooves are equally valid when scratches are considered.

Tammann[48] pointed out that (polycrystalline) metals contain internal boundaries and that their energy should be considered when calculating the area A which is being extended by the load W. At present, this idea is used above all to estimate the energies of grain boundaries[49].

If the wire after annealing looks like Fig. 3 above, if the grain boundaries aa, bb, etc. are all perpendicular to the length of the wire (i.e., also to the direction of the external force), and the extent of each boundary is πr^2 cm^2 (thus the above difference between r_0 and r_1 is disregarded), then the equation

$$W \cdot dl = \gamma_s dA + \gamma_1 dA_1 \tag{34}$$

is obtained instead of $W \cdot dl = \gamma_s dA$, see Eq. (30). Here γ_1 is the interfacial energy (per unit area) of grain boundaries, and A_1 is the combined area of all grain boundaries, equal to $n\pi r^2$, if n is the number of boundaries in the specimen. Obviously, $dA_1 = 2\pi nr \cdot dr = -2\pi n\nu(r^2/l) \, dl$. With Eq. (32), Eq. (34) is transformed into $W_0 = \gamma_s 2\pi r(1-\nu) - \gamma_1 2\pi n\nu(r_2/l)$. Assuming again that $\nu = 0.5$, we find

$$W_0 = \pi r \gamma_s - \pi n (r^2/l) \gamma_1 \tag{35}$$

Equation (35) contains two unknowns and thus appears useless, but a hypothesis permitting an experimental determination of the ratio γ_1/γ_s is presented in Section III.4. below. It is clear that all fundamental objections to Eq. (29) are equally valid in respect of Eq. (35).

The weight of these objections is such that the shrinkage method cannot be relied upon. Consequently, no table of the results obtained with it for the γ_s, mainly of metals, is given here. Those interested may be referred to the compilation[50] which includes 177 references; a more recent table will be found in Ref.[51].

Two remarkable results, still from the period covered in[50] and still deserving an independent evaluation, ought however to be described here. When a single crystal of silver[52] or copper[53] was heated under conditions usual for the shrinkage determination of the γ_s, neither extension nor contraction could be observed under a reasonable load W. Thus single crystals appeared either to possess no surface tension at all or, at least, to have an ability to resist the effects of γ_s.

The absence of deformation in single crystals was considered to be a consequence of the Nabarro – Herring (1948, 1950) theory of metal flow under small stresses. It is postulated that the "flow" is a result of diffusion of vacancies (or vacant lattice sites), that both sources and sinks of these vacancies are likely to be more common in the metal – vapor interface region and in grain boundaries than in the bulk, and that the resulting strain rate is directly proportional to the externally applied stress. The original theory paid no attention to surface effects[54], but it easily accounted for the fact that, in a single crystal, in which no grain boundary was present and only a few vacancies existed, no metal "flow" took place. On the other hand, if filament shrinkage and extension require a special mechanism, then they cannot be caused by surface tension which must be operating equally in metals (single crystals or polycrystalline), inorganic salts, organic fibers, and so on. This is another weakness of the zero-creep method.

Not many attempts on applying this procedure to organic filaments have been recorded in the literature. In a relatively recent paper[55], extruded fibers of paraffin wax were examined and the conclusion was reached that the method was unsuitable for determining the γ_s of paraffinic solids.

Some of the more recent work on the shrinkage method, mainly as applied to metals, is briefly reviewed here, to "complete" the story. It follows at once from Eq. (35) that

$$\frac{W_0}{\pi r} = \gamma_s - n \frac{r}{l} \gamma_1 \tag{36}$$

in other words, the ratio $W_0/\pi r$ is a linear function of r and from this function both γ_s and γ_1 can be determined[56]. This relation was applied to indium wires, the diameter of which was varied (unfortunately) only between 0.1 and 0.3 mm, at about 147 °C. The γ_s of In appeared to be between 600 and 750 dyne/cm, and γ_1 between 220 and 290 dyne/cm.

The modification introduced in[41] was employed[57] to find the γ_s for indium (142 °C) : 633 dyne/cm; for bismuth (239 °C) : 521; for tin (215°) : 673; for thallium (272°) : 562; for lead (309°) : 560; whereas for gold, at a temperature as low as 215 °C, γ_s seemed to be between 666 and 681 dyne/cm. In[41] the γ_s of a barium – aluminum silicate glass was found equal to 312 dyne/cm at 600 °C.

Lead wires gave[58] γ_s of 610 ± 20 in argon, and 515 ± 20 dyne/cm in nitrogen, both at 317 °C. This strong effect of a relatively inert gas on the creep of lead should be rechecked.

Silver wires at 1211 °K in hydrogen gave[59] γ_s values of 1520 ± 210 erg/cm^2 and γ_1 of 480 ± 110 erg/cm^2. Tubes of 25 μm gold foils were heated in air[60]; the

γ_s appeared to be 1480 ± 80 at $903\,^\circ$C and 1300 ± 80 erg/cm^2 at $1033\,^\circ$C if effects observed after a heating for over 50 hours were disregarded.

Some results for higher-melting metals are[61] : iron ($1480\,^\circ$C) : $\gamma_s = 2525 \pm 125$; $\gamma_1 = 985 \pm 50$; cobalt ($1455\,^\circ$C) $\gamma_s = 2595 \pm 125$, $\gamma_1 = 1010 \pm 50$; nickel ($1370\,^\circ$C) $\gamma_s = 2490 \pm 120$; $\gamma_1 = 930 \pm 50$ erg/cm^2. The value of 1970 ± 175 erg/cm^2 for cobalt (0.013 cm thick wires) at $1354\,^\circ$C in hydrogen[62] is not in a good agreement with these results. For titanium at $1600\,^\circ$C the value of approximately 1700 erg/cm^2 is given[63].

The γ_s of molybdenum at $2200\,^\circ$C in a vacuum was[64] as high as 8800 erg/cm^2, whereas in argon at 2350° it was only 1960. The author discounts the former value because the Mo wires (0.013 cm in diameter) lost 5–20% of their initial weight during the experiment in the vacuum; this is a striking example of the volatility referred to a few pages back. Chromium was tested in argon only and gave $\gamma_s = 5200$ erg/cm^2 at $1550\,^\circ$C. This tension also appears unusually high to the present reviewer.

Thin foils rather than wires were used in the two following investigations. The γ_s of zirconium at a temperature "near the melting point" appeared[65] to be 1850 ± 240 dyn/cm. For a copper containing traces of bismuth, $\gamma_s = 1350$ erg/cm^2 was found[66].

In some instances, difficulties of interpretation emerge when too much is expected from the zero-creep method. Two recent examples refer to the apparent γ_s of nickel[67] and copper[68] wires. For both metals, the rate of extension of the wires, which rate increased with the applied load W in the usual fashion as long as the load was not much greater than W_0, almost ceased to depend on W when this exceeded W_0 by a factor of, say, 3. Moreover, an increase in the (small) gas pressure of oxygen appeared to lower the γ_s in one region of pressures, and to raise it in another region. The preferred values of γ_s at extremely low oxygen concentrations were 2.12 J/m^2 for Ni at $1300\,^\circ$C, and 1.48 J/m^2 for Cu at $1027\,^\circ$C.

The authors venture to express the opinion that, under some conditions, the observed changes in the wire length are caused by effects different from surface tension. The present writer believes that these changes basically are volume rather surface effects. As long as the experimenter does not think of processes in the volume (stress relaxation, recrystallization, etc.) he cannot satisfactorily account for the phenomena he observes.

It may be repeated here that all the numerical values for γ_s, if correct, would represent a kind of average for different crystal planes exposed along the cylindrical surface of the wire. An additional drawback of the shrinkage method is indicated in Section III.2 below.

A method of "lying cylinder", similar in principle to that of zero-creep, was introduced and applied to tin[69]. However, two later teams[58, 70] could not reproduce the initial results: the cylinder simply refused to deform even at temperatures as high as $T_m - 0.7\,^\circ$K.

2. Small Crystals

If surface tension, analogous to that in liquids, really exists in solids, then also capillary pressure P_c must exist (Laplace 1805). The pressure at any point on the concave (convex) side of a curved interface would be by

$$P_c = \gamma_s \left(\frac{1}{R_1} + \frac{1}{R_2} \right) \tag{37}$$

greater (smaller) than in the gas phase across the boundary; R_1 and R_2 are the local radii of curvature counted positive when they are located in the solid, and negative when in the gas. When Eq. (37) is applied to a cylindrical wire of radius r, it follows that the pressure within the wire must be by γ_s/r greater than that outside; in this system $R_1 = r$, and R_2, that is the radius swinging along the length of the wire, is infinitely long if the wire is straight and surface roughness is disregarded.

The presumed effect of this pressure difference on the zero-creep measurements was rarely mentioned in the literature, but its existence would not be too difficult to test. If P_c were as important as believed[71], then the W_0 of a prismatic wire would have been twice that of a cylindrical wire of an identical cross section, and the W_0 of a wire with a flat horizontal end would be quite different from that of a wire ending in a hemisphere. Neither of these differences ever was observed.

On the other hand, P_c is believed to cause several other phenomena that may be used to measure or estimate γ_s. Two of these are reported on in this section.

1. It has been observed by some experimenters, but not by the others, that the experimental lattice constant a in crystals of ordinary size was different from that, $a + \Delta a$, found in extremely small crystals. A recent example[72] refers to vacuum-deposited copper grains whose "diameter" D (they were, of course, not spherical) varied from 24 to 240 angstroms. The lattice constants calculated from the (111) reflexions increased from 3.577 to 3.6143 angstroms when the grain volume decreased, but the particle size had no definite effect on the reflexions from the {220} plane.

The results presumably were distorted by the unavoidable heating of the microcrystals occurring in the electron beam and other side effects. However, if the Δa detected in the (111) reflexions is real, then the surface stress or rather surface tension γ_s causing the compression of the lattice can be calculated. If R_1 and R_2 in Eq. (37) are supposed to be identical and equal to 0.5 D, see above, then $P_c = 4\gamma_s/D$. Let k be the compressibility of the micro-crystal, i.e., $-dV/V.dp$. Then the relative compaction under the pressure, that is $-dV/V$, is $4\gamma_s k/D$. The ratio dV/V may be approximated as $3\Delta a/a$. Hence

$$\gamma_s = \frac{3\Delta aD}{4\,ak} \tag{38}$$

From this relation, the highest values of γ_s appeared to be 5325 ± 1700 dyne/cm, and the lowest (when the Δa was zero) : 0 ± 450 dyne/cm.

The former value is much greater, and the latter much smaller than those quoted in Section III.1. The main objection is, however, against Eq. (38) itself rather than its numerical conclusions. To derive Eq. (38) it is necessary to postulate that the "diameter" D is twice the radius of curvature. This relation is valid for spheres but is entirely wrong for micro-crystals surrounded by nearly flat faces. The two radii of curvature (R_1 and R_2) of a plane interface are each equal to infinity, independently of the actual dimensions of the microcrystal. As long as the surface of separation is plane, $P_c = 0$ everywhere, and, if small crystals appear to be denser than the large ones, the reason has to be looked for somewhere else.

It is realized that ultramicroscopic copper crystals deposited from vapor must have irregular surfaces of high degree of roughness, so that perfectly plane boundaries occupy only a small part of the vapor − solid interface. If surface tension γ_s really exists, then capillary pressure would be variable on a rough surface, see Fig. 4. It would be equal to zero along the stretch aa, reach perhaps 10^{10} g/cm · sec^2 along the corner b or corner c, and so on. Nowhere will it be equal to $4 \gamma_s/D$. The atoms at a, or at b, or at c are not affected by the average dimensions of the particle, *i.e.*, by D.

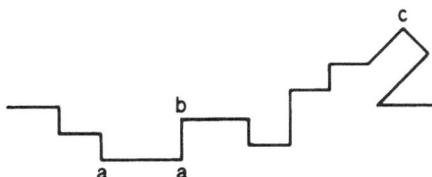

Fig. 4. Schematic provile of microscopic crystals deposited from vapor. The hypothetical capillary pressure is zero along the stretch aa, and is very strong at points b and c

In a nearly simultaneous study[73], the lattice spacing of thin gold platelets was determined (also by electron diffraction) as a function of the plate thickness. When this was about 390 angstroms, the above a was 4.0769 angstroms. In crystals as thin as 115 angstroms, a + Δa was 4.0728 angstroms. To account for these results, uniaxial compression of the platelets by an external force was postulated, and the most likely value of this force (per unit length) appeared to be 1070 ± 75 dyne/cm. Because the physical nature of this compressive force has not been elucidated, a more detailed criticism of the theory is not possible.

2. In the field − ion microscopy, the emitter usually is a thin single-crystalline wire with a hemispherical head, see Fig. 5. The diameter $2r$ of this head generally is between 10^{-5} and 10^{-4} cm. Thus capillary pressure $2\gamma_s/r$ reaching, say, 5×10^8 g/cm · sec^2 may be expected. When the emitter is at work, an intense electrostatic

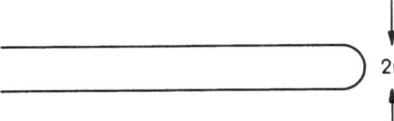

Fig. 5. The electrode in a field-ion microscope. $2r$ is the diameter of the tip

field E exists all along the gas − solid interface. The energy density of this field is $E^2/8\pi$ (because the dielectric constant of the gas is practically = 1).

According to a hypothesis[74], the hemispherical tip preserves its shape when the capillary pressure is equal and opposite to the electrostatic traction $E^2/8\pi$. A more detailed treatment leads to the equation

$$\frac{\gamma_s}{r} = \frac{E^2}{C\pi} \qquad (39)$$

in which C is a numerical constant which depends on the geometry of the tip. For tungsten at $1200\,°K$ the values $\gamma_s = 5200$ and 3800 g/sec^2 have been calculated.

Evidently, Eq. (39) cannot be valid whenever the tip does not possess an exact hemispherical symmetry; for instance, whenever traces of crystallographic planes are visible in the field-ion microscope, the value of the radius of curvature becomes nearer to infinity than to the r of Fig. 5. It is interesting to note also that the theory admits no direct effect of the electrostatic field on the value of γ_s. In liquids, the effect of E on γ gave rise to a whole branch of science usually known as electro-capillarity. An attempt to inaugurate an electrocapillarity of solids is mentioned in Section III.9.

The dimensions of a crystal affect also its breaking stress; this effect is referred to in Section III.3.

The melting point is supposed to depend on the crystal size also when the capillary pressure is not involved. It is assumed that the change in free energy taking place during solidification contains a term proportional to the volume of the prism formed, and another term equal to the sum $\gamma_{s1}A_1 + \gamma_{s2}A_2 + \gamma_{s3}A_3$, in which γ_{s1} is the specific surface energy of the first crystal face whose area is A_1, and the other symbols have the analogous meanings. If the prism is a very thin plate, then only one kind of crystal faces (that represented by the two large opposite bases) has to be counted. With this approximation, and some other hypotheses which cannot be scrutinized here, the equation

$$T_1 = T_m \left(1 - \frac{2\gamma_s}{\rho_3 \lambda l}\right) \tag{40}$$

is derived[76]. In it, T_1 is the melting point of a plate of thickness l cm, T_m is the bulk melting point, and λ is the heat of melting per unit mass (erg/g). The quantity ρ_3 (i.e., the density of the solid) was selected by the reviewer, because in the original it is not specified, whether ρ_1, ρ_2, or ρ_3 is meant ρ_1 is vapor density. This uncertainty should be kept in mind also when inspecting Eqs. (66) and (68).

Equation (40) was applied[77] to single crystals of polyethylene, about 130 angstroms thick. The γ_s of the {100} faces turned out to be 72, and that of the {110} faces: 64 erg/cm^2. It is obvious however, that in this instance not the tension at the vapor — solid boundary, but rather that (γ_{sl}) acting in the melt — solid interface is meant. The former tension would be expected to be much greater than the latter; perhaps 10 or 100 times as high. — It has been mentioned above, that thin plates are imagined to be under an uniaxial compression; thus the theories of Ref.[73] and Ref.[77] are not in agreement with each other.

From another series of measurements of the T_1 of polyethylene, when the lamella thickness varied from 250 to 1000 angstroms, the γ_s or γ_{sl}, still using Eq. (40), seemed to be[78] near 100 erg/cm^2 for the base faces of the crystals. A difficulty in applying Eq. (40) to polyethylene is the uncertain value of the temperature T_m. If the lowest probable T_m is used, the calculated γ_s (or γ_{sl}) rises[79] with the annealing temperature of the specimen from 29 to 72 erg/cm^2; and when the highest possible limit of T_m is used, the range of γ_s (or γ_{sl}) is from 49 to 104 erg/cm^2.

A value of 102 erg/cm^2 for the γ_s of the end surfaces of the low-melting form of poly-trans-1.4-isoprene was derived[80] in the same manner. The lamella thickness was supposed to be equal to the spacing shown by low-angle X-rays and varied between 150 and 200 angstroms.

3. Surface Energy of Rupture

As mentioned in the Introduction above, this method originated with Dupré[4]. The first task of the experimenter is to separate that part of the work of rupture which is spent on the mechanical deformation of the sample from that used up to create a new surface $2A$ (on both sides of the crack together).

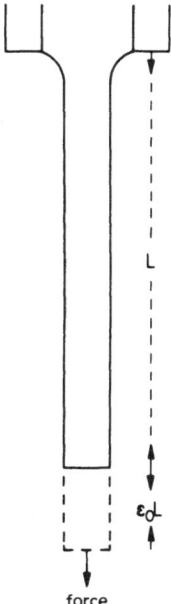

Fig. 6. Elongation of a bar, of length L, by a tensile force. The maximum absolute extension is $\epsilon_0 L$

Consider a rod of length L, see Fig. 6, and suppose that its maximum strain (or maximum relative elongation before rupture) is well reproducible and equal to ϵ_0. If the force causing rupture is W, this force performs the work

$$\mathfrak{W} = \epsilon_0 L W \tag{41}$$

before the rod snaps, taking Hooke's law to be valid. If the experimenter extends the rod by only 0.99 $\epsilon_0 L$ cm and then permits it to contract again, the work performed is 99% of the above $\mathfrak{W}$, and no fracture occurs. This is a simple example of the difference between the mechanical work and the surface energy gained, or a simple indication of the fact that Dupré's "γ" is only a small fraction of the total work spent on breaking the solid.

In principle, it should be possible to disentangle that part of $\mathfrak{W}$ needed to increase the gas — solid interface by a unit area in the following manner. Let the bar of Fig. 6

be extended almost but not quite to ϵ_0. Then a horizontal scratch is made on its vertical surface. If the scratch deepens and results in fracture, this proves that the decrease $d(SE)$ in the strain energy of the bar (this decrease is caused by the crack propagation and the resulting contraction of the solid) is greater than, or at least equal to, the free energy of the two nascent surfaces. In mathematical form:

$$- d\,(SE)/dx \geqq 2 \text{ ``}\gamma\text{''} \ w \tag{42}$$

if w is the width of the crack (normal to the direction of the crack advance, so that $dA/dx = w$) and x is the variable depth of the fissure.

Whether the scratch does or does not grow, depends on its sharpness (i.e., the radii of curvature of the gas — solid interface at the front of the crack) and on the initial depth $2c$ of the scratch. A relation between c and "γ" was derived by Griffith[81], see also Petch[82]. It is

$$\text{``}\gamma\text{''} = \frac{\pi c\, \sigma_m^2}{2 E} \tag{43}$$

σ_m is the average tensile stress causing fracture and E is the modulus of elasticity of the bar. A simplified derivation of a similar relation may not be out of place here.

If the Hooke's law is valid and the stretching of the bar is very slow, then the work of extending the bar from L to $L + dL$ cm is

$$\mathfrak{W} = \frac{\sigma_m^2 L}{2 E} \ \text{erg/cm}^2 \tag{44}$$

per unit area (normal to the load). When a flat crack, $2c$ cm in width and $2c$ cm in depth, perpendicular to the tensile force, is introduced, the solid can contract a little from the area $4c^2$ cm^2, that is in a volume roughly equal to $4c^2 L$. The relative degree of contraction, at a first approximation, may be set proportional to the ratio $2c/L$. Hence, the total strain energy of the bar is lowered (by the crack) by $8\,c^3\, \sigma_m^2\, K/E$, K being a (dimensionless) constant of proportionality. According to Eq. (42), this quantity must be at least as great as the surface energy $4c^2$ "γ" of the crack. Hence

$$\text{``}\gamma\text{''} \leqq \frac{2Kc\, \sigma_m^2}{E} \tag{45}$$

This relation becomes identical with (43) if K is set equal to $\pi/4$.

According to a newer theory[83], see also[84], the fracture energy "γ" calculated from Eq. (43) or a similar relation has no connection with true surface energy. Energy is needed not to create a new surface but to deform the solid to its maximum strain. This rule is unquestionably correct for liquids. As was proved by Plateau (1869), a liquid cylinder can be extended by an external force until its length exceeds its circumference; then it spontaneously splits into two spherical drops, and the combined area of the drops is smaller than that of the stretched cylinder. Thus, external work is required to stretch the cylinder, not to break it.

When Plateau's rule is applied to solids, it may be said that a crack advances whenever the corresponding decrease $d\,(SE)/dx$ in the strain energy is at least as great as the work necessary to extend the material (in front of the crack) to its maximum strain, that is, to rupture. Two semi-quantitative formulations of this idea are possible.

1. For a crack, w cm wide, to advance by dx cm in a solid, a ribbon (or curtain) of this solid, dx cm thick and w cm wide, must be extended to its greatest possible elongation. If the curtain of length L can extend freely, the work of extension is, see Eq. (44), $\xi^2\,Lw\cdot dx/2E$ ergs. The cohesion ξ is equal to the breaking stress σ_m (g/cm $\cdot$ sec^2) of the theoretical solid, otherwise identical with the real, but free of all defects. ξ rather than σ_m should be used here because dx is so small that no defect can find place in the curtain. The work per unit area corresponds to the hypothetical "γ". Hence,

$$\text{``}\gamma\text{''} = \frac{\xi^2 L}{2E} \tag{46}$$

It is clear that this quantity is not a surface property at all.

The relation between ξ and σ_m usually is approximated as

$$\xi = \sigma_m \left(\frac{2c}{R} \right)^{0.5} \tag{47}$$

if the crack is an elliptical hole of the length (perpendicular to the direction of the force) $2c$ cm and if R is the radius of curvature of the ellipse at the most stressed point of the latter. Introduction of Eq. (47) into Eq. (46) leads to

$$\text{``}\gamma\text{''} = \frac{2\,\sigma_m^2\,cL}{R\,E} \tag{48}$$

Usually it is assumed that $L = R$. Thus

$$\text{``}\gamma\text{''} = \frac{2\,\sigma_m^2\,c}{E} \tag{49}$$

This relation externally is very similar to Eq. (43) but includes only volume properties.

2. In the derivation of Eq. (46) it was assumed that the curtain in front of the growing crack could be stretched independently of the material around it. This

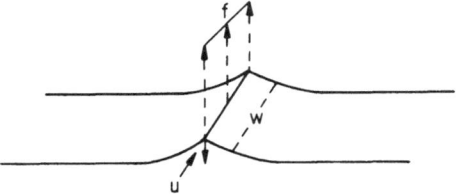

Fig. 7. Force f acting along length w raises a ridge whose greatest height is u

approximation, although employed in many theories of mechanics, is less than satisfactory. An approximation, in which the involvement of the whole solid near the crack is taken into consideration, is based on Boussinesque's analysis[85] of the effect of a force acting along a line.

If such a force f, uniformly spread over a length w, see Fig. 7, lifts a ridge on the solid, the height u of a point of the ridge (not too near to the line on which the force acts) is given by the relation

$$u = \frac{2f}{\pi wE} \ln \frac{R_\infty}{r} \tag{50}$$

r is the distance of the elementary volume considered from the above line, and R_∞ is that distance (from this line) at which the effect of f becomes negligible. Let us find the value of u at such a distance r that $\ln(R_\infty/r) = 10$. The work performed by the external force in lifting material from zero level to level u is fu, that is, $2f^2\ln(R_\infty/r)/\pi wE$. With $\ln(R_\infty/r) = 10$, the work becomes $\mathfrak{W} \approx 6 f^2/wE$. On the other hand, $\xi = f/w\Delta$, if Δ is the thickness of the "line" along which f acts; the dimension of f is g · cm/sec^2, and the dimension of ξ is g/cm · sec^2. Thus $\mathfrak{W} \approx 6 \xi^2 w\Delta^2/E$ ergs, and the work per unit area, *i.e.*, the work corresponding to the above "γ", is

$$\text{``}\gamma\text{''} \approx \frac{6 \xi^2 \Delta}{E} \tag{51}$$

This relation contains Δ instead of L as in Eq. (46).

In the foregoing analysis, the mathematical similarities between the new and the Griffith theories are emphasized. From the point of view of physics, the differences between the two approaches are more important. In the new approach, the energy of fracture (per unit area) is the work required to stretch a curtain of material, in front of the crack, to its maximum elongation or strain ϵ_0. This work may be formulated as $\epsilon_0\xi L$, L being a length the exact meaning of which depends on how rigidly the "curtain" is attached to the surrounding solid.

The three main advantages of the new theory are:

1. It can account for those experimental values of "γ" (a few of which are presented below) that are far too large to be accepted as true surface energies.

2. It gives a reasonable explanation for the observation that "γ" generally rises with the maximum strain ϵ_0.

3. It explains the heat evolution during fracture: the above work $\mathfrak{W}$ does not remain in the two fragments (resulting from fracture) as their surface energy; its major part is released as heat, and the rest remains as the frozen strain energy in the two deformed fragments.

Experimental Determinations. Several determinations of "γ" were based directly on Eq. (43). A crack of length c was made on the external surface, or a crack $2c$ long was produced in the middle of the specimen, and the breaking tensile stress σ_m was measured for the weakened specimen. A few typical results are quoted here.

A lead – alkali, a soda – lime, a borosilicate, and an aluminosilicate glass afforded[86] "γ" values of 1160, 1700, 3460, and 3340 erg/cm^2, respectively. A Pilkington float glass had[87] "γ" = 4.2 ± 0.5 J/m^2 or 4200 erg/cm^2. A soda – lime glass, tested by three different methods[88], appeared to have "γ" ranging from 5000 to 11000 erg/cm^2.

These data, when compared with the surface tension γ_s = 312, as determined by Khagabanov[41], see Section III.1, show that fracture energy in many instances is much greater that the energy determined by the shrinkage method. As stated above, an advantage of the new theory of "γ" is, that it can account for the unexpectedly high values of this quantity.

The ϵ_0 of usual glasses is much smaller than that of the common metals. Hence, in agreement with the new theory, the "γ" of metals greatly exceeds that of silicate glasses. Thus, the "γ" of a steel varied[88] between 11 x 10^6 and 24 x 10^6, and the "γ" of two aluminum alloys was in the range between 10^7 and 49 x 10^7 erg/cm^2. In these instances, the ratio of "γ" to γ_s (as given by the shrinkage method) exceeded 1000 or 10000; evidently, different properties are measured by the two procedures.

No convincing explanation is available at present for the observation[91] that the "γ" of single crystals of tungsten depended on their thickness l. As long as l was between 400 and 800 μm, "γ" was near 1.7 J/m^2, but at l = 200 μm it reached 30 J/m^2.

The values of "γ" obtained for polymers also are far greater than those for glasses, as are also their maximum strains ϵ_0. By a special method[90], the "γ" of a poly (methyl methacrylate) between 17° and 80 °C was found to vary irregularly between 3 x 10^5 and 9 x 10^5 erg/cm^2, in a reasonable agreement with Ref.[88] in which values in the range 3 x 10^5 to 7 x 10^5 are recorded.

Using Eq. (43) and defining the crack length in different manners, the fracture energy of another sample of poly(methyl methacrylate) appeared to be 1.6 to 2.6 x 10^5 erg/cm^2. In a later study[92] the effect of the molecular weight (MW), varied by irradiation, of this polymer on "γ" was determined. When the average MW was varied from 2 x 10^4 to 10^5, "γ" increased from 5.3 x 10^3 to 10^5 erg/cm^2, but a further rise of MW (to 11 x 10^5) caused "γ" to advance only to 1.7 or 2.4 x 10^5 erg/cm^2. If, as above, "γ" $\approx \xi \cdot \epsilon_0 L$, the $\epsilon_0 L$, $i.e.$, the absolute extension of the "curtain" in front of the advancing crack, must be between 10^{-6} and 10^{-5} cm for the polymers of a high MW.

For a phenol-formaldehyde resin at room temperature, the "γ" was 5 x 10^4 erg/cm^2; it decreased on heating ($e.g.$, to 175 °C) when also the modulus E of elasticity became very small[93].

In many instances, "γ" increases with temperature. The γ of pure liquids always decreases when temperature T rises[94]. Thus the temperature coefficient of "γ" is different from what would be expected from the true surface energy; on the other hand, positive values of d"γ"$/dT$ agree with the new theory of fracture energy, because ϵ_0 of many materials is greater the higher the temperature.

A striking example of this behavior was described[95] for vanadium carbide VC. In the temperature range of roughly 20° – 700 °C this material is brittle and its "γ" remains near 3000 erg/cm^2 (as for glasses). Considerably above 800°, VC may be classified as ductile, and its "γ" at 1000 °C is about 3 x 10^5 erg/cm^2. A positive

d"γ"$/dT$ was ascertained also in splitting tests (see below). Thus, the "γ" of a zinc
— cadmium alloy appeared[96] to be 780–1150 at −196 °C and 5620–5930 at 25 °C.

In many fracture tests, a splitting method was employed. Figures 8 to 10 schematically indicate three of the arrangements used. In Fig. 8, a slice, h cm thick, is separated from the rest of the solid by pushing in a wedge which may be as thin as 0.01 cm but may also be more robust and have one of various shapes; also the part broken off may be as thick as the remainder. In Fig. 9, a plate, $2h$ cm thick and w cm wide, is split in the middle to the depth c cm, and the force f required to extend the fissure is measured. Figure 10 shows peeling: a ribbon, h cm thick, is peeled off the main body of an identical solid.

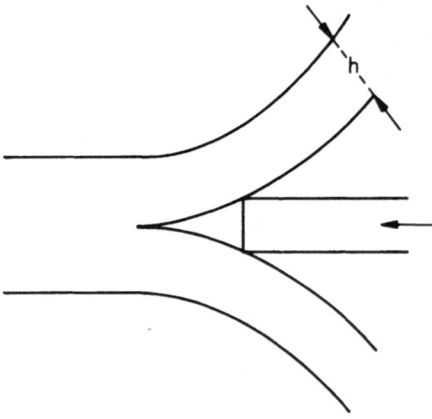

Fig. 8. A slice, h cm thick, is cleaved from a solid by pushing in a wedge

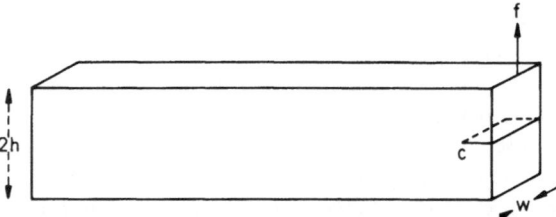

Fig. 9. A plate, $2h$ cm thick and w cm wide, is split by force f; the preformed crack is c cm deep

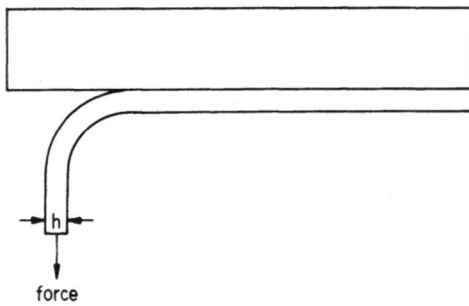

Fig. 10. Peeling. A ribbong, h cm thick, is peeled from a rigid solid by an external force

In principle, in the tests of the first type (Fig. 8), the force applied to the wedge need not be known. This force bends the two (partly separated) components of the solid, and the work of splitting can be calculated from the shape of the bent members. If x is the Cartesian co-ordinate along which the wedge moves forward, and y is the height of the bottom surface of the upper member above the middle of the slit, then[97, 98]

$$"\gamma" = \frac{Eh^3 y^2}{12 x^4} \tag{52}$$

E is still the modulus of elasticity. The E is assumed to be independent of force, rate of rupture, and so on. It has been pointed out recently[99] that the wedge, unless it advances extremely slowly, is likely to cause vibrations (flexure waves) in the specimen which would necessitate an alteration of Eq. (52).

At least one consequence of Eq. (52), namely the direct proportionality of the ratio x^4/y^2 to h^3 has been confirmed by experiments[100] in which pyrographite films, 0.2 to 0.77 mm thick, were sliced off. The "γ" appeared to be close to 1850 erg/cm^2.

If the wedge of Fig. 8 is a rigid plate of constant thickness Y and if its front is x cm to the left of the crack orifice (x being a variable), then a relation between Y and x can be used instead of that between y and x in Eq. (52). In this manner, the relation

$$"\gamma" = \frac{3 Eh^3 Y^2}{16 x^4} \left[1 + \frac{3}{2} \left(\frac{Y}{x} \right)^2 - \frac{9}{4} \left(\frac{Y}{x} \right)^4 \right] \tag{53}$$

was derived[101].

In the arrangements of Fig. 9 and Fig. 10, knowledge of the external force f is needed. When the point of application of this force advances by dc cm, the work $\mathfrak{W} = f.dc$ is spent. However, a part $\mathfrak{W}_m$ of it is used up on macroscopic deformation of the specimen and thus should be substracted from $\mathfrak{W}$ to obtain the fracture energy $\mathfrak{W}_f$, equal to 2 "γ" A, A being the newly formed area of the crack (on one side). Apparently, there is still no unanimity as far as the calculation of $\mathfrak{W}_m$ is concerned. Equation

$$"\gamma" = \frac{6f^2 c^2}{Ew^2 h^3} + \frac{3 \alpha f^2}{2 G w^2 h} \tag{54}$$

was advocated in Ref.[102]; in it, c is the length of the crack, w and h are defined above, G is the shear modulus normal to the cleavage plane, and α is a numerical constant between 0 and 0.5 "for elastic behavior within the constraints of elementary beam theory". In some calculations, e.g., Ref.[96], only the first term on the right-hand side of Eq. (54) was employed; for a criticism of this approximation see, for instance, Ref.[100]. Deviations from Eq. (54) occurred when the ratio h/c was too low[102]. A semi-empirical relation between "γ" and this ratio was derived by Wiederhorn[103]:

$$\text{``}\gamma\text{''} = \frac{6f^2 c^2}{E w^2 h^3} \left(1 + 1.34 \frac{h}{c} + 0.45 \frac{h^2}{c^2} \right) \tag{55}$$

In peeling tests, Fig. 10, the material in front of the crack is compressed in the direction normal to the crack propagation, extended in this direction a little further from the crack tip, and so on. This sinusoidal compression and extension, described, for instance, in Ref.[104], usually is disregarded when discussing measurements of the types illustrated in Fig. 8 and Fig. 9.

The quantitative results obtained in splitting and cleavage tests are similar to those afforded by Eq. (43). For instance, the "γ" of the {100} face of KCl appeared to be[102] 110 ± 5 erg/cm^2. Depending on the equation used, the "γ" of the {111} face of SrF$_2$ seemed[105] to amount to 413—484 or to 252—389 erg/cm^2. Three different glasses (soda — lime, borosilicate, and alumosilicate, respectively), all in dry nitrogen at 300 °C, had[103] "γ" of 3.9, 4.6, and 4.6 J/m^2.

Single crystals of zinc were broken[106] by two cleavage and one tensile method. The "γ" was about 100, 400, and 575 erg/cm^2, respectively, for these three arrangements. Also this wide spread of results is easier to reconcile with the new than with the Griffith theory.

A value for the "γ" of pyrographite is given above in this section[100]. In a later publication[107], the dependence of this "γ" on the mode of preparation of pyrographite is described. When the material was precipitated at 1800 °C, at 2100 °C slowly, or at 2100 °C rapidly, the "γ" was, respectively, 4000, 350, and 1500 erg/cm^2. After annealing at 2300 °C for 3 hours, the "γ" of these solids became 1600, 350, and 1100 erg/cm^2. It is not difficult to account for such a variability by Eqs. (46) or (51), but true surface energy would be expected to possess a more stable value. On the other hand, the "γ" of every specimen was identical in air and in a vacuum.

Considerable differences between the energies of cleavage of mica *in vacuo* and in air have been recorded by Obreimov[97], the initiator of the splitting technique. Unfortunately, the reproducibility of measurements on mica often is unsatisfactory. Figure 11, taken from Ref.[101] (and simplified) shows the variation of "γ" (the ordinate) with the depth x of penetration of the wedge (the abscissa), see Fig. 8, in a moderate vacuum, at a speed of 0.5 mm/sec. It is seen that "γ" changes by a factor of 3 or even 10, when the wedge advances by 1 or 2 mm. Fluctuations of such a magnitude cannot be attributed to true surface energy. An explanation can be searched in Eqs. (46) and (51) or in the flexure waves[99] referred to above.

An interesting observation[108] again emphasizes the difference between "γ" and the true surface energy. According to Gibbs, traces of an impurity can greatly lower, but cannot greatly raise the surface tension of a liquid (or a solid in a corresponding equilibrium). The cleavage energy "γ" of calcite was 3 times as great for crystals containing traces of Cu, S, and Fe than for particulary pure crystals of Iceland spar.

Many recent publications contain "γ" values calculated from Eqs. (43), (52), etc., but this *specific fracture energy* is not identified with any energy residing in a surface; rather it is treated as a material constant whose true nature is not specified. The derivative $-d(SE)/dc$ is viewed as "the force conjugate to time rate of crack extension and as a measure of elevation of stress near the end of a crack"[109]. The

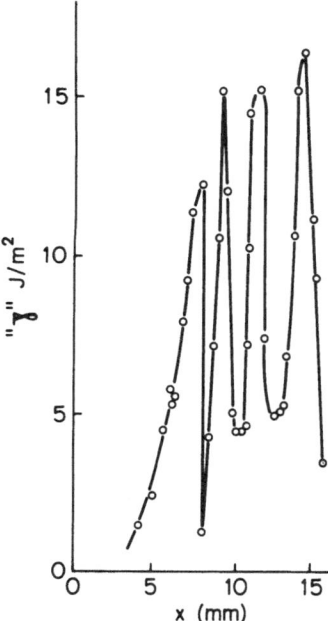

Fig. 11. The cleavage energy (J/m^2) of a mica specimen as a function of the distance travelled, x mm. Data of Ref.[101]

term *fracture mechanics* frequently is applied to the corresponding branch of science. Its findings are outside the scope of this review.

4. Grain Boundaries and Scratches

The energy associated with grain boundaries in polycrystalline solids, *i.e.*, γ_1 erg/cm^2, has been mentioned in Section III.1. It is reviewed here even more briefly than the surface energy (γ_s or "γ") with which this paper is primarily concerned.

In the most popular method of estimating γ_1, not this quantity itself but only its ratio to the surface energy, that is, γ_1/γ_s is determined. Let Fig. 12 be a highly

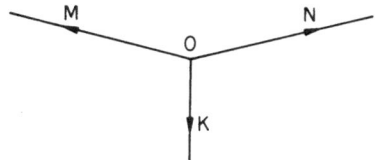

Fig. 12. A grain-boundary groove. The hypothetical tensions OM and ON balance the hypothetical tension acting along the grain boundary OK. The space above the line MON is occupied by a gas (vacuum)

magnified profile of a spot where the boundary KO between two identical crystals (or grains) A and B reaches the gas – solid interface. If it is assumed that tensions act along the surfaces MO and NO (both against the gas phase) and the interface KO, and if the upward continuation of the line KO divides the angle MON (the dihedral angle 2 ψ) in two equal halves, then the three tensions balance each other[110] as long as

$$\gamma_1 = 2\gamma_s \cos \psi. \tag{56}$$

Tension γ_1 acts downward along the line OK, and the two tensions γ_s act along MO and NO, respectively.

The latter two tensions are supposed equal because the two grains A and B are presumed identical. It is possible, of course, and often happens in real specimens, that the two gas — solid interfaces exposed on two adjacent grains belong to two different crystallographic planes. Then, in theory, also their surface energies should be different; thus, the tension along MO would differ from that along NO, and an equation more complex than (56) is needed. Several experimenters referred to this doubt, e.g., Allen[64], but a separate determination of γ_s of two neighboring grains is so difficult and unreliable that almost everyone was satisfied with Eq. (56) and blamed the discrepancies observed on the neglect of the difference between the two tensions (along MO and along NO).

An equation related to (56) is known in the capillarity of liquids; see, for instance, Ref.[94], p. 128. If, in Fig. 13, 1 means the gas phase, and 2 is a drop of an immiscible liquid floating on liquid 3, then, from the equilibrium of forces,

$$\frac{\gamma_{12}}{\cos 3} = \frac{\gamma_{13}}{\cos 2} = \frac{\gamma_{23}}{\cos 1} \tag{57}$$

γ_{12} is the tension in the interface between 1 and 2, angle 3 is that occupied by phase 3, and the other symbols have analogous meanings. This relation was applied[111] to an assembly of three aluminum crystals, all oriented along their [100] axes and grown together at 640 °C; naturally, only the ratios γ_{13}/γ_{12} and γ_{23}/γ_{12}, not the actual values of each tension, could be determined in this manner.

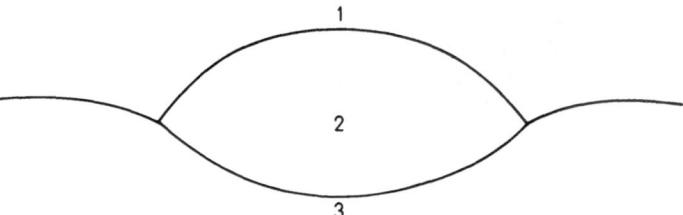

Fig. 13. Capillary equilibrium of a drop 2 floating on an immiscible liquid 3; 1 is the vapor

Doubts have been expressed, see Ref.[94], p. 128—129, regarding the validity of Eq. (57) for gas — liquid systems, but the validity of Eqs. (57) and (56), when applied to solids, is much less convincing. Typical liquids can attain the shape prescribed by capillary forces because their yield point is zero so that they can follow, in time, even the weakest deforming force (which does not significantly alter the density of the liquid). Marked deformations of solids are not so easy.

Moreover, most solids have a preferred crystalline shape and tend to preserve it or to revert to it. Hence, a solid, or an agglomerate of solids (grains, crystals) is not free to adjust to capillary forces, as postulated in Eqs. (56) and (57). The profile of a single crystal may look, for instance, like the line KLMN in Fig. 14, and the angle α usually is determined by the crystal structure. In a cubic crystal, frequently $\alpha = \pi/2$.

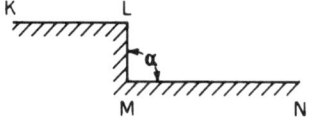

Fig. 14. Schematic profile of a single crystal. If tensions act along the surfaces KL, LM, and MN, they cannot balance each other

The surface tension, if it exists, acting along the face LM in principle is identical with that acting along MN. Obviously, these hypothetical tensions do not balance each other. Nevertheless the crystal is stable enough to survive several geological epochs. It is difficult to believe that capillary equilibrium, which is not needed for the stability of this crystal, becomes all poverful in a system of Fig. 12.

In addition to these fundamental doubts, experimental difficulties render the determination of cos ψ, Fig. 12, rather uncertain. The lines MO and NO of Fig. 12 in reality are not straight. Usually the profile is strongly curved as depicted in Fig. 15.

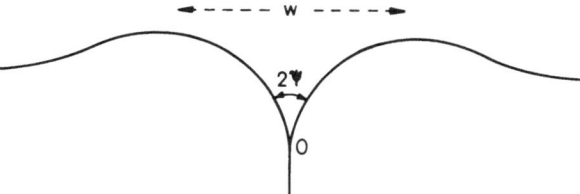

Fig. 15. Common profile of a grain boundary emerging on the metal surface (*i.e.,* interface with a gas). The angle at the groove bottom is 2 ψ. The distance between the summits of the two ridges on both sides of the groove is w

The angle 2 ψ of interest is that within, say, 10 angstroms from the star point 0. It is still impossible to determine this angle experimentally with a reasonable degree of precision. Some experimenters[112] used profilometers to measure ψ, but usual profilometer styluses have a diameter of 25 microns so that that part of the groove which is narrower than 25 μm is not explored at all. Also the bending of the interference lines (of light) across the grooves is not exact. In similar unfavorable circumstances, the contact angle between liquids also may be in error by ± 5°. In solid – solid systems the error presumably is at least as great, although more optimistic evaluations such as ± 4°, Ref.[113] or even ± 2° (Ref.[64]) for 2 ψ are found in the literature. On many metals, ψ appears to be near 80°. The values of cos (80° + 5°) and cos (80° − 5°) are respectively, 0.087 and 0.259; thus it is easy to commit a large error in cos ψ and, consequently, in the ratio γ_1/γ_s.

In some experiments, *e.g.,*[113], the metal surface was electroplated. Then the specimen was cut parallel to the plane of Fig. 12 and polished. The angle ψ was estimated on these metallographic sections. To rely on the data thus obtained, the experimenter must believe that

(a) the capillary forces which were all-important before the treatment (*i.e.,* electroplating etc.) remained dormant during the treatment, and

(b) that cutting and polishing did not alter the position and directions of the three interfaces within, say, 10 angstroms of the point 0, Fig. 12 and Fig. 15.

If the solid is not chemically pure, that is, practically always, some impurities tend to gather in the grain boundary regions; see[114] on bismuth accumulation in an alloy of Cu 99.8, Bi 0.02 weight-%. Thus, instead of an almost two-dimensional interface preferred by theoreticians, a grain boundary in most cases is three-dimensional. It differs from the adjoining grains in its composition and, we may be sure, in its state of stress. A groove, such as illustrated in Fig. 12, may be affected by stresses in its vicinity, but it is hazardous to attribute a true interfacial tension to such an interlayer.

In Section III.1, the upsetting influence of vaporization has been pointed out. Determination of the groove profile also is performed after a protracted heating (intended to achieve equilibrium) and thus may be hampered by evaporation. This effect was blamed for the observed[64] spread of the values of 2ψ in heated molybdenum foils: "preferential loss from the groove root was likely and would account for the low dihedral angles". This surmise reminds one of an explanation given over 60 years ago[115] to the high rate of vaporization of finely grained metals: these contain more grain boundaries than coarsely grained samples do; the crystal lattice near the boundaries is less perfect than in the bulk; and the looser the lattice, the higher the rate of its evaporation.

Experimental Determinations. The simplest test procedure consists in heating a polycrystalline solid to, say, $0.8\,T_m$ or $0.9\,T_m$, T_m being again the melting point on the absolute scale. In many cases, as in the shrinkage method of Section III.1, this treatment causes grooves to appear near the 3-phase lines in which two grains and the gas phase meet. If the heating, also known as *thermal etching,* was protracted, the experimenter usually assumes that the equilibrium state has been reached. Now he selects those spots in which the line KO, Fig. 12, seems perpendicular to the main gas — solid interface (far from the groove), and measures the angle 2ψ. Alternatively, many (say, 100) grooves are inspected, and the most common value of 2ψ is accepted as that belonging to grain boundaries perpendicular to the above main surface. It seems, however, to the present reviewer that the most common angle 2ψ on a polycrystalline solid should depend on the crystallographic orientations in it.

As soon as 2ψ is known, Eq. (56) affords the ratio γ_1/γ_s. In some papers, this is the only numerical result. In other studies, γ_s was determined by an independent method, so that the absolute value of γ_1 also was established.

Perhaps the most striking result of the measurements of the ratio γ_1/γ_s is that so many values of it are confined to the range 0.25—0.40. As a matter of fact, an earlier review[50] refers to "the general observation that the grain-boundary free energy is about one-third of the solid-surface free energy." Also in a latter study on UC, the ratio γ_1/γ_s was found to vary between 0.36 and 0.41[116]. Because so much of the usual treatment of surface effects in solids is based on the alleged analogy between them and the surface effects in liquids, it ought to be pointed out, that a corresponding ratio in liquids would be equal to zero.

Several attempts on theoretical prediction of γ_1 are found in the literature; according to some models, γ_1 is more closely related to the cuticular energy defined in Chapter V than to the true surface energy, but cuticular energy implies no contractile tendency and cannot pull along the line KO of Fig. 12.

The absolute value of γ_s can be found from a theory[117], which neglects the essential difference between solids and liquids, by measuring the gradual alteration in the shape of the groove region depicted in Fig. 15. The distance w between the two ridges is supposed to increase at a high temperature following the relation

$$w = 5 \, (\gamma_s/kT)^{1/3} \, (nD \, \Omega^2 \, t)^{1/3} \tag{58}$$

as long as the profile changes are caused by the diffusion of the material in the gas phase. In this equation, n cm^{-3} is the number of molecules in the unit volume of the gas phase, D cm^2/sec is the diffusion coefficient in the gas, Ω cm^3 is the volume of a molecule, t sec is time, and γ_s, k, and T have their usual meaning.

The above w was measured[118], for instance, for polycrystalline surfaces of UO_2 kept for various times in a helium atmosphere at various high temperatures. The slope of the curve $\ln w = f(\ln t)$ was between 0.33 and 0.37 so that the exponent $1/3$ in Eq. (58) seemed to be confirmed. When an approximate value for D was selected, the γ_s of UO_2 appeared to vary irregularly between 385 and 806 erg/cm^2; the annealing temperature ranged from 1660 °C to 1940 °C.

An analogous treatment is possible also for scratches of arbitrary depth and arbitrary position, independently of any grain boundary. The depth h of a scratch should decrease in time according to the relation

$$h = h_0 \, exp \, (-\pi \, \gamma_s t/\eta \, \lambda) \tag{59}$$

h_0 being the depth at that moment when the theory starts to be valid, t is time, η the viscosity, and λ the wave length of the periodic disturbance of the solid surface. This λ also is supposed to vary with time, and η is supposed to be constant (as in a Newtonian liquid!). When this theory was applied[112] to scratches made by a diamond stylus on a selenium film, 60 μm thick, deposited on aluminum, the γ_s at 39 °C appeared to be, for selected scratches, near 100 erg/cm^2. The disregarded scratches had irregular profiles "which may be attributed to crystallites a few hundredths of a micron high". The theory disregards crystallization and any crystal structure in solids.

In a similar investigation[119], from the rate of flattening of (artificial) microscopic ridges on nickel at 1219 °C in a vacuum, the γ_s was calculated to be 1820 ± 180 and 1900 ± 190 erg/cm^2 for the {100} and {110} planes.

In some instances, a thin metal film, freely hanging or deposited on a solid support, gives rise to holes on a heat treatment. This effect, noticed long ago, also is supposed to be influenced by surface energy. Silver films, about 400 to 2000 angstroms thick, on silica were heated[120] in the presence of oxygen and kept at 266 °C for several hours. After this treatment, a few holes were visible on each square micron of the film. The average area of each was supposed to be proportional to $\gamma_s^{0.8}$ but no attempt was made to derive the absolute value of γ_s.

Under some conditions of cooling from the melt, small bubbles near the interface of the metal and the metal oxide form spontaneously in tin and some other materials of a relatively low melting point. A subsequent moderate heating (e.g., to 223 °C for Sn) causes disappearance of these voids. If it is assumed that this effect is caused by the capillary pressure $2\,\gamma_s/r$, r being the radius of the void, then the equation

$$r^3 - r_0^3 = \frac{6\,D\,\Omega\,\gamma_s\,t}{kT} \tag{60}$$

results[121]; D again is the diffusion coefficient (in the solid!), Ω the volume of a tin atom, t time, and r_0 is the radius at $t = 0$. Two values for γ_s (about 350 and 990 erg/cm^2) were found in this manner, depending on the crystallographic direction along which the diffusion was believed to take place. Even more upsetting is the observation that the voids had tetragonal shapes so that the interfaces were plane; thus the values of r had no physical meaning and no capillary pressure was possible.

Equation (60) was used earlier[125] for bubbles (of uncertain composition) introduced in pure aluminum wires by plunging these from 650 °C into a saturated solution of calcium chloride kept at 100 °C. The bubbles were up to 500 angstroms across. They were characterized as octahedral but treated (in calculations) as spheres. The γ_s appeared to be at 150 °C, 175 °C, and 200 °C, respectively, 1160, 930, and 960 erg/cm^2.

5. The Bubble Method

The theory of this method is simple and directly based[122] on Laplace's law of capillary pressure. The experimental procedure, on the other hand, requires less common facilities than those employed in the shrinkage or the rupture methods. α-Particles[123] or ions of another noble gas are implanted in the surface layer of a metal. They remain very near to the irradiated surface and are invisible unless the specimen is heated (to e.g., 500 °C for copper). Then they assemble to ultramicroscopic bubbles. The number of these bubbles and their approximate radius r can be determined by isolating the damaged region and viewing it in an electron microscope. The total mass (n moles) of gas trapped can be determined by chromatography.

If the pressure inside the bubble is $2\,\gamma_s/r$ (Laplace's law) and the laws of ideal gases are applicable, then with the volume $V = (4/3)\,\pi r^3$ the relation

$$\frac{8}{3}\,\pi\,\gamma_s\,r^2 = nRT \tag{61}$$

is found, R being again the gas constant. Thus, as soon as r and n are known, γ_s can be calculated. However, the inside pressure P computed from n and r amounted, for instance, to 1500 atmospheres (1.5×10^9 g/cm · sec^2), so that the product PV was too great for ideal gas laws to be valid. Using the experimental values of $P = f(V)$ for the gas in question, estimates for the γ_s of Cu at 500 °C were between 1400 and 2000 erg/cm^2.

The method has been criticized[124] because it assumes zero stresses in the metal around the bubble. When a nail is driven into wood or a bullet is lodged in a stone wall, they are kept in place, as everyone agrees, by the compressive stresses created in the surrounding solid by the introduction of a foreign body. There is no reason for disregarding this explanation for the system of gas bubbles in metals. In some cases, bubble formation gave rise to visible deformations of the outside surface of the specimen; obviously, the high pressure in the bubbles was not balanced then by capillary pressure. In some other instances, the interface between the gas in the bubble and the surrounding metal appeared to be plane. That no capillary pressure acts across planes, is stressed already in Section III.4. The bubbles were, in general, too small to ascertain their shape, so that the sphericity implied in Eq. (61) remains a groundless hypothesis. An *experimentum crucis* would consist in producing bubbles in a transparent material of a high stress birefringence; the stress pattern in the vicinity of the bubble would then be directly visible; compare an analogous remark in Section III.1.

6. Capillary Pressure and Vapor Pressure

Numerous instructions on determining γ_s or the specific energy γ_{sl} of the melt – solid interface use the equation first derived by W. Thomson (Lord Kelvin) in 1871. One of its derivations follows.

When the vapor phase has volume V_1 and pressure p_1 and is in equilibrium with a liquid phase of volume V_2 and pressure p_2, then the differential of the free energy of the system is $dF = -p_1 dV_1 - p_2 dV_2$. The function $F + p_2 V_2$ also depends on the present state of the system only, so that $d(F + p_2 V_2) = -p_1 V_1 + V_2 dp_2$ is a complete differential. Hence

$$-\left(\frac{\partial p_1}{\partial p_2}\right)_{V_1} = \left(\frac{\partial V_2}{\partial V_1}\right)_{p_2}$$

When the volume dV_2 of the liquid evaporates, the volume of the vapor increases by dV_1; the two partial differentials refer to the same mass of substance. Thus $(\partial V_2/\partial V_1)_{p_2} = -\rho_1/\rho_2, \rho_1$ and ρ_2 being the densities of the two phases. Integration of the equation $(\partial p_1/\partial p_2)_{V_1} = \rho_1/\rho_2$ affords $p_1 - p_0 = (\rho_1/\rho_2)(p_2 - p_0)$. The pressure p_0 is that on both sides of a *plane* liquid surface. Pressure p_2 is different from p_0 whenever the liquid surface is curved. If its two principal radii of curvature are R_1 and R_2, then

$$p_2 - p_0 = \gamma\left(\frac{1}{R_1} + \frac{1}{R_2}\right) \tag{62}$$

Thus Thomson's equation

$$p_1 - p_0 = \frac{\rho_1}{\rho_2} \gamma \left(\frac{1}{R_1} + \frac{1}{R_2} \right) \tag{63}$$

is arrived at.

When the sum $(1/R_1) + (1/R_2)$ is positive, as in a drop, the vapor pressure is greater than above a plane surface; and when this sum is negative, as for a meniscus of a water column rising in a capillary tube, then $p_1 < p_0$.

Equation (63) was employed[126] to determine the γ_s of the end surface of potassium needles (whiskers), about 0.2 μm thick. These needles terminate in a "roof" of 3 inclined identical planes of the {110} indices. The planes are, of course, flat except for the unavoidable roughness. The author hypothesized, however, that one of the radii of curvature was equal to 1.22 l, if l is the length of each of the roof inclines. No credence can be given to this assumption, which is equivalent to stating that an atom (in the surface) situated, say, 5000 angstroms from the edge, vaporizes differently from one located, say, 6000 angstroms from it. With this unadmissible assumption, γ_s of K appeared to be about 170 erg/cm^2, when $p_1 - p_0$ was calculated from the rate of needle growth at different vapor pressures of potassium around it.

The derivation of Eq. (63), which can be found in textbooks, was repeated here to stress the fact that only curvature of the vapor — solid interface can cause differences between p_1 and p_0; no other length may arbitrarily be substituted for R_1 and R_2.

The triple point of a substance is reached when the vapor pressure of the solid phase is equal to that of the liquid phase. If both solid and liquid are subjected to external pressure (which may be caused by capillary forces), their curves of "vapor pressure versus temperature" lie above those for uncompressed phases and intersect at a temperature different from the triple point. The melting point T_m observed at atmospheric pressure, as a rule, is very near to the triple point. Thus the freezing temperature T_{mr} of a drop of radius r should be different from T_m.

The difference $T_m - T_{mr}$ is given by the Clapeyron-Clausius equation. Neglecting the variation of the latent heat of melting λ (erg/g) and of the derivative dp_1/dT in the narrow temperature range of interest,

$$T_{mr} - T_m = T_m \frac{v_2 - v_3}{\lambda} (p_1 - p_2); \tag{64}$$

v_2 and v_3 are the specific volumes of, respectively, the liquid and the solid. Hence,

$$T_{mr} - T_m = \frac{T_m \gamma_{sl}}{\lambda} \left(\frac{1}{R_1} + \frac{1}{R_2} \right) (v_2 - v_3) \tag{65}$$

Equation (65) is similar to Eq. (40). However, the former is, essentially, a consequence of the laws of thermodynamics, whereas Eq. (40) is a hypothesis and contains, instead of the radii of curvature, the thickness l which has no connection to curvature. Also, the difference between the densities of a liquid (ρ_2) and the corresponding solid (ρ_3) is small, so that the expression $(1/\rho_2) - (1/\rho_3)$ should have been used in Eq. (40) rather than $1/\rho_3$ alone. This error is rather common[127].

A modification of Eq. (65) was applied[128] to melt – solid interfaces. The treatment was restricted to interfaces of a single curvature, so that $1/R$ was used instead of the general term $(1/R_1) + (1/R_2)$. Let the interface profile be as indicated in Fig. 16, that is, at large values of x, the melt – solid boundary is flat and lies in the

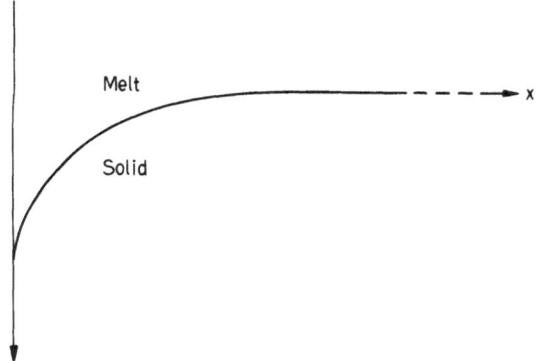

Fig. 16. Profile of a melt – solid interface. Far from the groove, the ordinate of the interface is $y = 0$

plane of $y = 0$. It is assumed further that T_{mr} is equal to Gy, G being an arbitrary constant. From this assumption, the radius R of curvature is a function of y such that

$$R = \frac{T_m \gamma_{sl}}{\rho_2 \lambda (T_m - Gy)} \tag{66}$$

The laws of capillarity lend no justification to this hypothesis.

As long as the shape of an interface is determined only by the capillary force, this shape can be an infinite plane, an infinite cylinder, a sphere, a catenoid, a nodoid, or an unduloid; this was found by Plateau over a century ago; see, *e.g.* Ref.[94], p. 8. Thus the shape assumed in Fig. 16 cannot exist unless other forces, such as gravitation, are operative, but no such force is considered in deriving Eq. (65). A second weakness of the theory lies in the use of $1/\rho_2$ instead of $(1/\rho_2) - (1/\rho_3)$; this error may raise the calculated difference $T_m - T_{mr}$ by a factor of, for instance, ten. A third weakness is, that the interface is implied to be perfectly smooth. Surface roughness can be detected even on precision – made ball bearings; thus the assumption of zero roughness on freely solidified solids is not convincing. A fourth defect is caused by the neglect of the crystalline nature of so many solids; a freely growing solid is likely to be bounded by crystallographic planes rather than by smoothly flowing curves. Finally, if the temperature T_{mr} really varies along the profile, then no thermodynamic equilibrium is possible; in this equilibrium, the temperature is identical in the whole system. Also no mechanical equilibrium is possible as long as the capillary pressure γ_{sl}/R is different in different points of the solid phase and, consequently, internal pressures are present. The material would tend to flow from the regions of a smaller R to those of a greater radius, and vacancies in the crystal lattice would move in the opposite direction.

Experimental Work. Experiments related to the theory of Ref.[128] are briefly reviewed here. Figure 17 is redrawn from Fig. 1(a) of Ref.[129]. A flat glass cell (or box), of only 0.25 mm inside thickness, is filled with water. A source of heat is attached to the right-hand wall, and a cooler to the left-hand wall. The ice — water interface would have been a vertical plane but, whenever a grain boundary (approximately perpendicular to this plane) is present, a groove, as shown in the sketch, forms. Assume, as a first approximation, that the temperature gradient G $(=dT/dy$, see Fig. 17) is constant in the box. Then the ice — water interface along the walls of the groove is at a lower temperature than that at, say, point P. If no supercooling

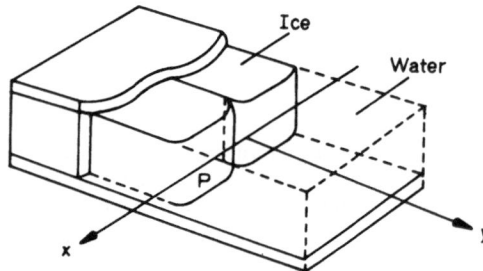

Fig. 17. An ice — water system with a grain boundary. Heat flows along the coordinate y from the right to the left. P is a point on the vertical interface of ice and water. Redrawn from Ref.[129]

occurs, the temperature at point $-y_1$ (situated in a groove wall) is by Gy_1 lower than that at P. Hence, $T_m - T_{mr} = Gy$. The radius of curvature of the groove walls can be determined from the curve $y = f(x)$, observed in a microscope. If the radius is R_1 at $-y_1$, then Eq. (66) can be used to calculate γ_{sl} because T_m, ρ_2 and λ are known from other sources.

The approximation of a constant G was dropped later[130], and different heat conductivities of ice and water were used to calculate the temperatures of the interface at various distances y; the convection in the liquid was, however, not taken into account. Moreover, the function $y = f(x)$ was believed to depend on the distance between two nearest grain-boundary grooves.

In the latest experimental work[131] not only the heat conductivities of ice and water, but also the latent heat of fusion were considered, but convection still disregarded. The importance of the refinements of the theory is clear from the comparison of the most recent value for the γ_{sl} of the ice — water interface, namely 29 erg/cm^2, with the early result[129] of $\gamma_{sl} = 41$ erg/cm^2. The probable limits of error were given as ± 9 erg/cm^2 in the early, and as ± 2 erg/cm^2 in the later paper; the former estimate appears to be too optimistic. For the interface of solid and liquid lead, $\gamma_{sl} = 76$ erg/cm^2 was calculated[130].

It is clear that all objections to Eq. (66) formulated above are applicable to these experiments, in spite of their ingenuity. Additional doubts are raised by the following details:

(a) The theory assumes that only one radius of curvature (R_1) is finite while the other (R_2) swinging in the vertical plane, is infinitely long. But the observation[129] illustrated in Fig. 18, redrawn from Fig. 1 (c) of the original, shows that also the vertical profile of the ice – water interface was curved, at least near the top and the bottom of the cell. It is difficult to believe, and was not proved, that R_2 was equal to infinity along the middle stretch of this profile; the vertical profile of every drop has finite radii of curvature at every point.

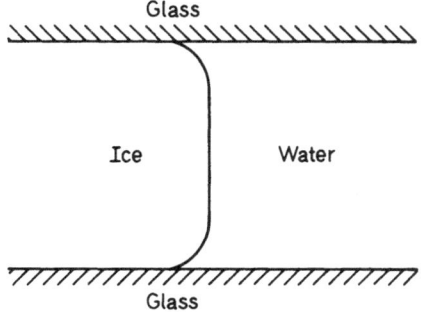

Glass

Ice   Water

Glass

Fig. 18. The vertical profile of the water – ice interface at a larger magnification. Redrawn from Ref.[129]

(b) Not only the above R_2 but also the curvature at the root of the groove is disregarded. Contrary to the roughly cylindrical interface, concave toward the solid, in the middle between the grooves or far from the only groove, the interface is convex on the ice side at the bottom of the groove, see point 0, Fig. 15. The horizontal radius of curvature at this point is much shorter than that near point P of Fig. 17; thus the root of the groove should have a much higher melting point than the T_m valid for a plane interface. The rapid evaporation from the groove bottom is mentioned in Section III.4; an analogous effect would be expected to occur also in the systems of this section but no reference to it could be found in the original publications.

(c) The radius of curvature R_1 in the deeper parts of the groove is at least as difficult to measure as the angle 2ψ discussed in Section III.4. In the work on lead, the result of which was mentioned above, the metal contained 0.04 weight-% Sb. The melt – solid system "was frozen rapidly". "The sudden accumulation of antimony adjacent to the interface at the start of solidification remains as a permanent 'marker' of the equilibrium shape"[130]. No experiments supporting this hypothesis are described by the authors.

The theory is simplified if the angle 2ψ is equal to zero[132]. This situation is almost impossible to prove and, from Eq. (56), the equality $2\psi = 0$ is possible only as long as

$$\gamma_1 = 2\gamma_{sl} \tag{67}$$

which can happen in exceptional circumstances only. If $\gamma_1 > 2\gamma_{sl}$, the depth h of the groove would indefinitely increase in time. Consequently, the relation

$$h^2 = \frac{2T_m\gamma_{sl}}{G\rho_2\lambda} \tag{68}$$

derived from (67) is not convincing; G is again the temperature gradient (degree/cm) across the phase boundary. However, succinonitrile ($T_m = 58.1°$) really gave values of h^2 proportional to $1/G$. From these observations, $\gamma_{sl} = 9 \pm 1$ erg/cm^2 was calculated.

If p_1 differs from p_0, as in Eq. (63), then the rate of evaporation of a small droplet should be different from that from a plane surface of equal extent. This reasoning was extended[133] to the vaporization of minute silver crystals (e.g., 600 angstroms "in diameter") in an electron microscope, that is in a good vacuum. During the evaporation the particles appeared faceted. Nevertheless, Eq. (63) was applied and a γ_s of about 1.2 J/m^2 at 1000 °K was obtained.

7. Capillary Pressure and Solubility

Solubility C g/cm^3 is fully analogous to vapor pressure, and the effect of surfaces on the C of solids has been pointed out long ago by Ostwald (1899) and Freundlich[134].

Suppose that a solution contains one small cube, of edge l_1 cm, and one large cube, of edge l_2 cm, both solids consisting of an identical material. When dm grams of the first dissolve and then precipitate on the second cube, the interfacial area of the first decreases by $12\,l_1\,(dl_1/dm)dm$, and that of the second increases by $12\,l_2\,(dl_2/dm)dm$ cm^2. The two derivatives are, respectively, $dl_1/dm = 1/3\,l_1^2\,\rho$ and $dl_2/dm = 1/3\,l_2^2\,\rho$, if ρ is the density of the solid. The total change in the interfacial area A associated with the above transfer is $12\,l_2\,(dl_2/dm)dm - 12\,l_1\,(dl_1/dm)dm$, that is

$$dA = \frac{4}{\rho}\left(\frac{1}{l_2} - \frac{1}{l_1}\right)dm$$

This dA is negative because l_2 is greater than l_1, thus, the free surface energy F_s of the system decreases when the isothermal transfer takes place: $dF_s = \gamma_{sl}dA$.

On the other hand, this event causes a decrease in the osmotic energy of the system. The osmotic pressure of the saturated solution around the small cube is $(RT/M)C_1$, M being the molecular weight of the solid substance, R the gas constant, and T the absolute temperature. Analogously, the pressure near the big cube is $(RT/M)C_2$, C_1 and C_2 being the two solubilities. When the amount dm is reversibly moved from the first to the second region, the decrease in the osmotic energy is $dF_0 = (RT/M) \ln (C_2/C_1)dm$. The two expressions dF_s and dF_0, referring to the same process, must be equal, so that the relation

$$\frac{RT}{M}\ln\frac{C_2}{C_1} = \frac{4\,\gamma_{sl}}{\rho}\left(\frac{1}{l_2} - \frac{1}{l_1}\right) \tag{69}$$

results. If correct, it would afford measurable values for the interfacial tension γ_{sl}.

Equation (69) is incorrect because, as stressed above, it is the curvature which affects properties such as vapor pressure and solubility, not the actual extent A of the surface (or interface) area. As long as the vapor – liquid interface is plane, these properties are identical for water in a shallow trough (large A) and in a narrow test tube (small A). Moreover, the argument used by Ostwald is based on the notion that all six faces of each cube take equal part in the dissolution and the deposition processes.

Imagine, on the contrary, that the two solids fill truncated cones, as illustrated in Fig. 19. Then, the exposed area A_1 of the small crystal increases when dm g dissolves from it, and the exposed area A_2 of the large crystal decreases when the dm g is deposited on it. The sum $dA = dA_1 + dA_2$ may easily be positive. It is true, that

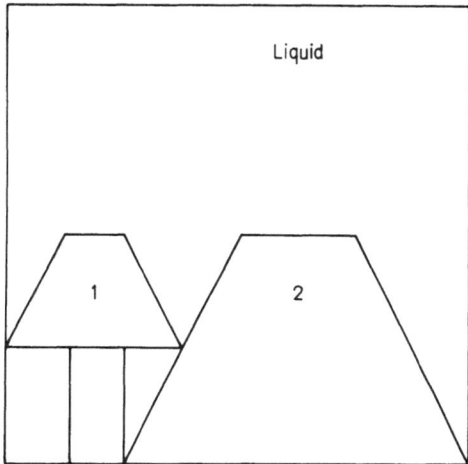

Fig. 19. The extent of the liquid – solid interface, as long as this is plane, has no effect on solubility. When the small solid 1, in a conical container, dissolves, its interface with the solvent increases. When a material is deposited on the large solid 2, the surface area of the latter decreases

also the interfacial area between the crystals and the container walls varies when the envisaged mass transfer takes place, but this is a small effect which, in general, cannot reverse the sign of the differential dA. Thus the dependence of solubility on surface area would appear to be a function of the shape of the solid; another improbable conclusion.

As soon as it is agreed that Eq. (69) is unjustified, two questions arise, namely: do phenomena predicted by it regularly occur in nature? If they do, how else can they be accounted for?

Coarsening of fine precipitates kept in the mother liquor for protracted time was observed in numerous instances: the number of very fine particless in the precipitate decreased in time, and the volume of several large particles simultaneously grew. This effect, after 1899, was frequently referred to as *Ostwald ripening*.

Quantitative studies of this ripening afforded a less clear picture. Many scientists attempted to determine the composition of slowly aging suspensions (containing particles of various sizes) by measuring the electric conductivity of the liquids.

The errors in this procedure, and the consequent uncertainty of the conclusions arrived at, were pointed out[135] in 1940. When a polarographic method was used instead, the ion concentrations proved to be identical in a solution in equilibrium with large crystals only, and in one containing both large and small particles. Precipitates of barium sulfate were observed for up to 14 years, but no change in the distribution of particle sizes could be detected[136]. Some other systems in which the coarsening was "extremely slow" are referred to in Ref.[137].

In spite of these findings, it cannot be denied that coarsening of precipitates is a common process. In all probability, it is caused by an equally common rule, according to which very small solid particles are less perfect (or even more defective) than larger particles of the same kind. When small crystals are obtained by crushing or grinding, the defects (dislocations, vacancies etc.) in them (per unit volume) are very numerous; cutting a metal foil can increase the frequency of dislocations in it to over 10^{11} per square centimeter[138]. When the method of preparation is by "precipitous" precipitation (for instance, by mixing two highly concentrated solutions), again very imperfect crystals are formed, contrary to large particles obtained by pouring together two dilute solutions. In the systems of the latter kind, only relatively few nuclei emerge and then grow by nearly regular capture of additional ions or molecules.

In disordered lattices, every ion or molecule is attracted to the rest of the particle less firmly than a corresponding unit belonging to a perfect lattice. Hence, the former escapes into the solution more easily than the latter. Careful experiments to test this explanation would be welcome.

Ostwald's method of estimating the γ_{sl} of liquid — solid interfaces is not popular at present. Consequently, only two relevant papers are reviewed here.

Different silica sols were prepared[139] by heating the initial colloidal solution of $SiO_2 + n\,H_2O$ at 80 °C for different lengths of time. The author points out that particles polymerized at 80–100 °C "consist almost entirely" of amorphous SiO_2, whereas those obtained at lower temperatures "appear to contain appreciable amounts of silanol groups" in their internal structure. Nevertheless, the whole difference in the solubilities of different particles was attributed solely to that of their specific surface areas. Thus the specific energy of the SiO_2 — water interfaces was calculated to be about 10^{-6} cal/cm^2, that is near 42 erg/cm^2.

Aescin, a saponin-related acid, was poorly soluble in water, but became readily soluble after thorough grinding[140]. Grinding had also other effects: the X-ray diffraction pattern of the particles changed from sharp (characteristic for good crystals) to diffuse (as from a distorted lattice). The melting point was lowered from 225 °C to 221 °C. The curves of the differential thermal analysis were grossly altered by comminution. Nevertheless, the change in solubility was relied upon to compute the γ_{sl} of the interface of aescin and its aqueous solution; the result was about 9 erg/cm^2. The γ_s of aescin (in air) was said to be near 45 erg/cm^2. — No experimental study of the Ostwald ripening after 1968 could be found by the present reviewer.

Coalescence of drops is caused by capillary pressure. Analogously it may be expected that sintering of metals, which is an important industrial process, also would be governed or at least influenced by the surface energy of the metal particles. An instance of the theoretical treatment of this problem can be found in Ref.[141].

8. Wetting

This group of methods is based on Young's equation of wetting, *i.e.*, Eq. (1) of the Introduction above. Two proofs of this relation are known, one dealing with forces, and the other, with energy.

(a) The force proof is indicated in Fig. 1. The two main assumptions, namely that γ_s and γ_{sl} exist as tensions, are rendered suspect by the existence of surface roughness on solids and its persistence for many millenia; if tensions, analogous to the surface tension in liquids, really were operative along surfaces and interfaces of solids, these would have been as smooth as liquids are. The positive capillary pressure acting on every hill on the surface would flatten the hill, and the corresponding negative P_c under every valley would raise the bottom of the latter.

It is mentioned in the Introduction that Young himself noticed that the component $\gamma \sin \theta$ remained unbalanced in his theory. This defect was re-discovered[142], and several experiments[143, 64, 144, 145] showed, that a ridge is raised along the 3-phase line, in which the vapor, the liquid, and the solid meet. This is true for mercury on gelatin, liquid tin on molybdenum, sodium borate melts on iron, and so on. Unfortunately, the precision of the experiments thus far (the height of the ridge usually is less than one micron) is insufficient to decide whether the operative force (per unit length) is equal to the component $\gamma \sin \theta$ or to the whole tension γ. The second alternative is preferred in Ref.[94], p. 242.

Whatever the force raising the ridge, the existence of the latter disproves Eq. (1). It is supposed in the "force proof" that γ_s and γ_{sl} both act along one straight line. In reality, the directions of γ_s and γ_{sl}, as indicated in Fig. 20, may form an angle estimated as $168°$ in one, and $177°$ in another system, rather than $180°$ postulated by Young.

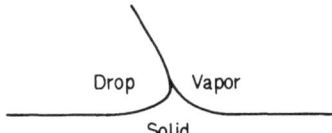

Drop Vapor

Solid

Fig. 20. Surface tension of a liquid drop raises a ridge on the supporting solid

(b) The energy proof starts from the equation

$$dF = \gamma dA + \gamma_s dA_s + \gamma_{sl} dA_{sl} \tag{70}$$

for the free energy F of the system in which A, A_s and A_{sl} are the areas of the vapor – liquid, vapor – solid, and liquid – solid interfaces. This expression is incomplete because the free capillary energy of a system depends, in general, also on the curvature of the liquid surface (and on the curvatures of the other two boundaries, if γ_s and γ_{sl} are real quantities). If, however, Eq. (70) is accepted as true, and if a movement of the 3-phase line along the solid surface is considered, such that $dA_s = -dA_{sl}$, then $dF = \gamma dA + (\gamma_{sl} - \gamma_s) dA_{sl}$. If it is further assumed that $dA = dA_{sl} \cos \theta$, then

$$\frac{dF}{dA_{sl}} = \gamma \cos \theta + \gamma_{sl} - \gamma_s \qquad (71)$$

In equilibrium, $dF/dA_{sl} = 0$, and Eq. (1) results.

Unfortunately, dA is equal to $dA_{sl} \cos \theta$ only in some, but not in all geometries, even when roughness is disregarded. When the drop is a part of a sphere (*i.e.*, the gravitational effects are neglected), and its perimeter expands or contracts a little, the changes in A and A_{sl} agree with the hypothesis. On the other hand, when the meniscus rises in a cylindrical capillary, the vapor − liquid area remains constant as long as the contact angle θ is constant, whereas A_s and A_{sl} continuously vary, and dA_s remains equal to $-dA_{sl}$.

Moreover, the formation of a ridge along the drop perimeter demonstrates that Eq. (70), which neglects the changes in curvature, neglects also the deformations of the solid, which often appear to be of significant dimensions.

Usually it is easy to determine the value of angle θ in a given system but too often this angle does not correspond to an equilibrium. This is caused by the effect known as hysteresis of wetting and which cannot be described here. However, an example of the error caused by this hysteresis, when calculating the γ_s of a metal, may be found in Ref.[146].

Thus Eq. (1), in spite of its venerable age and its wide popularity, must be considered a wrong guess. In addition, it contains two unknowns (γ_s and γ_{sl}) and, consequently, can be used only to calculate the difference $\gamma_s - \gamma_{sl}$ but not these two quantities separately. To find γ_s and γ_{sl}, each for itself, more hypotheses are needed.

In many instances, $\cos \theta$ appears to be a linear function of the γ of the liquid, so that

$$\cos \theta = 1 - b(\gamma - \gamma_c); \qquad (72)$$

b and γ_c are empirical constants, and γ_c is referred to as the "critical surface tension". Elimination of γ from Eqs. (1) and (72) leads[147] to the relation $\gamma_{sl} = \gamma_s - \gamma_c \cos \theta - (\cos \theta/b) + (\cos^2 \theta/b)$. If γ_{sl} is supposed to be a function of $\cos \theta$, whereas γ_s is supposed to remain constant, the $\partial \gamma_{sl}/\partial \cos \theta = -\gamma_c - (1/b) + (2 \cos \theta/b)$. This derivative is equal to zero when $2 \cos \theta = b \gamma_c + 1$. The value of γ_{sl} corresponding to this minimum of γ_{sl} (as a function of $\cos \theta$) is $\gamma_{sl} = \gamma_s - 0.5 \gamma_c (b \gamma_c + 1) - [(b \gamma_c + 1)/2b] + [(b \gamma_c + 1)^2/4b]$. If now it is assumed that $\gamma_{sl} = 0$ at this point, then γ_s can be calculated from the equation

$$\gamma_s = 0.5 \gamma_c (b \gamma_c + 1) + [(b \gamma_c + 1)/2 b] - [(b \gamma_c + 1)^2/4 b]$$

Thus $\gamma_s = 19.5$ erg/cm^2 was found for poly(tetrafluoroethylene). For further development of these hypotheses see Ref.[148].

An assumption of a comparable audacity was published in Ref.[149]. A drop covers the whole top surface of a truncated cone, as indicated in Fig. 21. Again, V, L and S refer to the vapor, the liquid, and the solid phases. It is proposed that, in this instance, the tension γ_s acting along the inclined slides of the cone does not affect the value of θ. Thus the latter is determined only by the two tensions operating in horizontal directions, that is, γ_{sl} (pulling point M to the right) and the horizon-

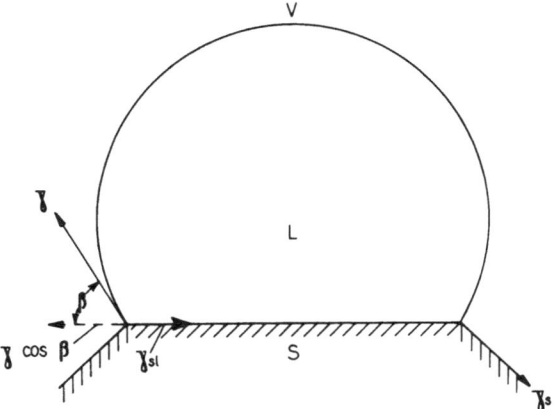

Fig. 21. If surface tension γ_s of the inclined planes of a solid (on which drop L rests) had no effect on the contact angle θ, then the value of θ would be determined by two tensions only, namely $\gamma \cos \theta$ and γ_{sl}

tal component $\gamma \cos \beta$ pulling it to the left. Thus $\gamma_{sl} = \gamma \cos \beta$ in the absence of motion. On the other hand, if an identical drop rests on a plane surface chemically identical with the above S, then Eq. (1) is said to be valid. Thus the rule $\gamma_s = \gamma (\cos \beta + \cos \theta)$ is arrived at, in which, in principle, all quantities except γ_s are directly measurable. Unfortunately, as already Laplace has noticed, the angle β, that is the contact angle at the edge of a solid, is indeterminate. At present, this uncertainty is attributed to the hysteresis of wetting. Consequently the values of γ_s calculated by the author (*e.g.*, 130 erg/cm^2 for the cube face of potassium chloride) cannot be considered trustworthy, even if the underlying theory were less objectionable.

Experimental Work. An experimental determination of γ_1 of ice was based[150] on measurements of the dihedral angles, see Section III.4. Figure 22 is a cross section of a joint between two ice crystals in their vapor; the dihedral angle again is 2ψ. Figure 23 is an analogous profile of this joint when the crystals were immersed in

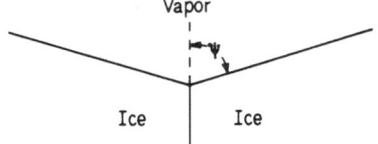

Fig. 22. The hypothetical tension between two ice grains is supposed to be balanced by the tensions acting along the vapor – ice boundaries. The dihedral angle is 2ψ

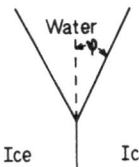

Fig. 23. The hypothetical tension between two ice grains is supposed to be balanced by the tensions acting along the water – ice interfaces. The dihedral angle is 2φ

49

liquid water; here the dihedral angle is 2φ. If we assume that γ_1, *i.e.*, the tension along the interface between the two grains, is not affected by the change in the medium from vapor to water, then $\gamma_s = \gamma_1/2\cos\psi$, see Eq. (56), and

$$\gamma_{sl} = \frac{\gamma_1}{2\cos\varphi} \tag{73}$$

If these two equations are introduced in Eq. (1), then the result is

$$\gamma_s = \gamma \frac{\cos\varphi, \cos\theta}{\cos\varphi - \cos\psi} \tag{74}$$

The accuracy of the numerical results obtained depends, in addition to that of the assumptions made, also on the correctness of the experimental values of the angles ψ and φ. For the determination of ψ, a formvar replica of the region pictured in Fig. 22 was prepared, then removed and coated with silver; the ψ was measured interferometrically on this coating. The reliability of the replica method recently was questioned[151]. Also the authors[150] state that the values obtained by them for ψ were considerably greater than the true values determined by cutting a razor blade at precisely 161° "and observing it under the interference microscope". It is regrettable than many other experimenters omitted checking their precision in this or another manner. After correction, ψ was said to be $72 \pm 1°$ so that only an error of $\pm 1°$ was admitted. The value of φ appeared to be $10 \pm 5°$, that is, the error conceded was much greater. The θ was estimated to be about 1°. The calculated γ_s was near 110 erg/cm², and γ_{sl} near 33 erg/cm². The effect of the local melting of ice and the local freezing of water on these results is impossible to estimate.

A similar scheme was resorted to later[152]. The measured angles were those of grain boundaries of UO_2 in argon (ψ), those of identical grain boundaries in molten nickel (φ), and the contact angle made by a drop of nickel on solid UO_2 in argon at 1500 °C. The spread of the ψ values was 149–158°, and of the φ angles 144–180°. The γ_s of uranium oxide in argon seemed to be 750 ± 150 erg/cm², and the ratio γ_1/γ_s was near 0.45.

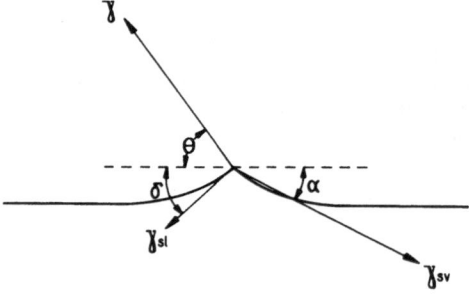

Fig. 24. Supposed equilibrium at a ridge raised by surface tension on a solid. The angles between the horizontal plane (parallel to the main solid surface) on one hand, and the directions of γ, γ_{sl}, and γ_{sv} on the other hand are, respectively, θ, δ, and α

A little more elaborate was the "multiphase equilibria" approach, adopted earlier in Ref.[64]. Equations (56) and (73) are still used, but Eq. (1) is not because direct observation proved that the drop (*e.g.*, of liquid tin) raised a ridge on the solid (*e.g.*, molybdenum) so that the vertical cross section of the system looked like Fig. 24, see also Fig. 20. Let the angles formed by the tension γ and the hypothetical tensions γ_s and γ_{sl} with the horizontal plane be, respectively, θ, α, and δ. As a correction, symbol γ_{sv} was substituted for γ_s because the energy of the vapor — solid interface presumably was influenced by the vapor of the liquid (*i.e.*, molten tin). Of course, the γ of tin melts presumably also was altered by dissolution of molybdenum, but this possibility was disregarded. For equilibrium of all horizontal forces, the equality

$$\gamma_{sv} \cos \alpha = \gamma_{sl} \cos \delta + \gamma \cos \theta \tag{75}$$

is required. On the other hand, from Eqs. (56) and (73) writing γ_{sv} for γ_s, the equation

$$\gamma_{sl} = \gamma_{sv} \frac{\cos \psi}{\cos \varphi} \tag{76}$$

results. Since γ and all angles are measurable, Eq. (75) and (76) permit calculation of the two unknowns (*i.e.*, γ_{sl} and γ_{sv}).

It is clear that the above mode of computation suffers from the defect pointed out at the start of this section, in the discussion of the "force proof" of the Young equation (1). If the forces are supposed to cancel each other along a vertical, rather than a horizontal, projection, then the relation obtained

$$\gamma \sin \theta = \gamma_{sl} \sin \delta + \gamma_{sv} \sin \alpha \tag{77}$$

is different from Eq. (75) but, on the face of it, is just as legitimate as the latter.

The values for the γ_s (or γ_{sv}) of Mo at 1600 °C, derived from Eqs. (75) and (76), ranged from 940 to 1990 erg/cm^2. For γ_1, supposed to be independent of whether the grain boundary ended at a vapor or a liquid phase, an even greater spread was found, namely 390 to 1100 erg/cm^2 in 8 experiments. In this series, the ratio γ_1/γ_s exceeded 0.5, that is, was unusually high, compare Section III.4.

The "multiphase equilibrium technique using silver as the liquid phase" was resorted to in Ref.[153], but the details were not given. The γ_s of stainless steel containing $0.001 - 0.006\%$ boron appeared to be near 880 erg/cm^2 at 1050 °C, and the γ_1 again was unusually high, namely about 620 erg/cm^2.

Four equations and 5 independent angles have been used[154] to estimate the γ_s of the uranium carbide of the composition near to UC. In addition to Eq. (56) supposedly valid for the point at which a boundary between two UC grains emerges toward an argon atmosphere, also the relation

$$\gamma_s^* = \frac{\gamma_1}{2 \cos \psi^*} \tag{78}$$

was employed, in which γ_s^* was the specific free energy of the interface between solid UC and saturated uranium vapor and ψ^* was the corresponding angle; γ_1, on the contrary, was not allowed to change when uranium vapor was added to argon. The third relation needed is Eq. (73): the carbide was immersed in liquid uranium metal. The fourth was a "compromise" between the Young equation (1) and Eq. (75) in that $\cos \alpha$ of the latter was implicitly equated to unity, so that

$$\gamma_s^* = \gamma_{sl} \cos \delta + \gamma \cos \theta \tag{79}$$

Here again, only the horizontal components of the tensions cancel each other, while the vertical components are left to shift for themselves. The 4 equations solved for the surface tension of UC afford

$$\gamma_s = \gamma \frac{\cos \theta \cdot \cos \varphi \cdot \cos \psi^*}{\cos \psi \, (\cos \varphi - \cos \delta \cdot \cos \psi^*)} \tag{80}$$

all quantities in the right-hand side of this equation are, in principle, directly measurable.

Unfortunately, the system selected was not in physico-chemical equilibrium: uranium carbide was gradually going into solution in liquid uranium, and "intergranular penetration and erosion of UC occurred to an extent which increased with temperature" that was varied between $1180°$ and $1720\ °C$. No wonder then, that the θ observed in different tests was as low as $37°$ and as high as $110°$. Also for the surface tension γ of liquid uranium values from 780 to 1510 dyne/cm have been obtained at a constant temperature of $1600\ °C$. The γ_s of UC in argon was estimated to be approximately 730 erg/cm^2 at $1325\ °C$, and γ_{sl} (at the boundary $U - UC$) appeared to be near 140 erg/cm^2 at $1100\ °C$.

The angle δ, see Fig. 23, was treated as negligibly small in a following paper[155] so that Eq. (79) reverted to Eq. (1). Thus only 4 angles ($\theta, \varphi, \psi, \psi^*$) were needed to solve the four equations appearing in[154]. The systems studied were those of UO$_2$ in argon (ψ), in argon + vapors of copper (ψ^*), in liquid copper (φ), and of UO$_2$ on which a drop of molten copper was resting. At temperatures below $1500\ °C$, the θ remained near $100°$, ψ and ψ^* were near $142°$, and φ was about $158°$. The γ_s and γ_s^* values were 350 erg/cm^2 at $1400\ °C$, and the γ_{sl} (i.e., the interfacial tension between UO$_2$ and liquid copper) was about 560 erg/cm^2.

9. Miscellaneous Methods

Nucleation. As explained in Section III.6, small droplets have a higher vapor pressure than plane surfaces of an identical composition at an identical temperature. The difference between the two pressures is given by Eq. (63). This effect renders condensation of vapors more difficult than it would have been at $\gamma = 0$.

A vapor mechanically separated from the liquid phase and rapidly cooled (for instance, by expansion in a cloud chamber) to a temperature T_1, at which its pressure is p' and greater than the saturated vapor pressure p_0 above a plane liquid surface at $T_1{}°$, in many instances still experiences no nucleation, that is, no formation of liquid phase occurs. Condensation would give rise to minute droplets; and it takes place spontaneously only as long as the pressure p' is not only greater than p_0 but also exceeds p_1, this p_1 being again the pressure of the vapor in equilibrium with the liquid nucleus formed.

When formation of condensation nuclei just starts, we may assume p' to be only a little greater than p_1. If the radius of the nucleus is known, the γ can be computed from Eq. (63). Usually, however, the γ in this equation is supposed to be equal to the surface tension of the liquid in bulk also for the smallest drops. This supposition was rejected by several scientists; see, e.g., Ref.[94], p. 63. Thus, application of W. Thomson's equation even to the formation of liquid droplets is beset with difficulties.

The situation is even more uncertain when the nucleus is solid. A solid particle of, perhaps, 10 or 100 atoms or molecules has no reason to be spherical or to possess definite radii of curvature (R_1 and R_2); hence, utilization of Eq. (63) in the vapor — solid systems is not feasible.

However, at least one attempt was made on calculating the free surface energy of a solid condensation nucleus[156]. The material was argon, and the interatomic potentials were calculated using the Lennard-Jones approximation. The above energy of a cluster containing n atoms came out to be $A \gamma_s n^{2/3} + Bn^{1/3}$. The values of the constants A and B were, but the magnitude of γ_s was not calculated.

As would be expected, it is even more difficult to asses the importance of γ_s for heterogeneous nucleation (i.e., formation of nuclei on other solids). Only one relevant reference is quoted here[157].

Apparently, the direct transition from vapor to solid is less common than the double transition vapor — liquid — solid, see, e.g., Refs.[158–160]. From the rate of solidification of metal droplets (average diameter near 0.005 cm) at temperatures $60°$ to $370°$ below their normal melting points, the γ_{sl} was concluded[158] to be, for instance, 24 for mercury, 54 for tin, and 177 erg/cm^2 for copper. For this calculation it was necessary to assume that each crystal nucleus was a perfect sphere embedded in the melt droplet; the improbability of this model was emphasized above.

When a similar theory (which appears objectionable to the present reviewer also on other grounds) was applied to the formation of ice in water droplets[160], "the critical nucleus < was > assumed to be a hexagonal prism of height equal to the short diameter". No capillary pressure acts across plane faces of a prism. Nevertheless the author found a value (for the γ_{sl} of water — ice) near 20 erg/cm^2 for drops of about 0.002 cm in diameter at $-37\,°C$.

Electric Effects. The science of electrocapillarity exists for over 100 years. When a liquid surface is electrically charged, its tension changes, and these changes can be followed quantitatively with the standard methods of measuring γ. It is tempting to apply the notions of electrocapillarity to solid surfaces, and a recent book[162] is a particularly rich example.

Figure 25 indicates the principle of one of the arrangements used. A metal ribbon 1 is loaded with a weight 2 and suspended from a membrane 3; it is surrounded by the electrolytic solution 4. When alternating current passes through the ribbon,

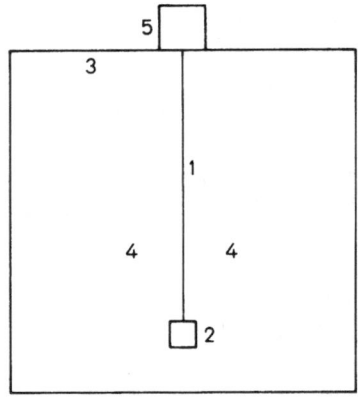

Fig. 25. One of the arrangements used to demonstrate the effect of electric charge on the γ_s. 1 is a metal ribbon loaded with weight 2, suspended on membrane 3, and immersed in an electrolyte solution 4. The vibrations are recorded by piezoelectric indicator 5. Simplified from Ref. [162]

the latter vibrates; the vibrations are picked up by the piezoelectric indicator 5 and suitably displayed on an oscillograph.

The author believes that these vibrations are caused by changes in the γ_{sl} of the electrode, produced by charging and discharging it, but no convincing proof of this belief could be found in the book by the reviewer. The method is not supposed to afford absolute values of the γ_{sl} but, unfortunately, also the derivatives $d\gamma_{sl}/dq$ and $d\gamma_{sl}/d\varphi$ are not given in customary units; q is the surface density of charge on the ribbon, and φ is the potential of the latter.

An older method may be mentioned here although it also never led to numerical values of γ_s. If a single crystal of zinc, cut to a wire, is axially extended, the force needed for this deformation is reduced when the wire is electrically charged [163].

IV. Methods of Measurement and the Basic Notions

A perusal of the foregoing chapter demonstrates, it is believed, that no procedure suggested so far for determining the γ_s or γ_{sl} is valid. It may be worth while to summarize here the main reasons for this failure.

The most common cause of it is the neglect of 3-dimensional effects as compared with those in two dimensions. Thus, all stresses in a loaded wire or ribbon are disregarded in the shrinkage method, Section III.1. The work of deformation leading to rupture is a bulk effect which does not receive its due consideration in the calculation of fracture energy, Section III.3. Bulk deformations associated with thermal etching, Section III.4, demand more attention than was alloted to them by many scientists. The method of bubbles, Section III.5, is invalid both because of the above neglect (that is, that of the volume stresses around the bubble) and because of another popular error, namely an erroneous treatment of capillary pressure P_c.

The local value of P_c depends on the local curvature and has no direct relation with the total surface area. Disregard of this rule invalidates the theories of Sections

III.2, III.6, and III.7. Solids tend to be crystalline; crystals in equilibrium tend to have plane surfaces; and P_c across these is always zero. The vapor pressure of perfect crystals is independent of their dimensions. What is true for vapor pressure, is true for solubility also.

Surface roughness is a third ubiquitous source of error of the approaches summarized in Chapter III. It will be recalled again in a subsequent paragraph.

The methods based on wetting, Section III.8, all are undermined by the proved faultiness of the fundamental Eq. (1). Some of them employ additional hypotheses unlikely to be true. It is difficult to avoid the impression that, the more elaborate the experimental work, the more doubtful hypotheses are needed to evaluate the results of the measurements.

The conclusion, that none of the known methods of measuring γ_s (or γ_{sl}) is satisfactory, in spite of the ingenuity and the industry spent, raises the question: perhaps our failure is explained simply by the fact that the quantity we are seeking does not exist? The preferred answer, be it *yes* or *no*, probably would be erroneous, as long as the physical meaning of γ_s (and γ_{sl}) is not defined more precisely. Chapter V reviews the theoretical and experimental studies concerning the *cuticular energy* of solids, which undoubtedly exists, so that the questions here may be formulated more exactly, namely: do solids have a surface tension analogous to that of liquids? do they possess a surface energy of the kind known in liquids?

In liquids, the work needed to extend the surface is the work required to pull the necessary number of particles from the symmetric into the asymmetric field. This reasoning appears to be immediately applicable to solids also, but difficulties arise as soon as the problem is examined with more diligence.

As has been repeatedly stressed in this review, all solid surfaces are rough. Let the irregular line in Fig. 26 represent the profile of a vapor — solid interface, greatly enlarged. An atom in position A is almost surrounded by the gas phase; consequently,

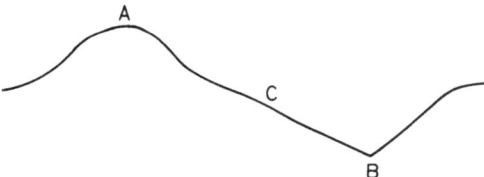

Fig. 26. Profile of a rough solid surface. The work of bringing a molecule from the bulk to position A is much greater than that valid for position B; and for position C the work has a middle value

the work of transporting it from the bulk to its present site was large. On the other hand, the work of moving an atom from an internal position to position B was small. Hence, the γ_s would be much smaller at B than at A, and have a middle value at, say, C. Equilibrium of these different tensions is impossible without involving stresses and strains in the volume of the solid; but then the idea of *surface* tension ceases to be useful or justifiable.

By mechanical necessity, surface tension along a curved surface entails capillary pressure P_c normal to the surface. The capillary pressure at A would push the hill around A down, and the capillary pressure at B would push the valley bottom up. No equilibrium is possible in a system of these pressures, and the persistence of roughness proves that the pressures, if they really exist, are counteracted by other forces which cannot be left out when discussing the γ_s.

Ideal crystals are perfectly smooth, but the above difficulty is present in them as well. Ideal crystals have extremely sharp edges, whose radius r of curvature (swinging in a plane perpendicular to the edge) may be as short as 1 or 2 angstroms (while the other radius is equal to infinity). Hence, P_c acting on the edge and equal to γ_s/r, would be $1000/10^{-8}$ g/cm $\cdot$ sec^2 if the value of 1000 g/sec^2 is accepted for the γ_s. Such a pressure (10^{11} g/cm $\cdot$ sec^2 or 10^5 atmospheres) would crush most crystals.

An additional difficulty was introduced many years ago by the hypothesis that different crystal faces possess different surface tensions. If, for instance, the prism face has a γ_s greater or smaller than the γ_s of the basis, then the edge between the two is pulled in two different directions by two unequal forces in the absence of any other force which could stabilize the system. It is usual, however, to hide this defect by stating that a crystal face has its characteristic surface *energy*, that is, admitting that γ_s is not analogous to γ.

The belief in a close similarity between solid and liquid surfaces is weakened also by the common observation that the surface layers on typical solids are thicker and less uniform than the corresponding layers in typical liquids. The exterior stratum (whose properties differ from those in the bulk) has no definite thickness τ in liquids, but the order of magnitude of τ usually is believed to be 10 angstroms for liquids ranging from argon to water. In solids, this τ may reach several microns. A glass exposed to moist air usually is covered with a degraded region of a low refractive index, whose τ may exceed 1 μm. The hardness of a machined metal is greater in the external 25 μm than in the deeper strata. Numerous additional examples are listed in Ref.[94], pp. 184, 187. If a metal plate is surrounded with an oxide of variable composition, it is impossible to allocate the (supposedly) measured γ_s to any of the substances present there.

V. Cuticular Energy

1. General

The surface layers of solids usually differ from the deeper zones of the same specimen in their chemical composition, their degree of lattice perfection (*e.g.*, the frequency of dislocations), their state of stress, and so on. This renders unpalatable the notion of a surface tension in solids, but suggests the existence of a kind of surface energy, unknown in liquids, which it was proposed to designate as cuticular energy.

A particularly simple example would be the energy of a sample of aluminum metal in air. The surface of the metal is covered with a continuous layer of aluminum oxide which may be less than 20 angstroms in thickness in a dry atmosphere but will be much thicker in moist air or after contact with hot water. Consequently, the energy of the specimen is the sum of two different terms, one referring to the metal-lic bulk, and the other to the oxide coating. The difference between cuticular and true surface energy will be clear after comparing Eqs. (81) and (82) with Eqs. (83) and (84).

A cube of aluminum, of 1 cm edge, is dissolved in a suitable solvent, and the heat Q_0 of solution is carefully measured. If the oxide coating was 10^{-7} cm thick, this cube consisted of 6×10^{-7} cm^3 of Al_2O_3 and 1.000 cm^3 of metal. An identical cube is cut into 10^{12} cubes, the edge of each being 10^{-4} cm long, and the powder obtained is dissolved in the above solvent. The heat of solution is Q_p. Its magnitude is different from that of Q_0 above all simply because the powder, prepared in air, contained $10^{12} \times 6 \times 10^{-8} \times 10^{-7}$ cm$^3 = 6 \times 10^{-3}$ cm^3 Al_2O_3 for each 0.995 cm^3 of metal (after accounting for the different densities of the two substances). Hence, if q_0 is the heat of dissolution of 1 cm^3 of Al, and q_1 is that of 1 cm^3 of Al_2O_3, then, with sufficient precision,

$$Q_0 = q_0 + 6 \times 10^{-7} q_1 \tag{81}$$

and

$$Q_p = 0.995 \, q_0 + 0.006 \, q_1 \tag{82}$$

The two measurable values are significantly different.

The above example suffices to formulate the 3 main differences between the true surface energy (as in liquids) and the cuticular energy.

(a) The specific cuticular energy (joules/m^2 or erg/cm^2) is a function of the specimen rather than of a substance as such. The thickness of the oxide film on aluminum metal varies from sample to sample. A liquid has a specific surface energy characteristic of the substance.

(b) Cuticular energy usually is not an equilibrium property. Oxide films exist on aluminum only as long as the equilibrium with oxygen of the air has not been attained; in equilibrium practically no metal is left. Surface energy of liquids exists, and commonly is measured, when the vapor – liquid system is in a stable equilibrium.

(c) The third distinction is perhaps the most important. Surface energy in liquids implies surface tension, that is, limitless tendency to contract. No such tendency is caused by cuticular energy. The oxide layer on aluminum may be in a state of tensile stress, but this stress would disappear after a very small contraction. Generally speak-ing, a compressive stress in the oxide film is just as probable as a tensile stress. If a cube of aluminum (with its oxide coating) is rolled into a thin sheet in the absence of oxygen, the area of the specimen would increase manifold but the total cuticular energy may, in principle, remain constant because the amount of Al_2O_3 did not change.

Usually, however, the other sources of cuticular energy (C.E.), referred to at the start of this section, also must be taken into account.

Now it is possible to discuss the question: is the calculated surface energy U_s or "γ", reviewed in the second chapter of this article, more similar to the true surface energy γ (as in liquids) or to the C.E.? It appears that the similarity between U_s and C.E. is the winner.

The energy excess possessed by a broken, as compared with an unbroken, crystal exists because the atoms (or ions, or molecules) at the rupture surface are attracted by the solid stronger than by the vacuum. This field of force causes re-arrangement of the particles but produces no surface tension.

It was shown in some calculations[164] that the external layer of ions in ionic crystals (or atoms in noble gases) is "unnaturally" spread or contracted over the fracture surface, so that its energy would be lowered by a little closer (or less close) mutual approach of the atoms or ions (in the plane parallel to the fracture surface). This effect, if real, still is completely different from those observed in liquids. The maximum lowering of the energy of the above distorted layer would be achieved by a small change (about 1%) in the average distance between the atoms (ions); further extension or contraction would raise the energy instead of lowering it. In liquids, surface tension tends to contract the surface area incessantly.

The first two peculiarities of the C.E. also seem to be encountered in the U_s. The excess energy of a real crystal fragment depends on how this was obtained; usually, the fracture surface is imagined to be perfectly smooth, but real surfaces are rough, have steps, and so on. The surfaces assumed in theoretical calculations are not in contact with any other substance. If a gas, however, is admitted, then the "unsaturated valencies" will be "saturated" with adsorbed gas molecules, and the asymmetry of the field will be reduced. Thus the structure calculated for the external layer may be in equilibrium only in excellent vacuum, and the duration of this equilibrium would depend on how rapidly gases and vapors leak into the evacuated vessel. This remark shows, by the way, how illogical are the attempts to correlate the experimental estimates of γ_s or "γ" with those calculated from the theories of Chapter II.

C.E. presumably affects also some experimental results on the γ_s. In particular, the effect of particle size on vapor pressure and solubilities, Sections III.6 and III.7, are related to C.E. more than to the true surface energy; as already pointed out, small particles usually have a less perfect crystal lattice than do larger crystals. A more direct estimate of C.E. is afforded by the measurements of the heat of dissolution.

2. Heat of Solution and Dispersity

In this section, mainly older results are quoted because the method seems to be out of fashion at present. The principle of the method is outlined in the foregoing section: the Q_0 of large, is compared with that, Q_p, of many minute particles. If the total, as distinct from the free, surface energy of the solid is U_s, see Eq. (9) in Section II.2, per unit area, then the corresponding energy of a large crystal is U_sA_0, and that of the n powder particles obtained from an identical specimen is nU_sA_p, A_0 being the surface area of the big crystal, and A_p the average surface area of a powder particle. Of course, nA_p is much greater than A_0, and in many systems A_0 may be disregarded in comparison with nA_p.

If, however, all terms are considered, then Q_0 is supposed to be

$$Q_0 = Q_\infty + U_sA_0 \tag{83}$$

if Q_∞ denotes the heat of solution of a specimen with zero surface. The analogous equation for Q_p is

$$Q_p = Q_\infty + nU_sA_p \tag{84}$$

The surfaces vanish during dissolution, and the energy which resided in them must be a part of the experimental heat. The contrast between the Eqs. (81) and (82) on one hand, and Eqs. (83) and (84) on the other hand renders the distinction between the two approaches (*i.e.*, C.E. versus surface energy) particularly evident.

In perhaps the earliest experimental work on the subject[165], fine-dispersed sodium chloride was prepared by sublimation and rapid cooling. In a microscope, the particles appeared to be, on the average, 1.3 μm "in diameter", corresponding to a specific surface area of 2.14 m^2/g. Unfortunately, the shape of the particles necessary to establish this correspondence is not specified in the original. The Q_p differed from the Q_0 by only about 1%, and the calculated value of U_s was approximately 400 erg/cm^2. In a subsequent paper[166], the product nA_p was estimated also from the number of particles in 1 g of powder, and the U_s appeared to be between 360 and 400 erg/cm^2. In the third publication[167], NaCl powder was obtained by grinding in an agate mortar, and the value of U_s for particles of 1.2−1.4 μm "in diameter" was roughly twice as great as that of the sublimed particles; presumably the crystal lattices of the former were even more defective than those of the latter.

The next group of relevant studies deals with metal hydroxides. These compounds will have a peculiar C.E. whenever their external region contains fewer or more hydroxyl groups than an equal volume far from the surface (*e.g.*, because of evaporation). Particles of $Mg(OH)_2$ were prepared[168] by wet precipitation, and their average shape and dimensions were estimated from the X-ray analysis. The finest particles were prisms, 90 angstroms long and 250 angstroms thick. The difference between Q_0 and Q_p amounted to roughly 3%. The calculated U_s was 330 erg/cm^2. Various ferric hydroxides seemed[169] to have $U_s \approx 350$ erg/cm^2.

Cadmium oxide prepared at 800 °C had coarser particles (about 2500 angstroms) than those obtained at 300 °C ($\approx$ 200 angstroms) and had a smaller heat of dissolution in a mixture of HF, HCl, and H_3PO_4; the difference $Q_p - Q_0$ was[170] about 0.015 $Q_0 \cdot$A U_s of roughly 500 erg/cm^2 was computed from the test data. A specimen of γ-Al_2O_3 consisted of "primary" particles whose average edge length was calculated from the width of the X-ray lines and varied between 24 and 41 angstroms[171]. The heat of dissolution of these samples in a mixture of HF and HCl was a linear function of the specific surface area, arrived at on the assumption that the particles were cubes. The U_s turned out to be approximately 560 erg/cm^2.

Highly dispersed gold sols evolved more heat during dissolution in ICl_3 than did coarse gold suspensions[172]. A value of 670 erg/cm^2 for the U_s of Au was derived from these data. The authors considered it to be too low for the "true" U_s but, if the difference between the two experimental heats was caused by the cuticular energy present in the gold particles and the quantity 670 erg/cm^2 belonged to the C.E. rather than to U_s, then no objection can be raised.

The area of a MgO powder was determined from its gas adsorption capacity and its heat of dissolution was taken from the literature[173]. The U_s appeared to be about 1000 erg/cm^2.

When oxides of cobalt and nickel were reduced with hydrogen[174] at various temperatures (300–420 °C), the dispersity of the resulting metal was greater the lower the temperature. Unfortunately, the method of determining the dispersity is not specified in the original. When the surface area per gram increased, for instance, from 5 m^2 to 17 m^2, the heat of solution in hydrochloric acid rose by over 3%. The values of U_s computed from the experimental data were 3700 for nickel and 3600 erg/cm^2 for cobalt. From the point of view of the new theory, these values indicate, perhaps, a high degree of disorder in the obtained microcrystals.

Round sodium chloride particles of 0.2 mm "in diameter" were prepared by rapid cooling of droplets of a NaCl melt[175]. Their heat of dissolution in water exceeded that of large crystals by about 0.3% only. Consequently, the precision of the U_s values was low; the results of 3 pairs of experiments ranged from 4000 to 30000 erg/cm^2. This high range may have been caused by the fact that the minute "spheres" were far from internal equilibrium.

Much smaller crystals, still of NaCl, obtained by electrostatic precipitation of smoke, gave smaller values of U_s. The dimensions of single particles have not been determined but the specific surface area nA_p of the powder was found from the isotherm of nitrogen adsorption and varied between 1 and 50 m^2/g. The difference $Q_p - Q_0$ was proportional to nA_p. The recommended value of U_s was 276 erg/cm^2 at 25 °C[176]. A larger energy was calculated in an earlier paper[177] but the sodium chloride used then was shown to contain sodium nitrate. The nitrate tended to concentrate in the external layers of the NaCl crystals and raised the cuticular energy of the specimens analogously to the oxides considered in the foregoing section and the water referred to above in this section.

Gas adsorption was used also for determining the specific surface area nA_p of the powder in several other investigations. The nA_p of CaO varied[178] from 0.3 to 8 m^2/g, and that of Ca(OH)$_2$ from 0.1 to 26 m^2/g. Each sample was dissolved in nitric acid. The preferred values of U_s were 1370 ± 250 for CaO and 1180 ± 100 erg/cm^2 for Ca(OH)$_2$ at 23 °C. When precipitated silica gels were dried at different temperatures for different times, the nA_p of the dry material obtained varied[179] from 382 to about 700 m^2/g. The heat of dissolution in a mixture of HF and HNO$_3$ was greater for the fine-dispersed than for the coarser powder. From the difference, U_s of 259 ± 3 erg/cm^2 was calculated, at 23 °C. The U_s of hydrated silica was estimated to be near 130 erg/cm^2, that is, not significantly greater than the total surface energy of liquid water (118 erg/cm^2 at 25 °C).

Other papers in this series dealt with silicates which, unfortunately, were not chemically uniform. The mineral tobermorite (Ca$_3$Si$_2$O$_7$, 2 H$_2$O) was made from Ca$_2$SiO$_4$ under various experimental conditions[180]. Strangely, the values of nA_p derived from the adsorption of nitrogen were in some instances only one-third or one-quarter those calculated from the adsorption of water vapor. When only the latter data were counted, then Q_p appeared to be a roughly linear function of nA_p, and U_s was 388 ± 30 erg/cm^2. The U_s for the trihydrate Ca$_3$Si$_2$O$_7$, 3 H$_2$O was said to be 320 ± 70 erg/cm^2.

When a trihydrate of an identical composition was prepared by grinding in a ball mill with water[181], the powder showed only one X-ray line near 3 angstroms, in contrast to the crystalline tobermorite which had two strong lines in this region.

The U_s of this powder was 415 ± 89 erg/cm^2. Presumably, these results are less trustworthy than those obtained for chemically homogeneous crystals. The most recent study[182] leads to 450 erg/cm^2 as the U_s of tobermorite.

Magnesium oxide samples resulting from dehydration of magnesium hydroxide had a greater heat of solution when the thermal treatment was performed at temperatures below 400 °C than had bulk specimens of MgO[183]. As the authors state, this difference "may be due to the powder having a higher specific surface or to the existence of strain energy in the lattice, or to both". Additional measurements confirmed, in the reviewer's opinion, the second alternative, that is, the importance of cuticular energy for the observed dependence of Q_p on nA_p. When the temperature T_t of the treatment was raised from 380° to 1380 °C, not only the Q_p decreased by over 3%, but also the water content of the samples was lowered from 3.7 to 0.12%, the density increased from 3.32 to 3.53 g/cm^3, and nA_p became 4.5 instead of 270 m^2/g. Nevertheless, the change of Q_p with T_t was attributed above all to the change in the nA_p. If the value quoted above of 1000 erg/cm^2 for the U_s of MgO was introduced in the equations, the calculated variation in Q_p was not very different from that observed.

Calcium carbonate crystals whose nA_p was 12–18 m^2/g (from gas adsorption) had a Q_p greater (by, *e.g.,* 10%) than those with a nA_p of 0.2–1 m^2/g, but the reproducibility of the results was too low to calculate the surface energy[184]. Perhaps, comparison of Q_p with the chemical composition, the degree of lattice imperfection, and so on, would have shown a better correlation. When the nA_p of goethite FeOOH and α-Fe$_2$O$_3$ varied from 0.003 to 3.6 m^2/g, Q_p increased[185] by about 30%. The calculated U_s was 1170 for goethite and 770 erg/cm^2 for Fe$_2$O$_3$.

When sodium trimetaphosphate Na$_3$P$_3$O$_9$ was ground in humid air, the relative volume of the amorphous ingredient (determined from the X-ray diffraction data) and the lattice distortion increased with the duration t of the grinding[186]. The authors concluded that the variation of Q_p was "mainly attributable to production of an amorphous material". It is hoped that comparisons similar to those in Motooka's work will be more common in the future.

A thermal method independent of the heat of solution was invented and tested in Ref.[187]. It is based on the differential thermal analysis. When a powder, such as a dried polystyrene latex (consisting of almost identical spheres) is gradually heated, the heat H evolved is the difference between the energy losses in the system, such as the decrease in the quantity nU_sA_p, caused by sintering, and the energy increases, such as the rise in the heat content of the bulk. The two ingredients of H can be separated by repeating the treatment on the already sintered material: the reversible components of H (related to, for instance, heat capacity) will be as prominent as during the first treatment, but the irreversible components will be absent (because nA_p has already reached its minimum value). The irreversible part of H is assumed to be equal to nU_sA_p, see Eq. (84). The U_s was 36 to 47 erg/cm^2 for 3 different polystyrene samples, and 83–90 erg/cm^2 for two samples of poly(vinylidene chloride).

VI. Conclusions

At least five conclusions may be formulated after perusing this review.

1. No method so far suggested for measuring the surface energy or surface tension of solids is satisfactory.

2. This failure may be caused, above all, by the fact that solids, contrary to liquids, cannot alter their shape without changing the strain energy in their volumes. The changes in strain energy are so much greater than those in surface energy that the latter remain unrecognized.

3. Solids possess an energy unknown in typical liquids. This cuticular energy exists because the surface region of innumerable solids has a chemical composition, a frequency of lattice defects, and so on, different from those in the bulk.

4. Small solid particles obtained by cooling of vapors, by grinding, or many other methods, usually have a less perfect lattice and more impurity than have bigger crystals of nominally identical composition. Hence, the cuticular energy of the former exceeds that of the latter.

5. Cuticular energy implies no tendency of the surface to contract, *i.e.,* no surface tension. The theoretical calculations of the difference in energy between a broken and an unbroken crystal, if correct, afford a quantity which is related to cuticular energy and, like this, causes no contractile tendency.

Note Added in Proof

To p. 11. According to N. H. Fletcher, J. Crystal Growth *28,* 375 (1975), the free energy of solid − melt interfaces in many systems (e.g., water − ice) is determined above all by the low entropy of the liquid layers adjacent to the solid surface. This loss of entropy occurs because the above layers are more ordered than the melt far from the solid.

To pp. 41−43. The criticism of the methods based on Eq. (65) was expanded by J. J. Bikerman in a paper submitted to a magazine. These methods are not justified also because the quantities p_1 and p_2 in Eq. (64) refer to two different systems, in each of which a uniform pressure (p_1 or p_2) acts. In the experiments of Fig. 17 and the analogous tests, two different pressures are supposed to act in one system. A detailed consideration of such systems shows that in them no reversible melting and solidification, fully depending on the local curvature, can take place. Moreover, the actual pressures in the containers used depended on the flexing of the container walls, mentioned in Ref.[131].

VII. References

[1] Young, Th.: Phil. Trans. Roy. Soc. *1805* I, 65

[2] Young, Th.: Suppl. to the 4th, 5th, and 6th eds. of the Encyclopaedia Britannica *3,* 211 (1824)

[3] Quincke, G.: Phil. Mag. [4] *36,* 267 (1868)

[4] Dupré, A.: Ann. chim. et phys. [4] *7,* 245 (1866)

[5] Gibbs, J. W.: Thermodynamics, p. 315. New York: Dover Publ. (first published in 1876)

[6] Benson, G. C., Yun, K. S. in: The solid − gas interface E.A. Flood (ed.), Vol. I, p. 203. New York: Marcel Dekker 1967

[7] Benson, G. C., Claxton, T. A.: J. Phys. Chem. Solids *25,* 367 (1964)

[8] Macmillan, N. H., Kelly, A.: Mater. Sci. Eng. *10,* 139 (1972)

[9] Schmit, J. N.: Surface Sci. *55*, 589 (1976)

[10] Allen, R. E., de Wette, F. W.: J. Chem. Phys. *51*, 4820 (1969)

[11] Erikson, W. D., Linnett, J. W.: Proc. Roy. Soc. London *A 331*, 347 (1972)

[12] Zadumkin, S. N., Khulamkhanov, V. Kh.: Fiz. Tverd. Tela *5*, 48 (1963)

[13] Chen, T. S., *et al.:* J. Chem. Phys. *55*, 3121 (1971)

[14] Zadumkin, S. N. in: Surface phenomena in melts and in solid phases formed from them [Russian] S. N. Zadumkin (ed.), p. 12. Nal'chik 1965

[15] Craig, R. A.: Phys. Rev. *B 6*, 1134 (1972)

[16] Schmit, J., Lucas, A. A.: Solid State Comm. *11*, 415 (1972)

[17] Paasch, G.: Phys. Status Solidi *65*, 221 (1974)

[18] Goldiner, M. G., Malinovskii, T. I., Yagubets, A. N.: Isv. Akad. Nauk Mold. S.S.R., Ser. Fiz., Tekh., Math. Nauk 1973, 31; Chem. Abstr. *80*, No. 18325

[19] Heinrichs, J.: Solid State Comm. *13*, 1599 (1973)

[20] Lang, N. D., Kohn, W.: Phys. Rev. *B 1*, 4555 (1970)

[21] Julg, A., et al.: Phys. Rev. *B 9*, 3248 (1974)

[22] Hietschold, M., Paasch, G., Ziesche, P.: Phys. Status Solidi *70*, 563 (1975)

[23] Burton, J. J., Jura, G.: J. Phys. Chem. *71*, 1937 (1967)

[24] Ferrante, J., Smith, J. R.: Solid State Comm. *20*, 393 (1976)

[25] Matsunaga, T., Tamai, Y.: Surface Sci. *57*, 431 (1976)

[26] Stefan, J.: Wied. Ann. Physik *29*, 655 (1886)

[27] Shcherbakov, L. M., Baibakov, V. S. in: Surface phenomena in metallurgical processes A. I. Belyaev (ed.), p. 172. Consultants Bureau: New York 1965

[28] Jones, H.: Metal Sci. J. *5*, 15 (1971)

[29] Fricke, R.: Kolloid-Z. *96*, 211 (1941)

[30] Gvozdev, A. G., Gvozdeva, L. I.: Fiz. Metal. Metalloved. *31*, 640 (1970)

[31] Ogorodnikov, V. V., Rogovoi, Yu. I.: Porosh. Metall. *1976*, No. 1, 70

[32] Oshcherin, B. N.: Phys. Status Solidi *25*, K 123 (1968); *31*, K 135 (1969); Phys. Chem. of Surface Phenomena at High Temperatures [Russian] V. N. Eremenko (ed.), p. 39. Kiev: Naukova Dumka 1971

[33] Wawra, H. H.: Materialprüfung *14*, 413 (1972)

[34] Reynolds, C. L., Couchman, P. R., Karasz, F. E.: Phil. Mag. *34*, 659 (1976)

[35] Benedek, C., *et al.:* Surface Sci. *48*, 561 (1975)

[36] Missol, W.: Phys. Status Solidi *B 58*, 767 (1973)

[37] Kashetov, A., Gorbatyi, N. A.: Fiz. Tverd. Tela *11*, 493 (1969)

[38] Zadumkin, S. N., Shebzukhova, I. G., Al'chagirov, B. B.: Fiz. Metal, Metalloved. *30*, 1313 (1970)

[39] Vingsbo, O.: Mater. Sci. Eng. *8*, 32 (1971)

[40] Berggren, B.: Ann. Physik [4] *44*, 61 (1914)

[41] Khagabanov, A. Kh., Zadumkin, S. N., Bartenev, G. M. in: Phys. Chem. of surface phenomena at high temperatures [Russian] V. N. Eremenko (ed.), p. 68. Kiev: Naukova Dumka 1971

[42] Maurakh, M. A., Orlov, A. S., Besshapova, M. R. in: Adhesion of melts [Russian] V. N. Eremenko, Yu. V. Naidich (eds.), p. 103. Kiev: Naukova Dumka 1974

[43] Sawai, I., Ueda, Y.: Z. anorg. allg. Chem. *180*, 287 (1929)

[44] Tammann, G., Boehme, W.: Ann. Physik [5] *12*, 820 (1932)

[45] Heumann, Th., Wulff, H.: Scripta Met. *10*, 1001 (1976)

[46] Bikerman, J. J.: Phys. Status Solidi *10*, 3 (1965)

[47] Bikerman, J. J.: Mater. Sci. Eng. *20*, 293 (1975)

[48] Tammann, G.: Nachr. Göttinger Ges. Wiss. *1912*, 557

[49] Udin, H.: Metals Trans. *191*, 63 (1951)

[50] Inman, M. C., Tipler, H. R.: Metallurg. Revs. *8*, 105 (1963)

[51] Tyson, W. R.: Can. Metall. Quart. *14*, 307 (1975)

[52] Greenough, A. P.: Phil. Mag. [7] *43*, 1075 (1952)

[53] Pranatis, A. L., Pound, G. M.: Metals Trans. *203*, 664 (1955)

[54] Greenough, A. P.: Phil. Mag. [8] *3*, 1032 (1958)

[55] Lowe, A. C., Riddiford, A. C.: J. Colloid Interface Sci. *32*, 292 (1970)

56) Heumann, Th., Wulff, H.: Z. Metallkunde 67, 87 (1975)
57) Shebzukhova, I. G., Khokonov, Kh. B., Zadumkin, S. N.: Fiz. Metal. Metalloved. 33, 1112 (1972)
58) Heumann, Th., Johannisson, J.: Acta Met. 20, 617 (1972)
59) Fiala, J., Čadek, J.: Phil. Mag. [8] 32, 251 (1975)
60) Hondros, E. D., Gladman, D.: Surface Sci. 9, 471 (1968)
61) Roth, T. A.: Mater. Sci. Eng. 18, 183 (1975)
62) Bryant, L. F., Speiser, R., Hirth, J. P.: Trans. Met. Soc. AIME 242, 1145 (1968)
63) Kostikov, V. I., Kharitonov, A. V., Savenko, V. I.: Fiz. Metal. Metalloved 26, 947 (1968)
64) Allen, B. C.: Trans. Met. Soc. AIME 236, 903 (1966)
65) Kostikov, V. I., Kharitonov, A. V.: Fiz. Metal. Metalloved. 35, 188 (1973)
66) Hondros, E. D., McLean, D.: Phil. Mag. [8] 29, 771 (1974)
67) Stickle, D. R. et al.: Metall. Trans. A 7, 71 (1976)
68) Bauer, C. E., Speiser, R., Hirth, J. P.: Metall. Trans. A 7, 75 (1976)
69) Rice, C. M., Eppelsheimer, D. S., McNeil, M. B.: J. Appl. Phys. 37, 4766 (1966)
70) Bruver, R. E., Glikman, E. E. in: Ref.[41], p. 86
71) Udin, H.: Metal interfaces, p. 114. Cleveland: Am. Soc. Metals 1952
72) Wasserman, H. J., Vermaak, J. S.: Surface Sci. 32, 168 (1972)
73) Smart, D. C., Boswell, F. W., Corbett, J. M.: J. Appl. Phys. 43, 4461 (1972)
74) Barbour, J. P. et al.: Phys. Rev. 117, 1452 (1960)
75) Dranova, Zh. I., Dyachenko, A. M., Mikhailovskii, I. M.: Fiz. Metal. Metalloved. 31, 1108 (1971)
76) Hoffman, J. D., Weeks, J. J.: J. Res. Natl. Bur. Stand. 66 A, 13 (1962)
77) Harrison, I. R.: J. Polym. Sci., Phys. Ed. 11, 991 (1973)
78) Zubov, Yu. A. et al.: Doklady Akad. Nauk SSSR 217, 1118 (1974)
79) Hinrichsen, G.: Polymer 10, 718 (1969)
80) Martuscelli, E.: Makromol. Chem. 151, 159 (1972)
81) Griffith, A.: Trans. Roy. Soc. A 221, 180 (1920)
82) Petch, H. in: Fracture H. Liebowitz (ed.). Vol. I, p. 353. New York: Academic Press 1968
83) Bikerman, J. J.: SPE Trans. 4, 290 (1964); Proc. 5th Intern. Congr. Rheology, Vol. I, p. 589. Tokyo 1969
84) Phillips, C. J. in: Fracture H. Liebowitz (ed.). Vol. VII, p. 28. New York: Academic Press 1972
85) Timoshenko, S., Goodier, J. N. in: Theory of elasticity, p. 90. New York: McGrow-Hill 1951
86) Shand, E. B.: J. Amer. Ceram. Soc. 44, 451 (1961)
87) Dobbs, H. S., Field, J. E., Maitland, A. H.: Phil. Mag. [8] 28, 33 (1973)
88) Davidge, R. W., Tappin, G.: J. Mater. Sci. 3, 165 (1968)
89) Irwin, G. R., Kies, J. A., Smith, H. L.: ASTM Proc. 58, 640 (1958)
90) Svensson, N. L.: Proc. Phys. Soc. 77, 876 (1961)
91) Cordwell, J. E., Hull, D.: Phil. Mag. [8] 27, 1183 (1973)
92) Kusy, R. P., Turner, D. T.: Polymer 17, 161 (1976)
93) Nelson, B. E.: J. Colloid Interface Sci. 47, 595 (1974)
94) see, for instance, Bikerman, J. J.: Physical surfaces, p. 48. New York: Academic Press 1970
95) Govila, R. K.: Acta Met. 20, 447 (1972); Scripta. Met. 6, 353 (1972)
96) Gilman, J. J.: J. Appl. Phys. 31, 2208 (1960)
97) Obreimoff, J. W.: Proc. Roy. Soc. London A 127, 290 (1930)
98) Orowan, E.: Z. Phys. 82, 235 (1933)
99) Burns, S. J.: Phil. Mag. [8] 25, 131 (1972)
100) Elyutin, V. P., Kostikov, V. I., Kharitonov, A. V.: Doklady Akad. Nauk SSSR 182, 376 (1968)
101) Metsik, M. S.: J. Adhesion 3, 307 (1972)
102) Westwood, A. R. C., Hitch, T. T.: J. Appl. Phys. 34, 3085 (1963)
103) Wiederhorn, S. M.: J. Amer. Ceram. Soc. 52, 99 (1969)

104) Bikerman, J. J.: The science of adhesive joints, 2nd edit., p. 182. New York: Academic Press 1968

105) Kraatz, P., Zoltai, T.: J. Appl. Phys. 45, 4741 (1974)

106) Maitland, A. H., Chadwick, G. A.: Phil. Mag. [8] 19, 645 (1969)

107) Elyutin, V. P., Kostikov, V. I., Kharitonov, A. V. in: Ref.41), p. 263

108) Gupta, Y. P., Santhanam, A. T.: Acta Met. 17, 419 (1969)

109) Irwin, G. R. in: Handbuch der Physik. S. Flügge (ed.), Vol. 6, p. 551 (1958)

110) Smith, C. S.: Amer. Inst. Mining Met. Engrs., Inst. Metals Div. 175, 15 (1948)

111) Hasson, G. et al. in: Grain boundaries and interfaces P. Chaudhuri, J. W. Mathews (eds.), p. 115. Amsterdam: North-Holland Publ. Co. 1972

112) Harris, L. B., Vernon, J. P.: J. Polymer Sci., Part A-2, 10, 499 (1972)

113) Greenough, A. P., King, R.: J. Inst. Metals 79, 415 (1951)

114) Powell, B. D., Woodruff, D. P.: Phil. Mag. 34, 169 (1976)

115) Rosenhain, W., Ewen, D.: J. Inst. Metals 8, 149 (1912)

116) Hodkin, E. N., Nicholas, M. G., Poole, D. M.: J. Nucl. Mater. 25, 284 (1968)

117) Mullins, W. W.: Trans. AIME 218, 354 (1960)

118) Maiya, P. S.: J. Nucl. Mater. 40, 57 (1971)

119) Maiya, P. S., Blakeley, J. M.: J. Appl. Phys. 38, 698 (1967)

120) Presland, A. E. B., Price, G. L., Trimm, D. L.: Surface Sci. 29, 435 (1972)

121) Pokrovskii, N. L., Khefni, A. Kh. in: Adhesion of melts [Russian] Yu. V. Naidich (ed.), p. 102. Kiev 1974

122) Lilburne, M. T.: J. Mater. Sci. 5, 351 (1970)

123) Barnes, B. S.: Phil. Mag. [8] 5, 635 (1960)

124) Bikerman, J. J.: J. Mater. Sci. 6, 176 (1971)

125) Westmacott, K. H., Smallman, R. E., Dobson, P. S.: Metal Sci. J. 2, 177 (1968)

126) Haubelt, R., Mennicke, S., Dittmar, W.: Z. physik. Chem. 95, 187 (1975)

127) Gladkikh, N. T., Larin, V. I., Khotkevich, V. I.: Fiz. Metal. Metalloved. 31, 786 (1971)

128) Bolling, G. F., Tiller, W. A.: J. Appl. Phys. 31, 1345 (1960)

129) Jones, D. R. H., Chadwick, G. A.: Phil. Mag. [8] 22, 291 (1970)

130) Nash, G. E., Glicksman, M. E.: Phil. Mag. [8] 24, 577 (1971)

131) Hardy, S. C.: Phil. Mag. 35, 471 (1977)

132) Schaefer, R. J., Glicksman, M. E., Ayers, J. D.: Phil. Mag. [8] 32, 725 (1975)

133) Sambles, J. R., Skinner, L. M., Lisgarten, N. D.: Proc. Roy. Soc. London 318 A, 507 (1970)

134) Freundlich, H.: Kapillarchemie. Vol. I, p. 218. Leipzig: Akad. Verlagsges. 1930

135) Cohen, E., Blekkingh, J. J. A.: Proc. Kon. Nederland Akad. Wetenschappen 43, 32, 189, 334 (1940)

136) Balarew, D.: Kolloid-Z. 96, 19 (1941)

137) Glazner, A.: Israel J. Chem. 13, 73 (1975)

138) Bikerman, J. J.: Ref.94), p. 192

139) Alexander, G. B.: J. Phys. Chem. 61, 1563 (1957)

140) Rosoff, M., Schulman, J. H.: Kolloid-Z. Z. Polym. 225, 46 (1968)

141) Kislyi, P. S., Kuzenkova, M. A.: Porosh. Metal. 1969, No. 11, 21

142) Bikerman, J. J.: Proc. 2nd Intern. Congr. Surface Activity, Vol. 3, 125. New York: Academic Press 1957; J. Phys. Chem. 63, 1658 (1959)

143) Bikerman, J. J.: Contributions to the thermodynamics of surfaces, p. 65. Cambridge 1961

144) Khlynov, V. V., Bokser, E. L., Esin, O. A.: Doklady Akad. Nauk SSSR 208, 820 (1973)

145) Khlynov, V. V., Bokser, E. L., Pastukhov, B. A. in: Adhesion of melts and joining of materials [Russian] Yu. Naidich (ed.), p. 39. Kiev: Naukova Dumka 1976

146) Bruver, R. E., Glickman, Ye. F., Tsarev, O. K.: Fiz. Metal. Metalloved. 26, 1136 (1968)

147) Rhee, S. K.: Mater. Sci. Eng. 11, 311 (1973); J. Colloid Interface Sci. 44, 173 (1973)

148) Rhee, S. K.: J. Mater. Sci. 12, 823 (1977)

149) Bakovets, V. V.: Doklady Akad. Nauk SSSR 228, 1132 (1976)

150) Ketcham, W. M., Hobbs, P. V.: Phil. Mag. [8] 19, 1161 (1969)

151) Bikerman, J. J.: Microscope 21, 183 (1973)

152) Bratton, R. J., Beck, C. W.: J. Amer. Ceram. Soc. 54, 379 (1971)

153) Mortimer, D. A., Nicholas, M. G.: Metal Sci. 10, 326 (1976)

154) Hodkin, E. N. et al.: J. Nuclear Mater. *39*, 59 (1971)

155) Hodkin, E. N., Nicholas, M. G.: J. Nuclear Mater. *47*, 23 (1973)

156) Briant, C. L., Burton, J. J.: J. Chem. Phys. *63*, 2045 (1975)

157) Lewis, B.: J. Appl. Phys. *41*, 30 (1970)

158) Turnbull, D.: J. Chem. Phys. *18*, 769 (1950)

159) Schaefer, V. J.: Ind. Eng. Chem. *44*, 1300 (1952)

160) Mason, B. J.: The physics of clouds, p. 168. Oxford Clarendon Press 1971

161) Turnbull, D.: J. Chem. Phys. *18*, 768 (1950)

162) Gokhstein, A. Ya.: Surface tension of solids and adsorption. [Russian] Moscow: Nauka 1976

163) Shchukin, E. D., Smirnova, N. V.: Fiz.-khim. mekh. materials *3*, No. 1, 90 (1967); Chem. Abstr. *67*, No. 76898 (1967)

164) Shuttleworth, R.: Proc. Phys. Soc. London *63 A*, 444 (1950)

165) Lipsett, S. G., Johnson, F. M. G., Maass, O.: J. Amer. Chem. Soc. *49*, 925 (1927)

166) Lipsett, S. G., Johnson, F. M. G., Maass, O.: J. Amer. Chem. Soc. *49*, 1940 (1927)

167) Lipsett, S. G., Johnson, F. M. G., Maass, O.: J. Amer. Chem. Soc. *50*, 2701 (1928)

168) Fricke, R., Schnabel, R., Beck, K.: Z. Elektrochem. *42*, 114 (1936)

169) Fricke, R., Zerrweck, W.: Z. Elektrochem. *43*, 52 (1937)

170) Fricke, R., Blaschke, F.: Z. Elektrochem. *46*, 46 (1940)

171) Fricke, R., Niermann, F., Feichtner, C.: Ber. deut. chem. Ges. *70*, 2318 (1937)

172) Fricke, R., Meyer, F. R.: Z. phys. Chem. *A 181*, 409 (1938)

173) Jura, G., Garland, C. W.: J. Amer. Chem. Soc. *74*, 6033 (1952)

174) Schubert-Birckenstaedt, M.: Z. anorg. allg. Chem. *276*, 227 (1954)

175) Jones, E. D., Burgess, D. S., Amis, E. S.: Z. phys. Chem. (Frankfurt) *4*, 220 (1955)

176) Benson, G. C., Schreiber, H. P., van Zeggeren, F.: Can J. Chem. *34*, 1553 (1956)

177) Benson, G. C., Benson, G. W.: Can, J. Chem. *33*, 232 (1955)

178) Brunauer, S., Kantro, D. L., Weise, C. H.: Can, J. Chem. *34*, 729 (1956)

179) Brunauer, S., Kantro, D. L., Weise, C. H.: Can. J. Chem. *34*, 1483 (1956)

180) Brunauer, S., Kantro, D. L., Weise, C. H.: Can. J. Chem. *37*, 714 (1959)

181) Kantro, D. L., Brunauer, S., Weise, C. H.: J. Colloid Sci. *14*, 363 (1959)

182) Brunauer, S.: J. Colloid Interface Sci. *59*, 433 (1977)

183) Livey, D. T. et al.: Trans. Brit. Ceram. Soc. *56*, 217 (1957)

184) Antolini, R., Gravelle, P. C., Trambouze, Y.: J. Chim. Phys. *59*, 715 (1962)

185) Ferrier, A.: Compt. rend. Acad. sci. Paris *261*, 410 (1965)

186) Motooka, I., Hashizume, K., Kobàyashi, M.: J. Phys. Chem. *73*, 3012 (1969)

187) Mahr, T. G.: J. Phys. Chem. *74*, 2160 (1970)

Received October 18, 1977

Trends and Applications of Thermogravimetry

Hans-Georg Wiedemann

Mettler Instrumente AG, 8606 Greifensee-Zürich, Switzerland

Gerhard Bayer

Institut für Kristallographie und Petrographie ETH-Zürich, Switzerland

Table of Contents

1. Introduction

Thermogravimetry (TG) is one of the classical methods of thermal analysis (TA). It measures the loss or gain in weight of a sample as a function of temperature, time, and also of the atmosphere. In agreement with the general definition it is a typical TA-technique, because it follows the changes of a physical parameter as a function of temperature (Fig. 1). Compared to DTA (Differential Thermal Analysis), which probably is still the most important thermal analytical method, TG was not used so broadly however. This limitation had some historical reasons and has been due originally to the lack of sensitivity and reproducibility, to the time-consuming operation and later on to the emphasis on thermochemical information which is directly available from DTA-data. However, continuing modifications in equipment design have resulted in great improvements in measuring capability and sensitivity and have led to broader applicability. Especially the combination of TG and DTA in one system has proved to be very useful since it allows one to measure simultaneously and quantitatively various physical and chemical transformations which occur during heating up of the material to be studied.

The development of apparatus for thermogravimetric analysis, the so-called "thermo-balances". from Wagner's spring balance[1] to the modern semi-automatic recording instruments, is an interesting example of the rapid progress made by modern technology in the field of chemical apparatus design. The expression "thermo-balance" was first used by K. Honda[2] in 1915 for a balance designed to measure the loss or gain in weight of a heated sample as a function of temperature (Fig. 2). These balances were originally applied to the solution of analytical problems, some years later they were used also in physico-chemical work. The following developments were concentrated on the recording and registration of the data, that is to say on the simultaneous and automatic recording of the weight against time and temperature. The leading workers in this field were M. Guichard[3] and P. Chevenard[4]. After this period much of the fundamental work was done by C. Duval[5]. During the last two decades it was the application of electronics which made it possible to design recording balances without retroactive effects, either by using the principle of magnetic force compensation or by using an electromechanical transducer to follow the position of the balance beam. Modern methods of temperature measurement

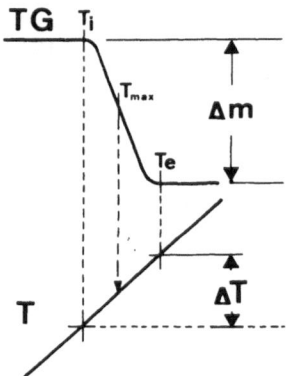

Fig. 1. Typical TG-curve with corresponding temperature curve

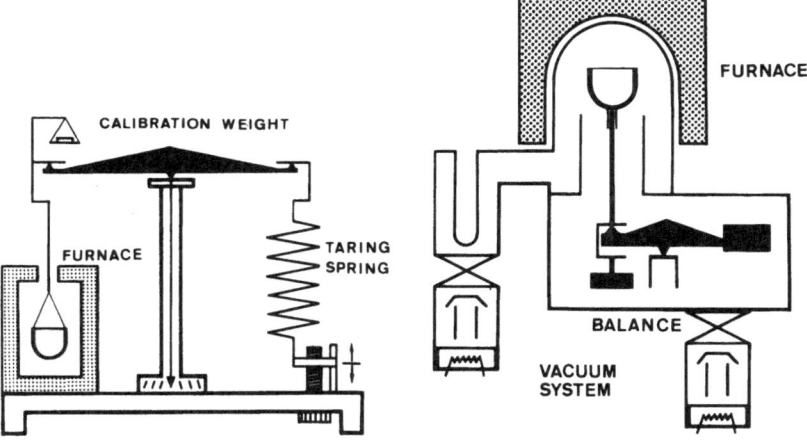

Fig. 2. Comparison of the classical Honda thermobalance with present, frequently used equipment

and control allowed linear heating programs which are of great importance in all thermal analyses[6-11]. This is true also for the control of atmosphere which has a pronounced effect on the reactions taking place in a material (Fig. 3). All modern thermobalances therefore are equipped with a high vacuum system which not only allows one to work in vacuum (down to 10^{-6} mm Hg) but also to change the gas atmosphere in the furnace chamber very rapidly e.g. from oxidizing to neutral, reducing or even reactive. The attachment of apparatus like gas chromatograph or mass spectrometer to the thermobalance made it possible to analyze continuously the gaseous decomposition products evolved during the heating of a sample, and to control the reaction atmosphere.

The information available from TG-data can be increased further by deriving and recording simultaneously the differentiated weight curve, dW/dt. This first derivative of the weight change is commonly referred to as derivative thermogravimetry (DTG). It is very useful for kinetic studies, for the determination of reaction rates and for interpreting complex thermogravimetric curves. Such curves may be caused

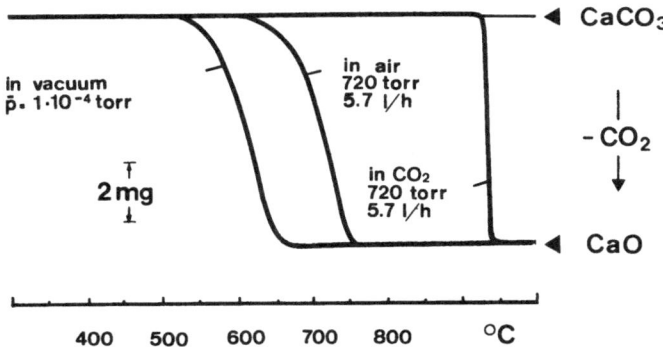

Fig. 3. Example for thermal decomposition in different atmospheres

H. G. Wiedemann and G. Bayer

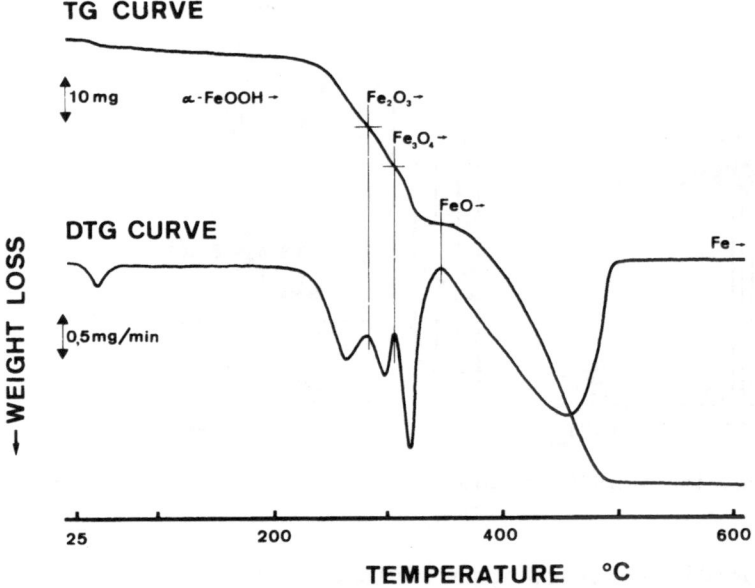

Fig. 4. TG-curve and its first derivative (DTG) for the reduction of α-FeO(OH) to Fe

Table 1. Thermogravimetric methods, simultaneous techniques and experimental

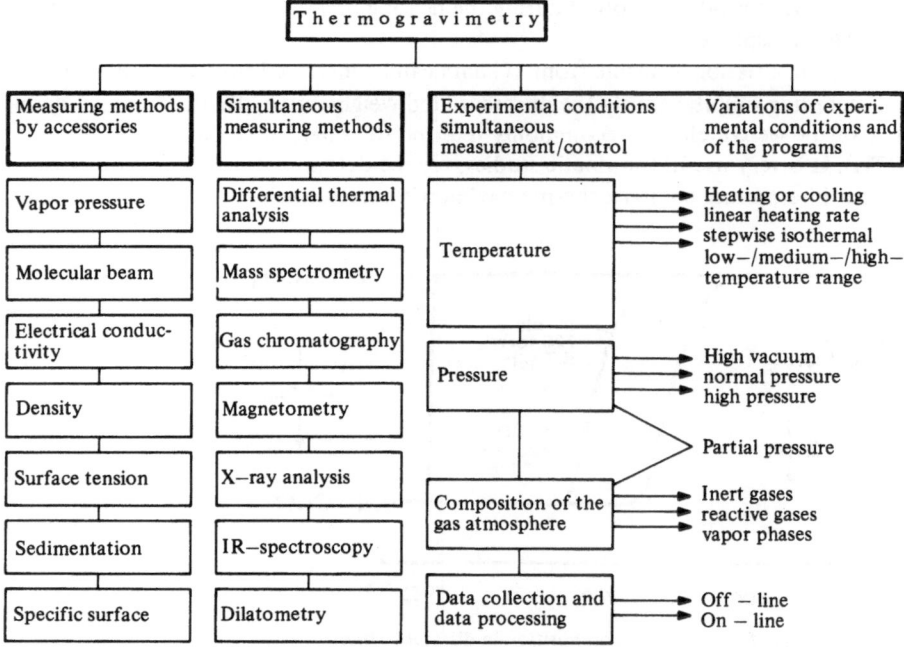

by changes in the order and rate of reaction or by overlapping of several reaction steps (Fig. 4).

Thermogravimetric instruments are important in many fields of research and industry[12]. They are particularly suitable for studies on reaction kinetics in the presence of a special gas atmosphere. In metallurgical work, thermobalances are used to investigate the oxidation and corrosion of metals and alloys. Furthermore, it is possible to measure the vapor pressure of substances at low and at high temperatures by Knudsen's effusion method, thereby following the vaporisation process thermogravimetrically. Adsorption equilibria, which are important in determination of specific surface area, may be investigated also. The use of TG in a magnetic field allows thermomagnetic measurements to be made, a. g. determination of Curie temperatures. One combination which is especially important both in basic and in applied research, apart from TG-DTA, is that of TG and X-ray diffraction. It will be obvious that many other combinations and applications of thermogravimetric methods are possible (Table 1).

For demonstrating the versatility and general usefulness of thermogravimetric methods, a large number of experiments was carried out in different areas of physical and inorganic chemistry and in various fields of applied research. The results of some of these investigations will be described in what follows.

2. Apparatus, Accessories and Experimental Technique

The different types of thermobalances which are used today in research laboratories and in industry are usually bound to and designed for specific applications. They may either be commercially available instruments, or assembled from individual components or completely homemade[6, 7, 13–15]. As an example, the set-up of a modern thermobalance is shown schematically in Fig. 5. This type of instrument with its

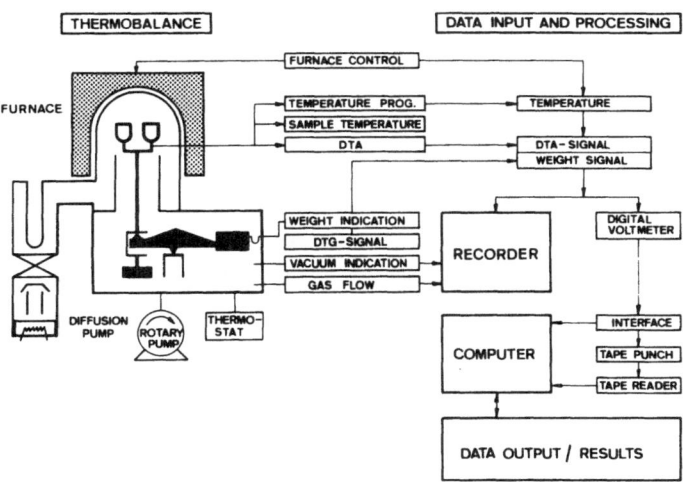

Fig. 5. Instrumental set-up and important components of a modern thermobalance

modular construction principle, allows different measurement techniques which are adapted to the special problem. In making specific measurements, a simpler system can be used which is still capable of being expanded to cover the most diverse variants of a multi-purpose instrument.

All measured values are normally registered by means of a recorder. In addition, an improved method for data collection and processing is possible today by use of a computer. This has the advantage of automatic, safer data collection in an easy-to-read form. Comparison with standard values, correction of the buoyancy effects, control of linearity, control of standard deviations and peak integrations are thus possible.

When the fields of application of thermo balances are considered, a compromise between an analytical balance and a micro balance in regard to sample capacity and weighing accuracy has to be found for various experimental methods. The ratio of loading capacity to sensitivity is of importance since thermo gravimetric curves should be measured with adequate sensitivity even when small samples are used, when, for example, the amounts available are small, and for reasons of uniform heat distribution. It is also possible to perform experiments with additional instrumention, *e.g.*, a measuring head for differential thermal analysis, or a high temperature Knudsen cell for vapor pressure determination. Toploading balances are very suitable for high temperature in investigations because of better temperature distribution in the reaction chamber and because of the smaller effects of temperature on the balance. The modern thermo balance is so designed that weighing can be made above or also below the balance, depending on the temperature range and auxiliary apparatus required. It is usually provided with measuring instrumentation protected against external influences such as shocks, corrosion, or temperature variation. Modern, well-equipped thermo balances facilitate carrying out precise physical and chemical experiments on the balance pan within the following limits: temperatures from $-160\,^{\circ}$C to $+2400\,^{\circ}$C, pressures from approx. 10^{-6} to 760 torr, stationary or flowing gaseous atmosphere of varied composition, including corrosive gases.

2.1. Balances

The balances used in thermogravimetric apparatus are distinguished by the different types of weight compensation, elastic forces in the case of the spring balance and counterpoise for the beam balance. Spring balances are rarely used in commercially available equipment, but rather in home-made, special thermogravimetric apparatus. They may show fatigue problems depending on the material of the spring, however, they are quite sensitive and applicable in corrosive atmosphere. The predominant role of the beam balance is due mainly to its high sensitivity also at high load. Fur-

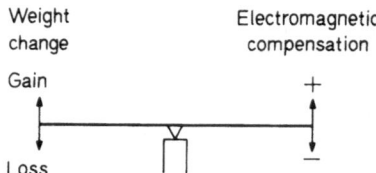

Fig. 6. Null-type balance system

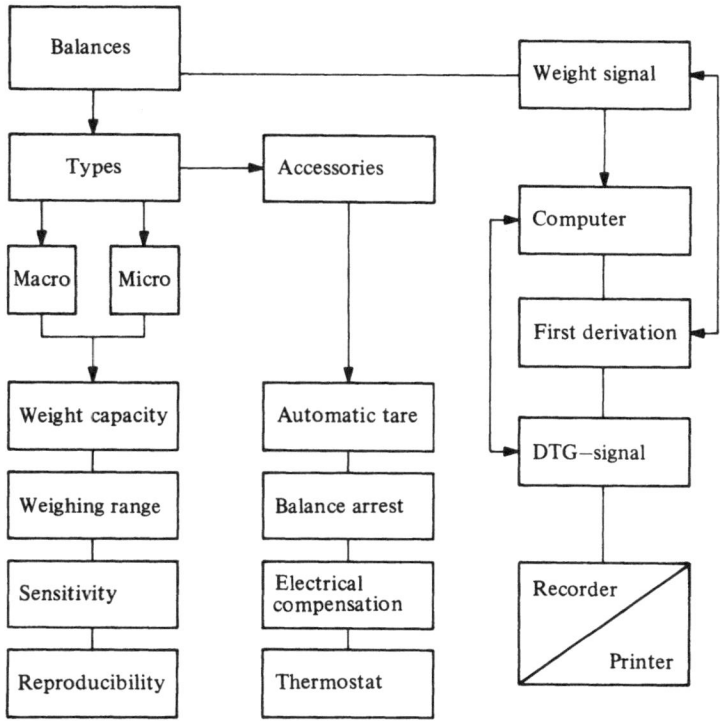

thermore, they have a solid construction and better mechanical resistance which also means better reproducibility. Most of these balances used in TG operate on the substitution principle and are of the null-point type. The construction is similar to that of a normal analytical balance but adapted, in respect to material and design, to the operating conditions of thermogravimetry. The balances are equipped usually with electro-magnetic compensation for continuous readout of results (Fig. 6). The weight changes are converted into an electrical signal which replaces the optical scale readout. The beam position is sensed here by photo-transistors and set by a control circuit. This proportional differential control also performs the balance damping which, in contrast to air damping, is effective both under normal conditions and in high vacuum. Built-in weights compensate for weight variations which exceed the electrical range. Calibration weights permit the sensitivity of the balance to be checked from the outside. Adjustment, if required, is made electrically. Crucible holders are provided instead of a balance pan which also accommodate the thermocouples. The measuring points of the thermocouples are in direct contact with the sample or its container.

Balances

As mentioned above, mainly two types of balances are used: beam balances and spring or torsion balances. Spring or torsion balances show a special relation between sample weight and sensitivity: small sample weights give high sensitivity and large sample weight means small sensitivity. The sensitivity and sample weight capacity for beam balances depends on their construction.

Macro Balances

These too can be very sensitive in the case of large sample weights when the adapted electrical system for weight indication is based on "force compensation". The advantage of this system is that it can be used for both, small and large samples. The larger weight capacity especially offers the possibility of applying special crucibles, *e.g.* Knudsen cells, DTA cells, ball value cells etc. Other advantages are the built-in thermocouples with direct contact to the crucible/sample, long-time stability of the weight signal also at very high temperatures, larger sample quantities in case of of inhomogenous samples.

Micro Balances

These are used for relatively small sample weights (0.1–20 mg) and for low but also for very high heating rates. For many technical products this thermogravimetric method is not so time consuming and gives results with adequate reproducibility. For all equilibrium measurements one can only use slow heating rates, and sample weights in the order of 10 mg.

Weight Capacity

Large weight capacities are required when sample holders with a high dead load are used. The same is also true for samples that are very inhomogeneous, or when very large pieces must be investigated in bulk form.

Depending on the instrument (thermobalance), the weight of the sample is followed and recorded within a specific sensitivity range. This means that one obtains the weight change expressed in a single weight curve. Some balances however allow

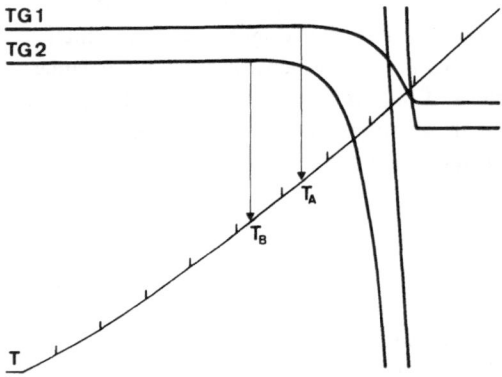

Fig. 7. Simultaneous registration of weight curves with different sensitivity (TG 1:1x, TG 2:10x)

the registration of the weight change simultaneously in two curves with different sensitivities. The advantage of this method is that the overall weight change is evident from one curve, while the other, with a 10 times higher sensitivity, allows very high resolution in case of very small weight effects. Figure 7 shows that the weight curve 1 gives the whole weight change of a sample during a linear heating period in one piece. The sensitivity in weight curve 2 is 10 times higher and therefore the evaluation of the onset temperature of the decomposition, T_B, is more accurate here than in the case of weight curve 1 where T_A is derived. In conclusion, one can generally state that the selection of the sensitivity depends on the material to be investigated, on the kind of reaction and on the type of experimental equipment.

Reproducibility

The number of influences in a thermobalance, which is a "continuously weighing balance". is relatively large. To a great extent these are dependent on the construction of the balance. Normal macro balances are not affected by small vibrations if they are properly installed, regardless of whether this is on an upper or lower floor. Important for good reproducibility are:
   good thermostated housing
   reproducible gas flow conditions
   constant humidity of the atmosphere in the balance (null point/drift)
   efficient transport of the condensable reaction products away from the balance parts
   If these conditions are fullfilled, a precision or *standard deviation* of ±0.5 mg in the 1 g range, ±0.05 mg in the 100 mg range,
±0.015 mg in the 10 mg range, can be reached.

The *accuracy – readability* in the various ranges is as follows

Range	1 g	100 mg	10 mg	1 mg
Accuracy	5 mg	0.1 mg	0.05 mg	0.01 mg
Readability	2 mg	0.2 mg	0.02 mg	0.005 mg

Accessories

Automatic tare is an important accessory apart from the mechanical tare so as to be able to start on the chart strip at any place one wishes.
   As with any normal analytical balance, it is necessary to arrest the balance for all operations between the measurements. Also in case of unexpected, explosion-like reactions which can displace the sample holder from its normal position it is better to arrest the balance for a moment and then continue the measurement.
   In balances which give two TG-curves with different sensitivity, an electrical compensation device switches automatically the expanded curve to the next passage in the chart strip (see Fig. 7 with the two recorded TG-curves).
   The thermostat is necessary for holding the surrounding temperature of the balance constant; it is of special importance for "high temperature long-time runs".

The weight signal is normally recorded in the manner indicated above. The first derivation can be performed with a derivative computer and is recorded simultaneously. Another way today is the use of a small or large programmable desk calculator which is connected to the thermobalance with an interface. The print-out of weight change per time unit gives the same information as the DTG-curve.

For quantitative evaluation of the results, both the TG-curve and the output of the printer of the computer are important. The recorded curve is important at every point during the experiment for handling the whole run. The first outprint is the starting point for further calculations.

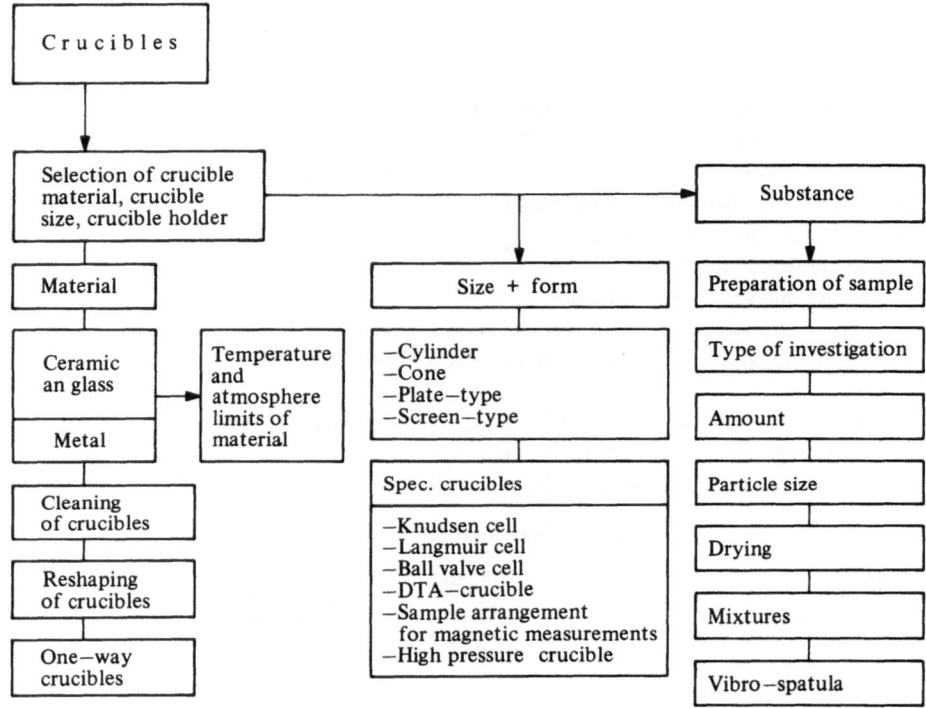

2.2. Crucibles

Crucibles must be suitable for the required experimental conditions with respect to their material, capacity and shape. As in the techniques of the chemical laboratory, crucible materials are selected to avoid the possibility of reaction between crucible and sample material. The main materials used are the precious metals, oxide ceramics, quartz and graphite. The size of the crucible is determined by the volume to be weighed.

Table 2 shows a compilation of different crucible materials, the working temperatures, atmospheres and some important physical data. Metal crucibles are used more for the investigation of clays, oxides, ceramics, glasses, inorganic materials as

Table 2. Crucible materials

Crucible material	Usable in air to	Usable in vacuum to	Fusion point	Density	Gas tight-ness	Resistance against temperature changes	Chemical properties
	$^{\circ}$C	$^{\circ}$C	[$^{\circ}$C]	[g · cm^{-3}]			

Ceramic and glass

Crucible material	Usable in air to	Usable in vacuum to	Fusion point	Density	Gas tightness	Resistance against temperature changes	Chemical properties
Pythagoras-material	1550	–	1820	2,9	Excellent	Satisfactory	Better resistance than porcelain
SiO_2	1200[1]	1200[1]	1750	2,1	Good	Excellent	Above 1200 $^{\circ}$C devitrification. Acid-resistant except concentrated H_3PO_4 above 300 $^{\circ}$C and HF
MgO	2200	1600	2700	2,8	Porous	Good	High temperature resistant
Al_2O_3	1900	1900	2050	3,4 to 3,9	Good	Good	Resists alkaline, alkali- and other metals
Magnesium-spinel	1950	1950	2135	3,8	–	Poor	Intermediary to MgO and Al_2O_3
BeO	2200	2100	2550	2,9	Good	Excellent	Resists alkaline substances as well as reducing influences of metals.
ZrO_2	2500	2300	2700	5,4	Low Porous	Poor	Acid- and basic resistant in high temperatures
ThO_2	2700	2200	3000	9,2	Good	Very poor	Resists basic substances
BN	~1000	1700	Subl.	2,25	Good	Excellent	Sensitive to water vapor
C	550	2300	Practically unmeltable	ca. 1,5	Porous	Excellent	Resists nonoxidizing acid- and basic fluxing agents
Pyrex Glass	400	400	500–600[2]	2,3	Good	Good	

Table 2. (continued)

Crucible material	Usable in air to	Usable in vacuum to	Fusion point	Density	Gas tight-ness	Resistance against temperature changes	Chemical properties
	°C	°C	[°C]	[g · cm^{-3}]			
Metals							
Pt	1600	1600[3])	1679	21,4	Good	Excellent	Vaporization takes place
Au	900	900	1063	19,3	Good	Excellent	Practically non-reactive with non-metallic substance
Ag	700	700	960.8	10,5	Good	Excellent	
Al	550	550	660	2,7	Good	Excellent	Not usable for reactive substance and atmospheres
Ni	800	800	1453	8,9	Good	Excellent	
Nb	250	2000 to 2100	2497	8,55	Good	Good	Acid resistant except HF and concentrated hot sulphuric acid
Mo	400	2300	2610	10,2	Good	Good	Resists nonoxidizing acids
Ta	250	2200 to 2500	2997	16,6	Good	Good	Similar to Nb
W	500	2800	3380	19,3	Good	Good	Similar to Mo

[1]) For short periods up to 1450 °C.
[2]) Annealing point.
[3]) Vaporization takes place.

salts etc. The ceramic materials are used mostly for metals, alloys and chemical compounds which are not inert towards metal crucibles. As already mentioned above, the crucible material must be suitable for the investigated material, for the expected highest test temperature and the surrounding atmosphere. Organic samples are investigated frequently in aluminium cells (in open or sealed capsules). For those organic compounds which attack aluminium, gold or platinum crucibles are preferred.

Cleaning of Crucibles

Depending on the material of the crucible, various cleaning agents may be used. Whenever possible, water and mild abrasives like sand are preferred. Cleaning with acids is usually applicable to platinum crucibles, but should be avoided in the case of alumina crucibles. Platinum crucibles can be cleaned especially thoroughly in a melt of potassium pyrosulfate. In all such operations it is important for the crucibles to retain their smooth surface, any scratches in the crucible wall may act as starting points for cracks and corrosion.

The Fig. 8 a shows tools for reshaping crucibles after use. During physical and chemical treatment crucibles are often deformed, especially noble metal. This results in a loss of good heat transfer to the thermocouple which records the temperature during experiments. Plastic tools are available for reshaping crucibles.

The points discussed before do not apply to the one-way crucible. Especially for organic compounds in the temperature range up to 500–600 °C, aluminium crucibles are used. Also for other compounds which are compatible with aluminium this type of crucible can be used.

Normally the size of the crucible is chosen according to the amount of sample to be studied. The type of thermoanalytical investigation is also important for the construction and form of the crucible (see special crucibles).

The most common crucible form in the laboratory is the cylindrical form (see Fig. 8 b). The size with respect to volume depends mainly on the expected weight change and on the homogeneity of the studied sample. For these types of crucibles lids are often used which do not close the liner hermetically, but rather influence the temperature homogeneity and the equilibrium of reaction by the self-generated atmosphere.

Cone. This type of crucible shape (Fig. 8 c) provides an increased contact area between crucible and its counterpart on the crucible holder. Especially in the combination TG-DTA this decreases the thermal resistance and allows a better resolution of the measured DTA-effect.

The flat crucible (Fig. 8 d) facilitates the spreading of the sample in the form of a thin layer. This kind of preparation is especially important for equilibrium studies and for reactions between the sample and the surrounding atmosphere. It also avoids any loss of substance during spontaneous decomposition reactions in high vacuum. The horizontal temperature gradient in these rather large crucibles (*e.g.* 20 mm diameter) must be taken into account and can be determined by a second thermocouple.

For larger samples, for achieving equilibrium conditions and for improved contact with the atmosphere the e-type of has proved to be very useful. When covered with a lid it also avoids any loss of substance during decomposition reactions.

Special Crucibles

Knudsen cells (effusion cells) are exclusively used for vapor pressure measurements (see vapor pressure) in the pressure range from 1 torr to 10^{-6} torr. In the low temperature range ($-20°$ – $+400$ °C) pyrex glass cells are applicable. Especially the vapor pressures of dyes, organic compounds can be measured in such cells, because metal cells may sometimes cause catalytic decompositions of the investigated materials. The Fig. 8 f shows a glass cell which is positioned on a four-hole capillary. The diameter of the cells is 20 mm. Cells with an orifice diameter of 0.5–5 mm are needed depending on the vapor pressure of the sample. The position of the thermocouple is so arranged, that the hot junction is in the center of the substance without direct contact. In the same temperature range metal cells (Al or stainless steel) can also be used when the above mentioned effect of catalytic dissociation does not occur. Alu-

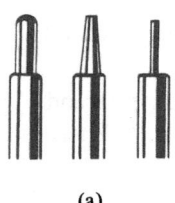

(a)

(b)

(c)

(d)

(e)

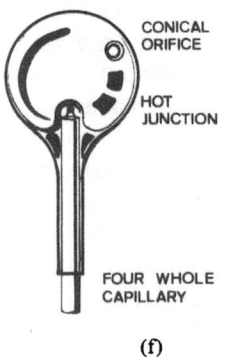

CONICAL ORIFICE

HOT JUNCTION

FOUR WHOLE CAPILLARY

(f)

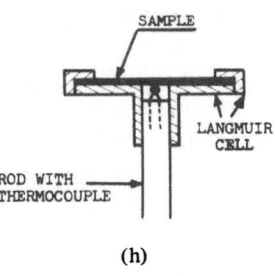

SAMPLE

LANGMUIR CELL

ROD WITH THERMOCOUPLE

(h)

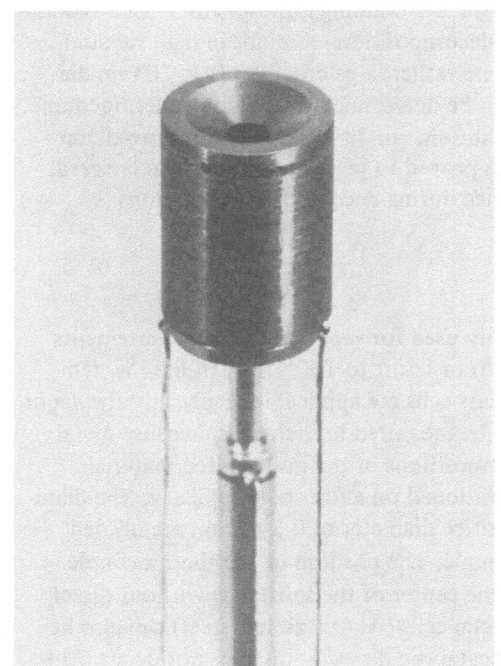

(g)

(i)

mina cells are used from low to high temperatures (400–1600 °C) especially for metals, alloys, salts ect.

Another high-temperature cell (Fig. 8 g, up to 2400 °C, can be produced from tungsten. Tungsten Knudsen cells are used primarily for high-temperature vapor pressure measurements, *e.g.* for metal oxides. They are suitable also for metals when graphite linings are applied to the inner surface. The vapor pressure can be determined from the effusion rate $\frac{\Delta m}{\Delta t}$ of the vapor emerging from the orifice by using the Knudsen equation $p = \frac{\Delta m}{\Delta t} \cdot \frac{1}{\pi \cdot r^2} \cdot \sqrt{\frac{2\pi \cdot R \cdot T}{M}}$ [dyn/cm^2]. The calibration of such cells will be discussed in the following Sects.

Langmuir cells (Fig. 8 h) are very similar to Knudsen cells. Here the vaporization of a solid from its uncovered surface can be measured in a similar way by thermogravimetric methods.

The ball-value is a special type of vaporization cell (Fig. 8 i) which is used for the high pressure range from 10 to 760 torr. Depending on the temperature range, the material of the cell and of the ball may be Al, steel, gold or sapphire. The ball is situated on top of the cell so as to form a valve which opens when the inside pressure is higher than the outside pressure plus the weight of the ball.

In the combination TG-DTA normally two crucibles of identical shape and size are used, one for the sample and the other one for the inert reference material. The crucibles may be *e.g.* cylindrical, conical, plate-type or any other form which favors a good contact between the sample, thermocouple and sample holder (Fig. 8 j). The thermocouples are placed either on the bottom of the crucible or in the center of a cylindrical crucible. The latter type allows the element to be brought protected or unprotected into the center of the substance.

DTA compares the sample temperature (T_s), with the temperature of a thermally inert material (T_r), as both are heated or cooled at a uniform rate. Temperature changes in the sample are due to endothermic or exothermic reactions which are caused by phase changes (fusion, structural transformations, boiling, sublimation, and vaporization), dehydration, dissociation or decomposition, oxidation and reduction, breakdown of crystalline structure, and various chemical reactions. Generally, phase transition, dehydration, reduction, and some decomposition reactions show endothermic effects, whereas crystallization, oxidation, and some decomposition reactions produce exothermic effects. When a reaction occurs in the sample the $T_s - T_r$ curve deviates from a horizontal position to form a peak in either the upward or downward direction, depending upon the enthalpic change. These peaks may be used as a first qualitative identification of the substance under investigation. Since the area under the peak is proportional to the heat-change involved the technique is useful for semiquantitative or quantitative determination of the heat of reaction. As ΔH is proportional to the amount of reacting substance, DTA can also be used to evaluate quantitatively the amount of substance present if the molar ΔH is known.

Sample holders for determination of the Curie point exist basically as two different types. A compact sample, *e.g.* a metal block can be investigated by attachment on a normal, rod-type TG-sample holder (Fig. 8 k). The sample must be provided with a hole for the TG-rod, the built-in thermocouple should not have any contact

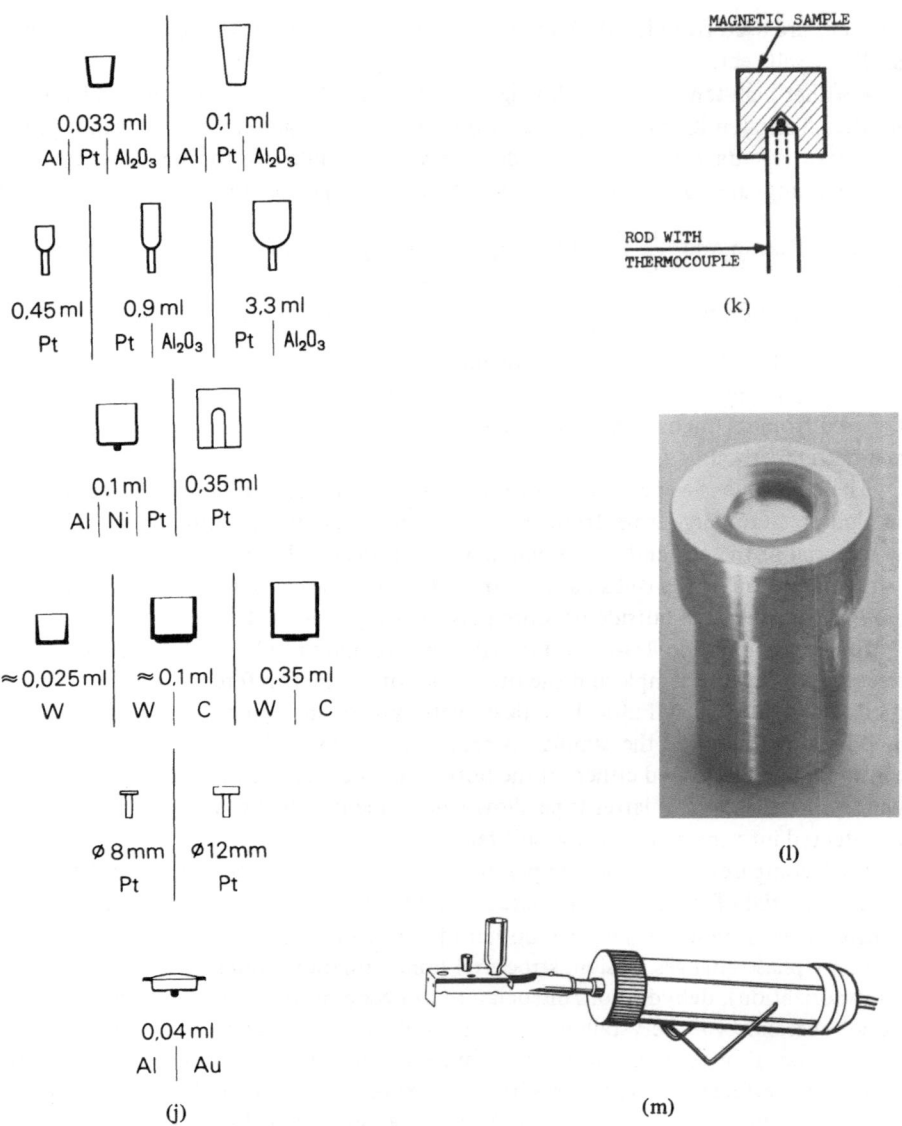

Fig. 8. Types of crucibles and auxiliary tools (a – m)

with the sample. The second type of holder is a closed crucible which protects the powdered sample from jumping out, due to the magnetic forces of the magnet during measurement. The dimension of the crucible is similar to that of the metal block.

An optional high-pressure crucible (Fig. 8 l) may be used for chemical reactions in solutions with an inherent vapor pressure of up to 100 atmospheres. The crucible is equipped with a centering pin for exact positioning on the measuring sensor.

Preparation of Samples

Instructions on how to prepare samples for thermal analyses can be given only very generally since the variety and complexity of the problems involved makes it impossible to specify strict rules of procedure. In carrying out series of tests on any particular material, it is advisable to make this up in sufficient quantity and also to test it by other analytical methods such as X-ray and IR. All samples should be as far as possible homogeneous. Sample conditioning, *i.e.* adjustment to a predetermined degree of moisture content, may be carried out either before or after samples are weighed and loaded on the thermobalance.

The sample weight is dependent on the homogeneity of the material and on the expected weight change. For very pure substances mostly sample weights from 10–20 mg are adequate for TG-investigations, smaller samples will show a more homogenous temperature distribution and shorter times for driving out the gaseous decomposition products.

Apart from tests carried out on single crystals, also fractions of uniform particle size should be tested in order to obtain precise readings and to ensure reproducibility of results. For runs in vacuo the use of particle sizes under 60 (0.25 mm) mesh should be avoided, since the gaseous products in the case of thermal decomposition may eject particles of the test material from the crucible.

In TG and simultaneous DTA procedures it is advantageous to free the samples of any physically adsorbed water and similar matter. Desorption can be effected by adopting the following procedures:
a) Pre-drying of the sample in a desiccator, drying cabinet or the like.
b) Placing the sample on the thermobalance and drying it for at least one hour under a vacuum of approx. 10^{-5} torr (any change in weight can be recorded in the course of this drying process), followed by flushing of the balance housing with the appropriate dry testing gas, and finally conducting the test in the usual manner. Samples where the reactions take place above 500 °C, can also be heated slightly during evacuation for faster drying.

When mixtures of substances are investigated, *e.g.* in solid state reactions, the mixture should be completely homogeneous and of uniform particle size. Smaller particle size, *i.e.* higher ratio surface area/volume, are important for such types of reaction:

A spatula is used not only for fast and precise filling of powdered substances, but also for vibration packing of samples in crucibles. A special holder allows the simultaneous vibration of crucibles (Fig. 8 m).

Crucibles of various forms and sizes can be vibrated. Uniform packing of the sample and reference is thus possible.

The spatula with a funnel and flow regulator can be used for weighing into very small crucibles or narrow-necked containers. The flow regulator is adjustable and holds back coarse granules assuring an even flow of fine substance.

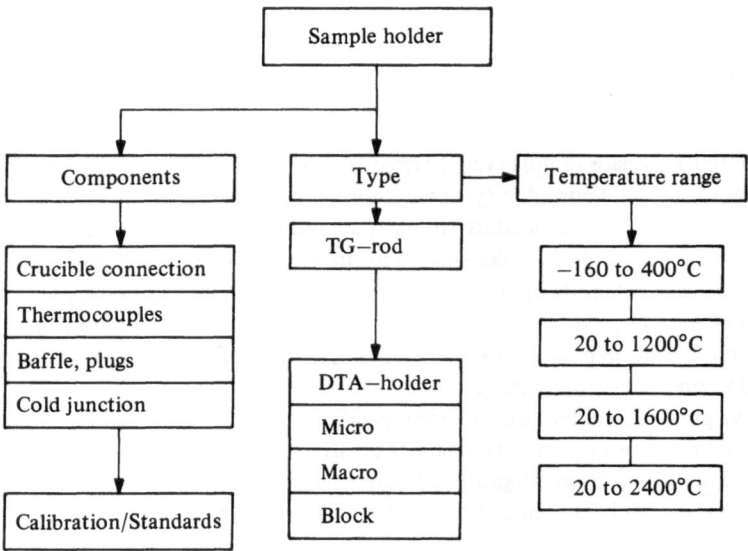

2.3. Sample Holders

TG and DTA are thermoanalytical methods which are very suitable for simultaneous investigations. Experimental parameters for both methods are almost identical. Many thermobalances combine these two methods. The construction of the special sample holders required is such that for DTA, as well as for the simultaneously measured TG, optimum conditions are possible. Low-mass sample holders assure that temperature gradients within the system are reduced to a minimum. Each sample holder is equipped with two thermocouples for ΔT measurements and for measurements of the furnace temperature. The elements for the ΔT measurement are designed so that large surface contact with sample and reference is achieved, thus reducing thermal resistance and increasing reproducibility of measurements. The thermocouples are placed either on the bottom of the crucible or in the center of a cylindrical crucible. The latter type allows the elements to be brought protected or unprotected into the center of the substance. The interchangeability of all sample holders and easy exchange of crucibles are important features.

TG – Rod

The basic components of a TG-rod and of a DTA-holder are very similar (Fig. 9).
All special type crucibles are compatible with the TG-rod, only for different temperature ranges are changes necessary.

The different types of TG-crucibles which can be used with this kind of rod were shown in Sect. 2.2.

A thermocouple is built into the rod, the "hot junction" is positioned at the end of the rod. The crucible bottom is in direct contact with this measuring point.

The built-in thermocouples can vary depending on the temperature range

platinum – platinum/rhodium (10% Rh)   25 – 1600 °C
nickel – chrom/nickel   25 – 800 °C
tungsten – tungsten/rhenium   25 – 2400°C
gold – nickel   25 – 550 °C

The baffle mounted on the rod is primarily designed to catch any substance which may jump out of the crucible during measurement. In addition, it provides the balance with protection against heat radiation. When the crucible support is inserted, the baffle rest in a socket and thus separates the reaction chamber or furnace chamber from the balance housing.

The plug permits the sample holder to be exchanged rapidly, *e.g.* for substitution of a DTA-sample holder by a holder with thermocouples fitting another temperature range.

It is important to know the temperature of the "cold junction" for evaluation of the exact temperature in a measured curve. In the case where the temperature is

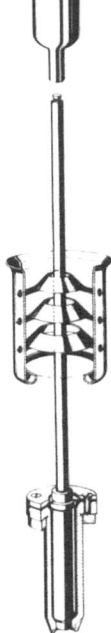

Fig. 9. TG sample holder with individual components

0 °C, no correction is required. For higher temperatures at the "cold junction" corresponding corrections have to be made.

Temperature Standards

For exact temperature evaluation the thermocouple must be calibrated with temperature standards. Today commercial temperature standards are available: Nat. Bureau of Standards (NBS) Washington offers standard reference materials (No 759, 758, 760) as DTA-temperature standards[16a-e] for 3 temperature ranges: 125–435 °C/295–675 °C/570–940 °C.

The materials are listed in Table 3.

Table 3. DTA-Temperature Standards
NBS-ICTA standard reference material

Transition Temperature Data (°C)

Material	Equilibrium Value[1]	DTA Mean Values[2]	
		Extrapolated Onsel T_A	Peak T_B
KNO_3	127.7	128	135
In (metal)	157	154	159
Sn (metal)	231.9	230	237
$KClO_4$	299.5	299	309
Ag_2SO_4		424	433
SiO_2	573	571	574
K_2SO_4	583	582	588
K_2CrO_4	665	665	673
$BaCO_3$	810	808	819
$SrCO_3$	925	928	938

[1]) NBS Circ. 500 (1952).
[2]) NBS Special Publication 260–40 (1972).

The procedure is to heat up a sample of these materials in the thermobalance. Figure 10 shows the DTA-curve and the evaluation of point A and B. These can be

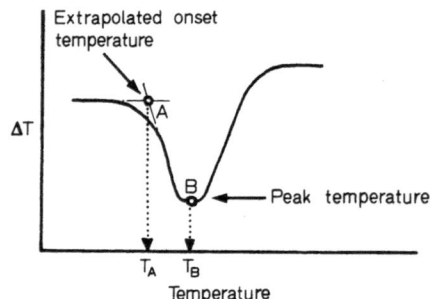

Fig. 10. Evaluation of onset- and peak temperature of a DTA-peak

compared with the data given for the NBS standards, which have resulted from measurement in 34 cooperating laboratories with 28 different types of commercial or custom-made instruments.

Manual Evaluation of the Calorimetrie Sensitivity

The following diagram (Fig. 11) shows the original DTA melting curve of indium and the inserted calibration rectangle which is required for the manual evaluation. The experimental parameters which were used for the calculation of ΔH are given below, and also the evaluation of $\mu V \cdot sec$ for the peak area.

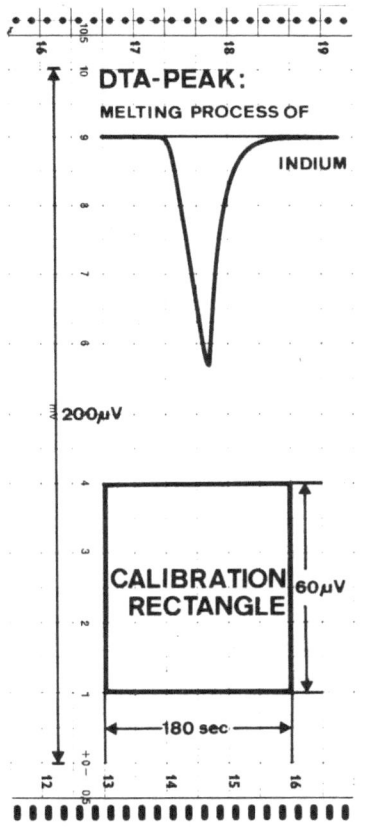

Experimental Parameters	
Sample weight of indium:	4.81 mg
Heating rate:	1 °C/min
Chart speed:	2 cm/min

Weight of the peak area: <u>70,57 mg</u>

↑

(cut out from chart paper)

↓

Weight of the rectangle area: <u>373,62 mg</u>

Calibration rectangle: 60 μV x 180 sec = 10 800 $\mu V \cdot sec$

Fig. 11. Quantitative evaluation of DTA-peaks

Peak area evaluated in $\mu V \cdot sec$:

$$\boxed{\begin{array}{c}\text{Peak area} \\ \text{in } \mu V \cdot sec\end{array}} = \frac{\mu V \cdot sec_\square \times \text{weight }^\triangledown}{\text{weight}_\square} = \frac{10800 \times 70.57}{373,62} = 2039,92 \ \mu V \cdot sec$$

Calorimetric Sensitivity:

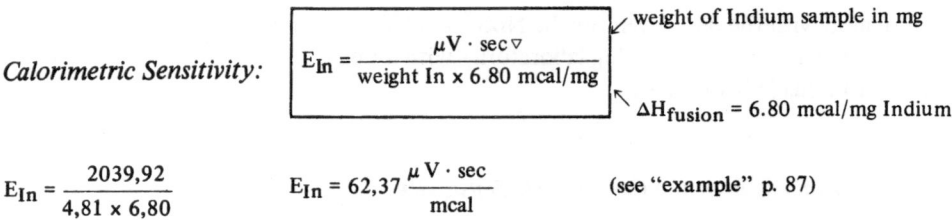

$$E_{In} = \frac{\mu V \cdot sec \, \triangledown}{\text{weight In} \times 6.80 \text{ mcal/mg}}$$

weight of Indium sample in mg

$\Delta H_{fusion} = 6.80$ mcal/mg Indium

$$E_{In} = \frac{2039,92}{4,81 \times 6,80} \qquad E_{In} = 62,37 \frac{\mu V \cdot sec}{mcal} \qquad \text{(see "example" p. 87)}$$

The calorimetric sensitivity E_{In} is then used for calculating an unknown ΔH of an investigated substance:

$$\Delta H = \frac{\text{peak area } (\mu V \cdot sec)}{\text{cal sensitivity } E \; \dfrac{\mu V \cdot sec \cdot mg}{mcal}} \qquad \Delta H = \ldots . \text{mcal/mg}$$

Calorimetric Sensitivity in Function of Temperature

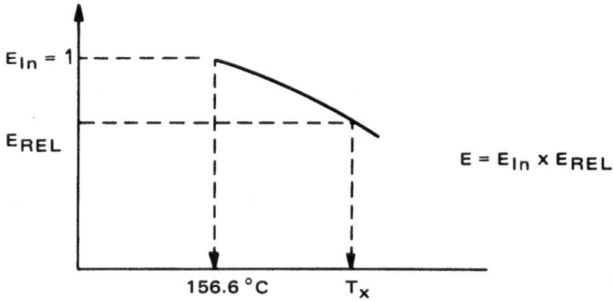

$E = E_{In} \times E_{REL}$

The calibration curve, where the sensitivity E_{REL} can be determined at other temperatures T_x, depends on the type of used thermocouples.

DTA-Holders

The components of the DTA-holder (Fig. 12) are similar to those of the TG-rod.

The crucibles of DTA-holders are chosen according to the sample weight. That means that for small samples or large samples one has to decide between macro and micro holders.

Some of the "hot junctions" are so formed that the crucible fits in it. Others are similar to the arrangement in the TG-rod.

Different, built-in thermocouples are used for covering the whole temperature range from -160 up $+2400$ °C. The selection of the thermocouple-type depends on the investigation which is to be carried out.

Baffle: See TG-rod.

Plug: See TG-rod. Also for "cold junction", calibration

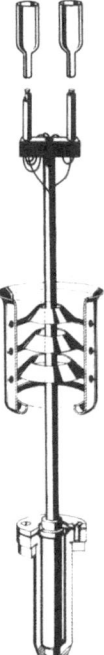

Fig. 12. Typical DTA sample holder

DTA-Microholders

Microsample holders are made from four-hole alumina capillaries. Small receptacles for the crucibles containing reference and sample are welded to the two thermocouples of the DTA holder at the hot junction points (Fig. 13 a).

The other end of the capillary tube carries a plug type connector which allows the crucible holder to be rapidly removed and replaced. The connector is adjustable with a centering device.

Technical Data: Temperature Range, 25–1600 °C. Crucible Material, Aluminum, for 25–580 °C range, Platinum, for 25–1600 °C range. Sensitivity ranges and maximum heating rates, for 500 μV 15 °C per minute, for 100 μV 10 °C per minute, for 20 μV 4 °C per minute. This crucible holder may be used in oxidizing and in inert gaseous atmospheres (e.g. air, N_2, Ar, He, CO_2 and the like), and also in reducing atmospheres, but in this latter case only up to a temperature limit of 1400 °C (Fig. 13 b).

A special type of micro DTA-holder (Fig. 13 c) uses large-surface, vapor-deposited thermopiles as the ΔT sensor which assure excellent heat contact with the crucibles. No undesirable heat exchange occurs at the sensor leads which are also vapor-deposited. This results in high sensitivity and precision.

The ΔT sensor is resistant to oxidation and to many corrosive gases over the temperature range up to 600 °C and is easily replaced.

The standard aluminum crucibles can be sealed vapor-tight (Fig. 13 d).

Another type of Micro TG/DTA sample holder is shown in (Fig. 13 e). It can be equipped with Al, Ni, or Pt crucibles with a miximum volume of 0.15 cm³. For use with the low-temperature furnace, or for sensitive DTA in the middle-range temperature furnace up to a maximum of 600 °C.

For very high temperatures up to 2400 °C a different Micro TG/DTA sample holder (Fig. 13 f) must be used which usually is equipped with tungsten crucibles. The thermocouple in this case is a combination of W − W 26% Re. This type can be applied for thermoanalytical measurements in high vacuum and in inert atmosphere (noble gases).

DTA Macroholders

The Fig. 13 g shows the sample holder with the two hot junctions of the DTA- and temperature control thermocouple. The two crucibles (sample and reference) must be placed in position so as to ensure satisfactory contact between hot junction and base of the crucible. After the experiment, any adhering crucible should be loosened carefully by heating, and the holder freed of impurities by annealing at approx. 1000 °C. Technical Data: Temperature Range, 25−1600 °C. Crucible Material, Aluminum, for the 25−580 °C range, Platinum, for 25−1600 °C, Alumina, for 25−1600 °C. The crucible holder may be used in all oxidizing and inert gaseous atmospheres, in high vacuum and in reducing atmospheres up to a working temperature of 1400 °C.

The sample block cell 13 h consists of an aluminum oxide block which can be removed from the base plate. The thermocouples serving for DTA and temperature measurement are reinforced up to the hot junction by ceramic oxide capillary tubes. The sample cell can be mounted in either of two alternative positions, according to the quantity of test substance used. The DTA thermocouples extend into either the large or the small hole.

The usual thermocouple materials are Pt and Pt-alloys. For increasing the rather low thermovoltage, thermoelements with "pallaplat" are very suitable, only however up to a maximum temperature of 1200 °C.

Whereas alumina ceramics are the usual materials for sample holders in the range up to 1600 °C, a special sample holder made of tungsten is necessary for ultra high temperature ranges. The thermocouple consists of W-W 26% Re. Such sample holders can be used only in high vacuum or in inert atmosphere.

(a)

(b)

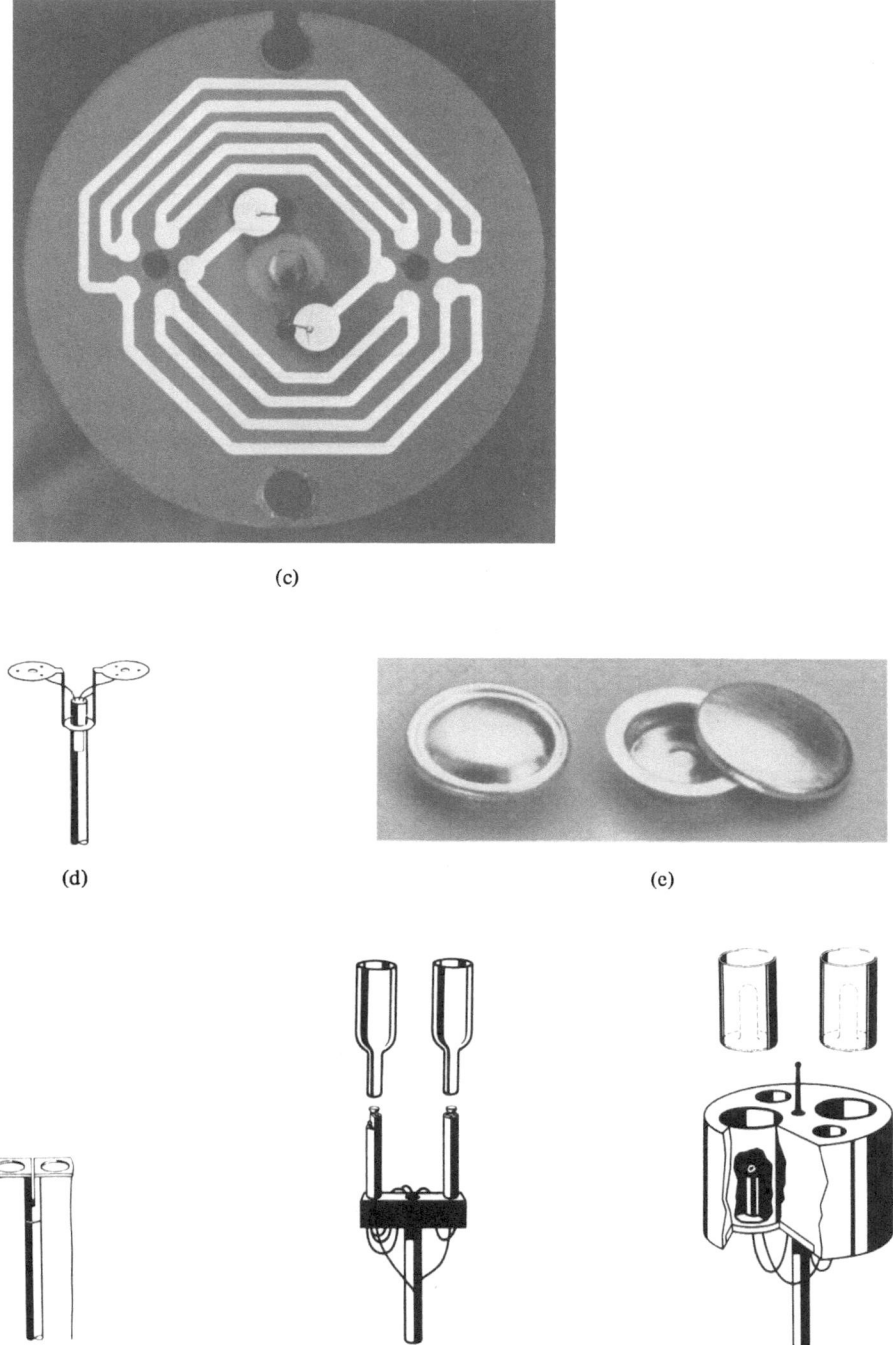

Fig. 13. Special types of DTA micro- and macro-sample holders (a – h)

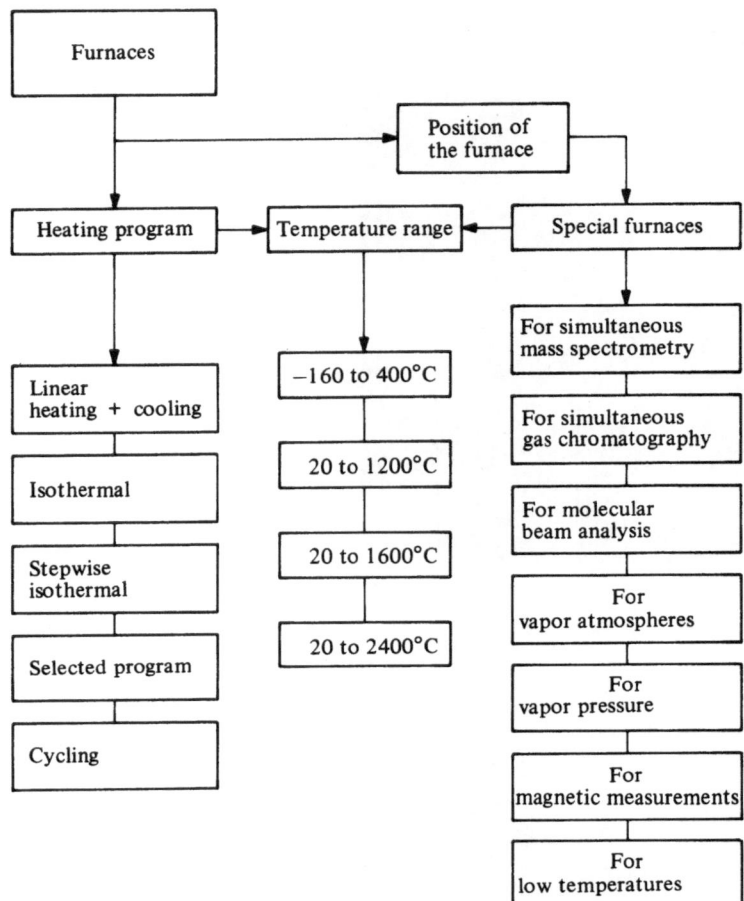

2.4. Furnaces

Heating Program

Of main importance is the linear heating and cooling of a sample during a thermo-analytical investigation: (Fig. 14 A).

Heating rate: 0.2 °C/min → 100 °C/min
Reproducibility of the used heating rate for comparison with other investigations. Selection of small and large heating rates depending on the type of investigation. Equilibrium measurements (*e.g.* 0.2 °C/min) or stepwise isothermal. Exploratory runs (standard rate 4−6 °C/min).

Some investigations require isothermal conditions (Fig. 14 B). The main point is usually to reach the temperature rapidly, so that one can follow the whole process under isothermal conditions.

There are two recommended procedures:

1) Separate heating of the furnace up to the required temperature and positioning of the furnace over the sample.

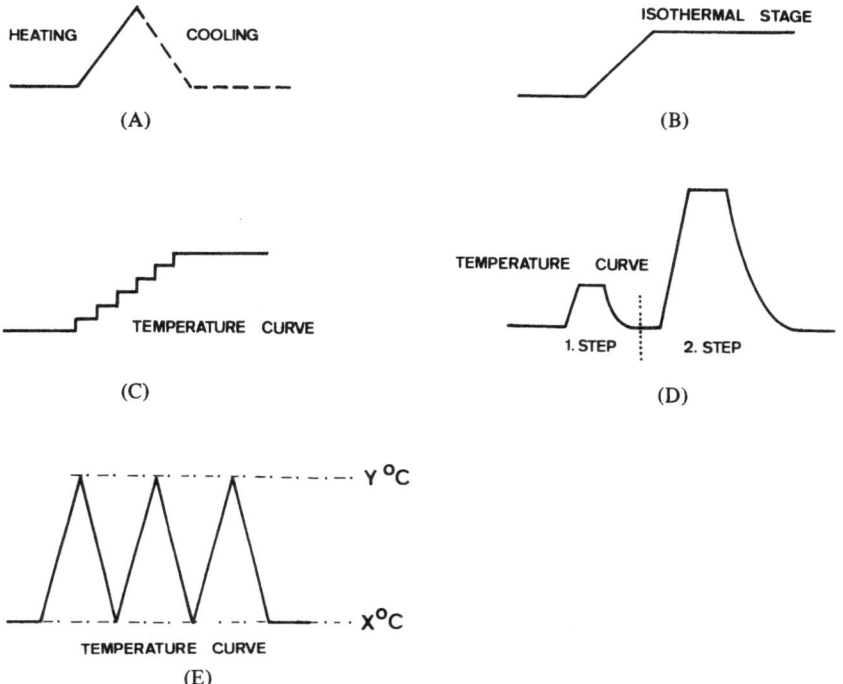

Fig. 14. Heating programs: linear heating and cooling (A), isothermal (B), stepwise isothermal (C), multistep program (D), cycling (E)

2) Heating up of the sample to the selected temperature with a high heating rate (*e.g.* 100 °C/min).

In such cases where even slow linear heating is too fast, stepwise isothermal heating is recommended (Fig. 14 C). Some apparatus have a built-in device or, alternatively, the realization is very simple made possible with two electrical switch clocks (see Sect. 4.1).

Selected Program

Some investigations require a preselected temperature program. For example, a substance should be dried at a special temperature, and afterwards one wants to know the complete heating behavior from room temperature to the selected temperature, with an isothermal step and uncontrolled cooling (Fig. 14 D). For such problems some of the commercial instruments are prepared either by computer, or by simple switching clock-programs which allow one to follow a preprogrammed course.

Most of the thermobalances have built-in, simple heating programs which allow *e.g.* linear heating and cooling with different rates, change to isothermal conditions, or to cycle the temperature between two preselected values (Fig. 14 E). The application of the latter is to check the reversibility of certain decompositions, or the reproducibility of DTA-peaks.

H. G. Wiedemann and G. Bayer

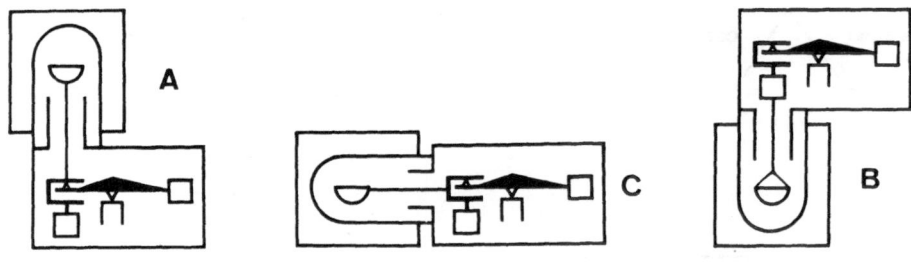

Fig. 15. Positions of the furnace with respect to the balance (A, B, C)

Position of the Furnace

There are basically three possibilities for arranging the furnace which used is with
a thermobalance. Above, below and beside the balance housing. The common arrangement (Fig. 15 A) today is placing the furnace above the balance. The advantages of this combination of balance and furnace are:

Very easy to change samples and sample holders and to carry out other operations; simple compatibility with other apparatus for simultaneous measurements;
convection goes the natural way, no additional shielding is required for protection of the balance against radiation from the furnace. The other two arrangements 15 B and 15 C which are used besides the first variant are equal in the distribution of use. The advantages of variant C are similar to that of A:
Very good temperature homogeny over a wide temperature range. Convection
in the furnace has only a very small influence on the balance without necessity
for special shielding. The disadvantages of this construction are the position of
the sample holder and its lever arm, its sensitivity to vibrations.
The arrangement B shows only few advantages and special applications in experiments which require the positioning of the sample below the balance e.g. in
thermomagnetic measurements.

The disadvantages are: Strong shielding for protection of the balance is required,
poor temperature homogenity, change of samples is complicated and time consuming.

The horizontal temperature distribution in various furnaces was measured around
a circle of 15 mm ϕ diameter in the height of the crucible position. Heat shadows
between the heating elements could be seen in the temperature distribution curves
of the high temperature furnaces which is connected with geometry. At 1000 °C the
differences are within the order of several degrees.

Special Furnaces: For Simultaneous Mass Spectrometry

Figure 16 shows a schematic sketch of a thermobalance for TG, DTA and mass spectrometer
measurements. – In the center is the thermobalance, enclosed in a vacuum-tight tank, with a
thermostatically controlled water jacket. The reaction chamber (R) is surrounded by the furnace and is clearly separated from the balance housing by a diffusion baffle. Diffusion pumps
(K) evacuate the balance housing and the reaction chamber.

94

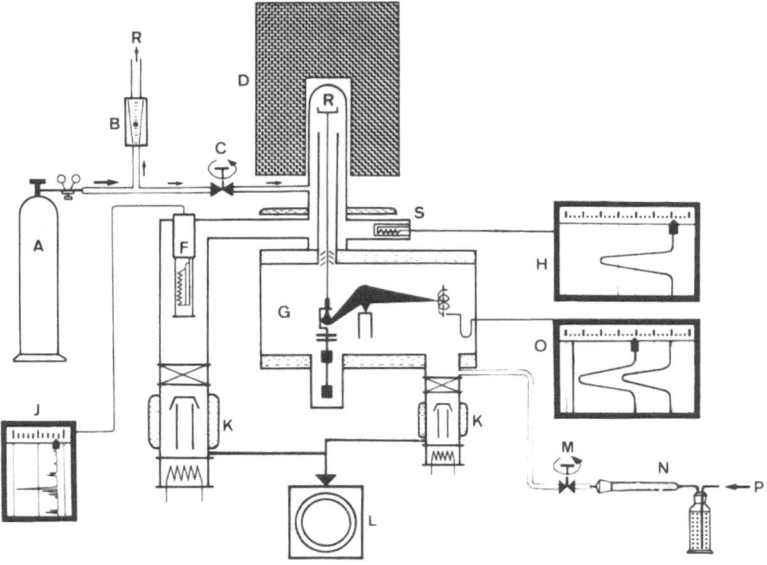

Fig. 16. Thermobalance combined with mass spectrometer

The decomposition gases immediately pass the mass analyzer (F) which, in turn, is connected to a recorder (J) through the mass spectrometer control panel.

Total pressure, required for detailed interpretation of the mass spectra, is determined with an ionization gauge (S). The gas inlet system (A, B, C) is used for calibration purposes. The relation between measured total pressure and the ion current of an injected specific gas permits calibration of the mass spectrometer in absolute partial pressure units or amps/torr.

A second gas inlet system (M, N, P) serves to set starting conditions and to vent the system after a test.

Operating data:

pressure range for the combination described – 10^{-4} to 10^{-6} torr

mass range – $1 - 100$ and $10 - 400$ amu

scan speed per mass unit, for line recorders – 1, 3, 10 sec. for oscilloscope recording – down to 0.001 sec.

This means that even very rapid reactions (detonations) may be followed on the thermobalance. Range, or single-mass recording is possible; thus, the pressure behavior of a specific mass, take for instance water, M/e18, or CO, M/e28, can be traced all over the complete decomposition range simultaneously with the thermal decomposition.

If the same chart speed is selected for TGA, DTA, total and partial pressure curves, the curves can be interpreted simultaneously as a function of temperature by making a simple overlay.

Working at Different Atmospheres

The Fig. 17 shows an experimental set-up for investigations of thermal decomposition accomplished in different atmospheres and atmospheric pressure.

In experiments which are carried out under normal pressure, the gaseous products are taken from a modified, commercially available low temperature furnace. The gas flows viscously in the specially designed capillary through an intermediate volume to the peristaltic pump. Hereby the gas pressure in the intermediate volume is reduced to approx. 1 mm Hg. By means of a variable

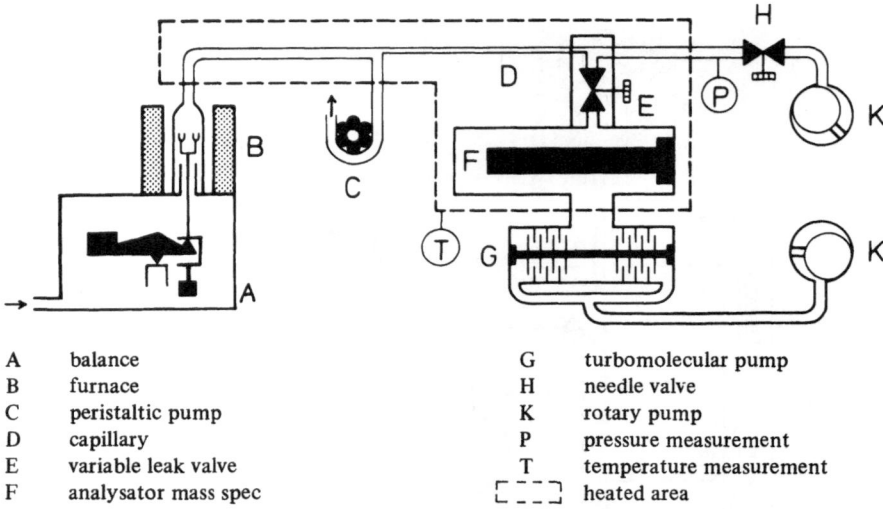

A	balance	G	turbomolecular pump
B	furnace	H	needle valve
C	peristaltic pump	K	rotary pump
D	capillary	P	pressure measurement
E	variable leak valve	T	temperature measurement
F	analysator mass spec	⌐ ⌐ ⌐	heated area

Fig. 17. Thermobalance for investigations in different atmospheres

leak valve, part of the gas enters under molecular conditions into the mass spectrometer located in the high vacuum region of the system.

Simultaneous Gas Chromatography

Compared to mass spectrometric analysis, gas chromatography is slower. As far as the quality of the results is concerned, both methods may be considered to be equal.

The Fig. 18 shows a schematic layout for simultaneous TG and GC measurements.

An interesting device is the furnace tube with a volume of only 35 cm^3, made of quartz. A small volume is essential, in order to keep the time delay low during gas transportation.

The decomposition gases are taken to the sampling valve which, according to the gas to be analyzed and the retarding column used, can be operated at specific intervals. A helium gas stream flushes the reaction gases into the gas chromatograph.

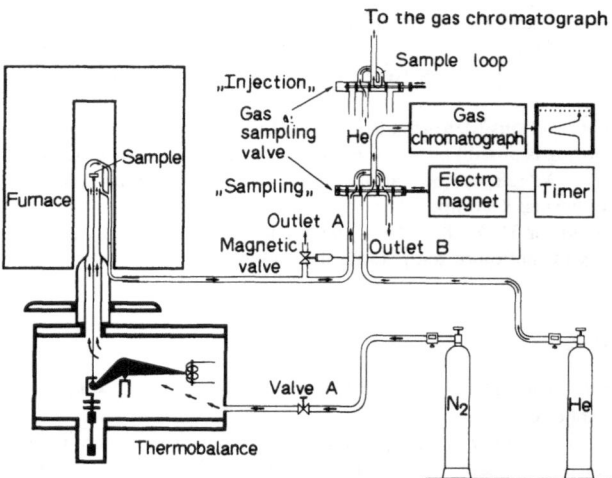

Fig. 18. Thermobalance with simultaneous gas chromatography

Molecular Beam Analysis

The combination of a heating X-ray camera with a conventional thermogravimetric apparatus is difficult, mainly because of geometrical and focussing problems. Therefore a different, new approach was taken in which the weight change of the sample is followed *via* its gaseous decomposition products. This method has the advantage that the sample can be mounted as usual in the sample holder of the X-ray camera and be heated up continously during exposure of the film without changing its position. The gaseous reaction products are then carried from the camera to the thermoanalyzer through a vacuum-tight link. They form a molecular beam which is directed to the balance pan. Due to the special, flat shape of the inlet orifice and of the balance pan, the arriving molecules bounce back and forth several times before they are removed through the vacuum system. The force of the impacting molecules is proportional to the first derivative of the weight change with time. Therefore the peak area of such recorded curves is directly proportional to the total change in weight of the sample. This new thermoanalytical method is called "Thermo Molecular Beam Analysis" (TMBA).

The Fig. 19 shows schematically a typical instrumental set-up in which a thermobalance is combined with a high temperature X-ray camera and with a quadrupole mass spectrometer.

The camera is mounted directly above the thermoanalyzer, its furnace is connected to the heated system of the impact plates through the heated vacuum-tight link. The vacuum in the balance is maintained by means of a shut-off valve, only the camera has to be opened during the sample change and evacuated afterwards.

The sample is used in finely powdered form and spread on a fine platinum wire mesh which is mounted in the sample holder frame (Pt). The thermocouple (PtRh 10%) for temperature measurement is directly in contact with the sample holder.

A special film holder allows transportation of the film with various rates. Time marks are printed automatically on the film for correlating the X-ray patterns to specific times and temperatures in the TMBA curves. The temperature program of the X-ray camera furnace is regulated by the thermobalance heating control system. Up to the maximum temperature of 1200 °C usual heating rates can be varied from 0.2 to 4 °C/min. The temperature of the impact plates can be held constant between room temperature and 450 °C and is recorded during the

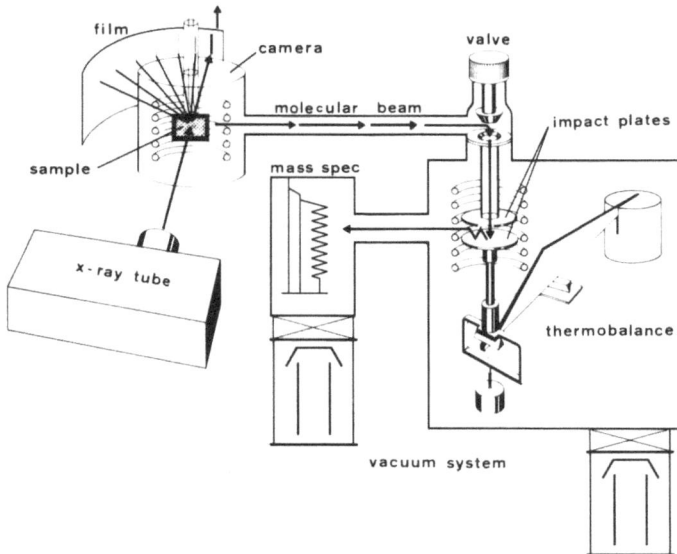

Fig. 19. Instrumental set-up of thermobalance, X-ray high temperature camera and mass spectrometer

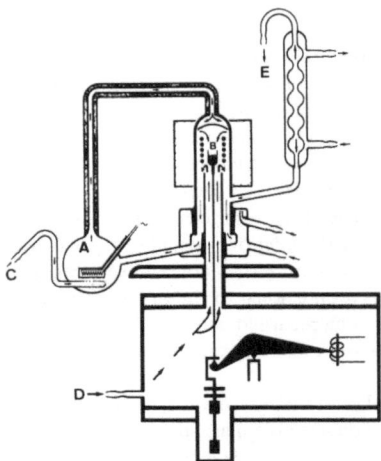

Fig. 20. Thermobalance with furnace for special vapor atmospheres

experiment. The distance of the impact plates is variable from 0.5 to 4 mm ± 0.01 mm by means of a micrometer.

Vapor Atmospheres

The Fig. 20 illustrates schematically the system for tests in special vapor atmospheres.

The crucible "B" containing the substance is placed on top of a sample holder in the center of the furnace heating zone. The sample holder is attached to the balance stirrup. In the flask "A" water vapor is generated, which is carried by the gas entering at "C" into the furnace through the insulated connecting tubes. The condensed water runs back to the flask "A". The gas entering at "D" flows up through balance and inner gas flow tube and prevents water vapor from the furnace to enter the balance housing. The gas leaves at "E" after passing through the back flow condenser.

Measuring the amount of water vapor passed through the furnace is possible only by feeding the water condensate into a graduated measuring cylinder instead of draining it back into blask "A". To compensate for the evaporated water, flask "A" has to be continuously replenished by a drop feeder.

If, for isothermal conditions, a constant gas flow through the flask "A" is used, the amount of water vapor carried by the gas into the furnace is also constant. The amount of water vapor

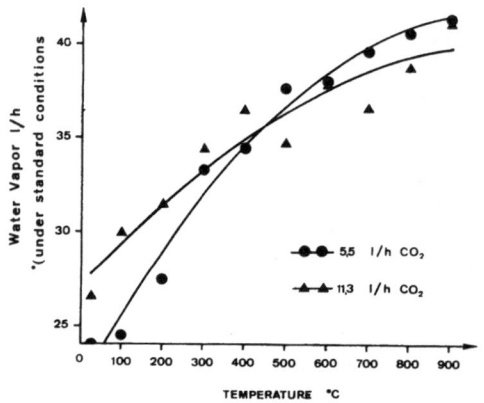

Fig. 21. Diagram showing the water vapor transport for two gas flow rates of CO_2 through the reaction chamber (furnace of the thermobalance) as a function of temperature under standard conditions* (20 °C and 760 torr)

carried along by the gas will constantly change with the temperature, when a linear heating rate is used for dynamic studies.

Tests were therefore made to establish the amount of water vapor that flows through the reaction chamber of the thermobalance at different temperatures. Results are shown in the Fig. 21 for two different CO_2[a] gas flow rates. The curves show the water vapor flow measured as condensate, as a function of temperature.

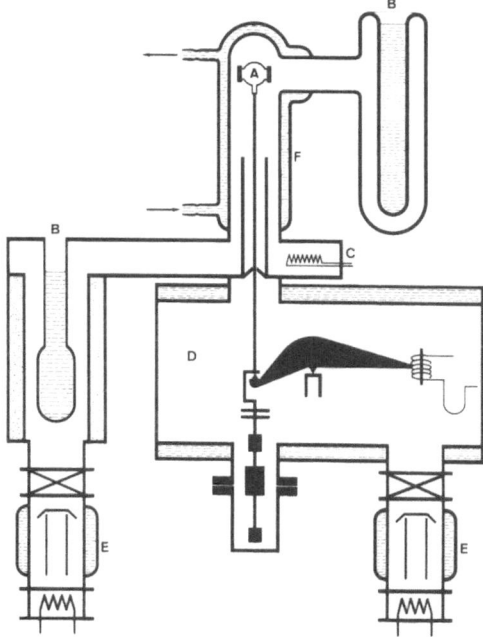

Fig. 22. Thermobalance for vapor pressure measurements. Schematic drawing of experimental equipment. A-Knudsen cell; B-cold trap; C-Ionization gauge; D-Balance and housing; E-Diffusion pumps; F-Thermostatically controlled reaction chamber

Vapor Pressure Measurements

The apparatus and the special accessories necessary for this work are schematically illustrated in the Fig. 22. The reaction chamber used for the Knudsen effusion method is positioned above the balance. The reaction chamber is thermostatically controlled and connected with a cold trap. Both of them are protected from outside temperature effects by an insulating material. This protection leads to a more constant temperature and a straight line in the recorded loss in weight.

Magnetic Measurements

For magnetic measurements a special furnace is not required; only some details must be considered (Fig. 23):

1) The furnace which is used together with the magnet should not shield too strongly the magnetic forces from the sample.

2) The magnet must be protected against heat radiation from the furnace. This means that during a long-time run at higher temperatures the magnet must be cooled. Only a magnet of constant temperature gives the same homogeneous magnetic field for the sample to be examined.

---

[a] Carrier gas in this case CO_2 (see Application 4.9, p. 130) for other gases is a special caliberatron required.

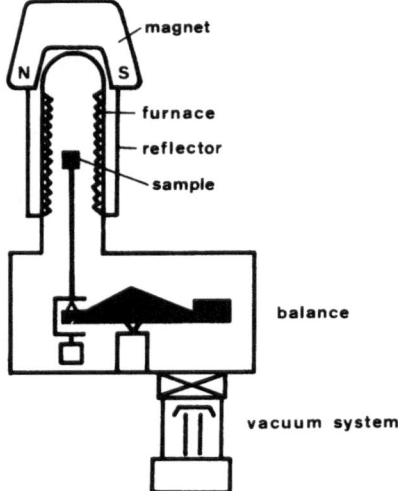

Fig. 23. Thermobalance for magnetic measurements

3) Measurements of metals and alloys in a inert atmosphere show only the magnetic transformation ferromagnetic ↔ paramagnetic. In oxidizing atmospheres the weight change of oxidation overlaps the effect of the magnetic transformation and cannot in this case be clearly identified.

For many ferrites the measurements of the ferrimagnetic ↔ paramagnetic transformation can be carried out in air.

Low Temperature Furnace

This glass furnace (Fig. 24) can be operated with non-corrosive gas atmospheres, *e.g.* helium, air, nitrogen, etc. Nitrogen is used as a coolant. The heating device F/E allows the speed of vapori-

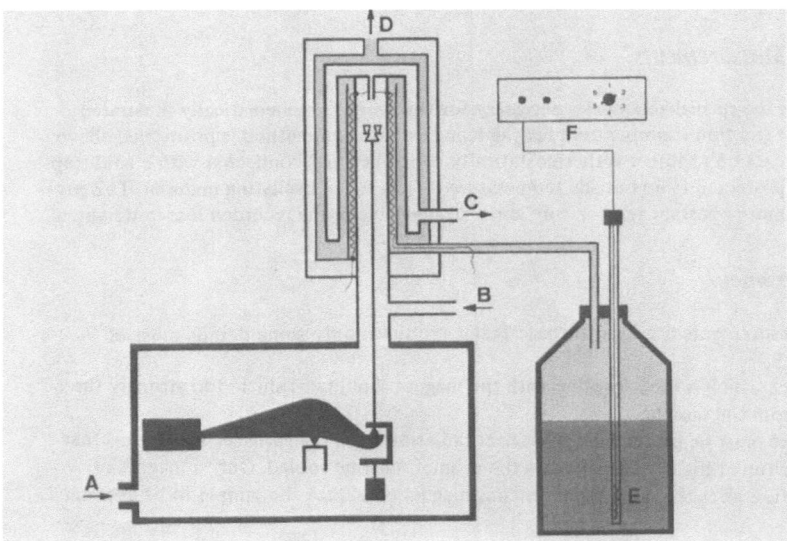

Fig. 24. Low temperature furnace for thermogravimetric measurements

zation of the liquid nitrogen to be modified and to cool down the furnace. D is the separate outlet of the vaporized coolant. The built-in bifilar heating coil permits heating up samples from $-140\,°C$ to $400\,°C$. The temperature is measured in this range with NiCr/Ni thermocouples. Two carrier gas inlets A and B permit operation in defined atmospheres. The fastest heating rates which guarantee a linear heating with respect to cooling are not higher than $4\,°C/min$.

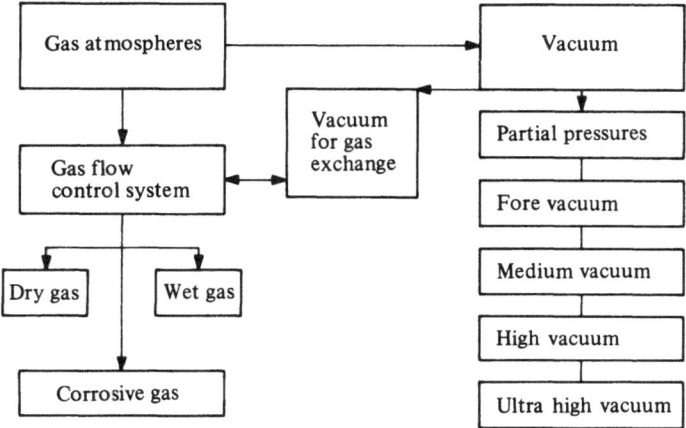

2.5. Gas Atmospheres

The fields in which the thermobalance can be applied depend on the possibilities of varying the conditions which affect the sample. The course of the reaction is particularly dependent on the ambient atmosphere and pressure. In the case of the balance design shown, it is possible to operate not only in a flow of air, but also in other defined atmospheres such as hydrogen, nitrogen, oxygen or gas mixtures.

Figure 25 shows a section of the entire vacuum system and its arrangement in

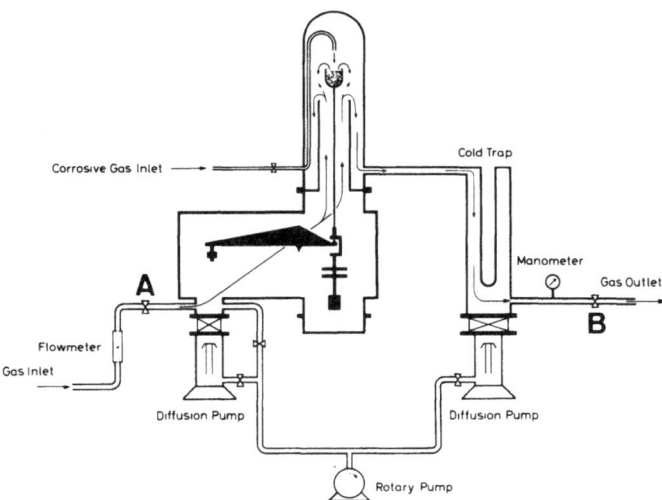

Fig. 25. Gas flow and vacuum system of a thermobalance

relation to the thermobalance. When the atmosphere is changed, the balance must be evacuated and then refilled with the appropriate purified gas by means of the needle valve A. When atmospheric pressure is reached again, the outlet valve H is opened. From that point on it is possible to work in a gas flow without affecting the weighing result. The path of the gas stream is indicated in Fig. 25. The labyrinth arrangement shields the balance from corrosive decomposition products and condensable substances. For measurements in corrosive gases, an inert gas must be pumped in through the gas inlet and a separate inlet used for the corrosive gas.

Because the balance housing is separated from the reaction chamber (cf. Fig. 25), one diffusion pump evacuates only the balance housing whereas the other evacuates the reaction chamber. The reaction chamber is connected to the diffusion pump by a cold trap. The decomposition products can be taken from the gas outlet for analysis or condensed on the cold trap by means of a coolant.

In the case of operations ranging between 760 and 10^{-3} torr, the two-stage rotary pump is adequate. Ultimate vacuum, using the diffusion pump installation, is better than 10^{-6} torr.

A precision aneroid manometer is used for measurements in the $760 - 1$ torr range. Thermocouple gauges are used in the $1 - 1 \times 10^{-3}$ range. A cold cathode ionization gauge is used in the high vacuum range down to 10^{-6} torr.

Gas Flow Control

The gas flow should be in the order of $0 - 10$ l/h and is normally kept constant during the measurement. If higher rates are necessary their effect on the weight curve should be determined by calibration runs. Usually noncalibrated flow meters are used. The amount of the gas, that means the flow rate through the balance, depends on the density and viscosity of the gas. Such values are either listed in tables or must be taken from calibration experiments.

For the reproducibility of all measurements it is very important to use a gas of constant composition, this is true especially for the humidity. The control of gas composition is possible either by special measuring cells (*e.g.* for H_2O, SO_2, CO_2) or by gas chromatography. Normally thermogravimetric measurements are carried out in dry gas atmosphere, this is important for kinetic studies.

Such special atmosphere is often necessary in decomposition and rehydration studies of hydroxides, hydrates etc. Furthermore also the catalytic effect of water vapor on certain reactions is of interest. For such studies the gas is saturated with water, or other vapors *e.g.* D_2O, alcohol, CS_2, etc. When higher water vapor concentrations are required special furnaces are available (see Sect. 2.4).

For investigations on dry and wet corrosion of metals and other materials additional components must be used. Shielding and protection of the balance parts against the corrosive gas is necessary.

For changing the gas atmosphere, the entire balance housing and the sample chamber must be evacuated down to approx. 10^{-3} torr. In cases where evacuation cannot be applied, the flowing exchange of the gases is recommended. Depending on the volume of the sample chamber and on the flow rate, this procedure may take around 3 hours for a total volume of 15 liters and a flow rate of 5 l/h.

Vacuum

The operating steps are very similar as for a gas exchange, except that a defined partial pressure is maintained during the whole experiment. Care must be taken that any gaseous reaction products generated do not change the pressure set. If necessary the pressure must be controlled and kept constant by means of a manostat.

The combination of a vacuum system and a thermobalance facilitates carrying out measurements at various pressures (vacua). Figure 25 shows the vacuum system connected to the thermobalance. Normally the vacuum system consists of a rotary pump, diffusion pumps, valves, cold trap, the actual recipients and the balance housing with reaction chamber. The balance housing may be evacuated separately by one diffusion pump while the other evacuates the reaction chamber which is connected to it by a cold trap. Reaction products may be collected for analysis at the gas outlet or condensed into the cold trap using a coolant. For work in the 760 to 1×10^{-3} torr range, a two-stage rotary pump is adequate. The final vacuum attainable with the diffusion pumps may be in the order of 1×10^{-6} torr. An aneroid fine vacuum gauge is provided for measurements in the fore vacuum range; for the medium vacuum range, thermocouple type pressure gauges are used and for the high vacuum range a cold-cathode ionization vacuum gauge. During measurements under vacuum, most substances give off their adsorption water even without heating. This is evident in the "running away" of the weight curve. The attainable final vacuum sets is only after the water has evaporated or been pumped out.

Table 4. Typical pumping times for reaching vacuum (system without sample)

Pumping time	From normal pressure To medium vacuum approx. 1×10^{-3} torr	To high vacuum approx. 1×10^{-6} torr
With frequent vacuum measurements	5 min.	25 min.
After long break	10 min.	45 min.

There are many applications for high vacuum thermogravimetry, *e.g.* determination of vapor pressures, degradation and degassing of materials in vacuum, reactions without oxidizing or reducing atmosphere. Of special importance are mass spectrometric gas analyses in combination with thermogravimetry. These require at least a vacuum of 10^{-3} to 10^{-6} torr. There exist also thermobalances for the ultra high vacuum range 10^{-6} to 10^{-9} torr. Special vacuum seals and turbomolecular vacuum pumps are necessary in this case.

3. Effects of Sample Properties and of Experimental Parameters

The information which may be derived from thermoanalytical measurements generally depends on a number of experimental conditions[17]. Parameters such as heating rate, geometry of sample holder and of furnace, crucible size and shape, but also the amount, volume, particle size, packing density and thermal conductivity of the sample are to some extent difficult to define and to reproduce. This is true especially also for the atmosphere around the sample which has to be controlled carefully. The factors which are of pronounced influence on the thermogravimetric curves and also in most of the thermoanalytical investigations will be discussed in some detail, referring to specific examples. For better clarity they will be subdivided into two categories, sample properties and experimental parameters.

3.1. Sample Properties

The *nature of the material* to be studied, which means its degree of crystallinity and perfectness of crystal structure, may have a significant effect on the thermoanalytical behavior. In spite of identical chemical composition of a certain material the variations with respect to structure, imperfections, grain boundaries, etc. are almost infinite. Of course many of these will not show in normal thermogravimetric analysis, with very sensitive apparatus characteristically different TG curves[18, 19] may be obtained however. As an example Fig. 26 shows the thermal decomposition of hydrozincite, $Zn_5(OH)_6(CO_3)_2$, whereby equal amounts of samples from natural origin and synthetic preparations are compared.

The *weight of the sample* and also its volume should preferably be small since this minimizes such effects as inhomogenous temperature distribution, retention of gaseous decomposition products etc. In most cases sample weights of the order of

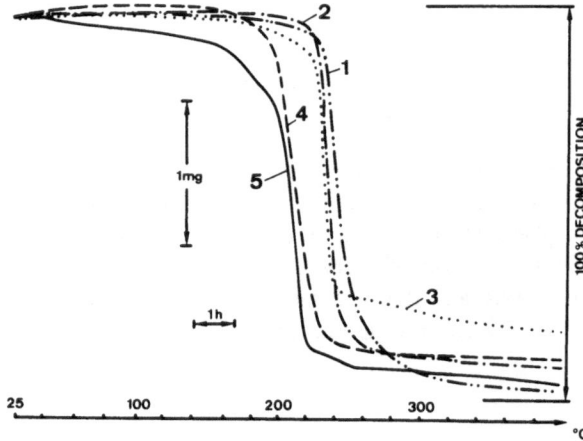

Fig. 26. Thermal decomposition of natural and synthetic hydrozincite. 1) Good Springs, USA; 2) Durango, Mexico; 3) Santander, Spain; 4) Synthetic from zinc-hydroxide; 5) Merck-product

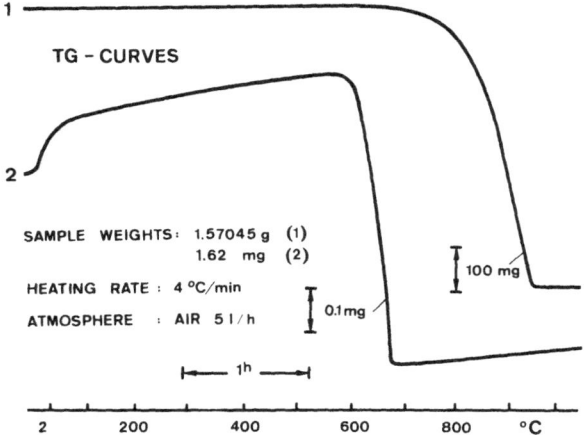

Fig. 27. Effect of sample weight and volume on the thermal decomposition of calcite

10 mg will be adequate. The effect of sample weight and volume is shown in Fig. 27 for the thermal decomposition of calcite.

The *homogeneity* in a sample during the course of a decomposition reaction is another test for the effects of sample weight but also of crucible shape and of the surrounding atmosphere. As an example the decomposition of zinc-hydroxide-carbonate (Merck, 250 mg) was investigated in flowing, dry air using a semi-spherical Pt/Rh-crucible, and 0.5 °C/min linear heating rate[18]. After reaching 185 °C, the sample was held at this temperature for 5 hours. This resulted in 50% decomposition (α = 0.5) to ZnO. After cooling to room temperature, concentric zones of the sample material were taken out of the crucible (see Fig. 28 and investigated by X-rays. The photometer tracings of the diffractograms proved the differences in composition of the various zones. The upper layer A had the highest proportion of ZnO ($\sim$70%), the middle parts B and C contained about 50% ZnO, and the bottom layer D showed only approximately 25% decomposition.

Another important factor is the *shape and particle size* of the sample. It is known that some materials *e.g.* clays, hydrates and carbonates may show changes in composition and in crystal structure after very fine grinding. Very generally speaking,

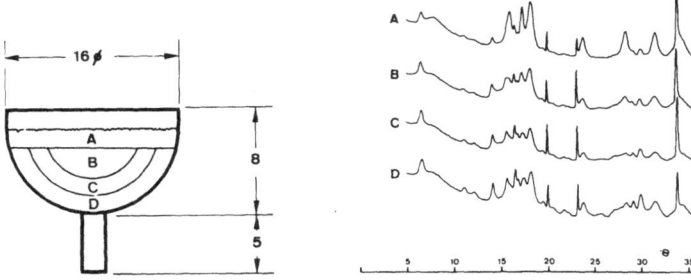

Fig. 28. X-ray diagrams showing differences in composition of various zones during heating of zinc-hydroxide-carbonate

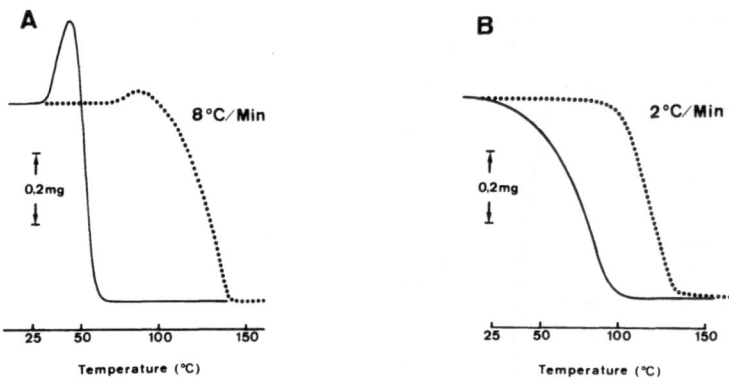

Fig. 29. Effect of heating rate and particle size on the dehydration of whevellite. A: 8 °C/min, B: 2 °C/min. Solid line: powdered material (150 mesh), dotted line: single crystal

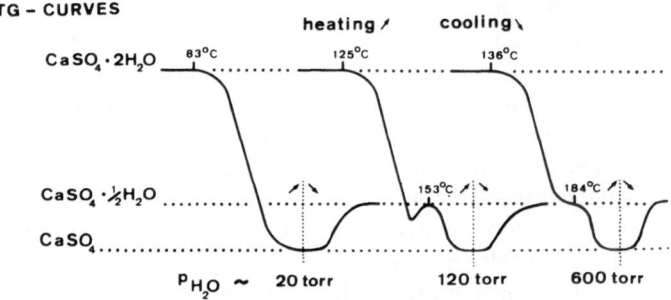

Fig. 30. Decomposition of natural gypsum under various partial pressures of water vapor

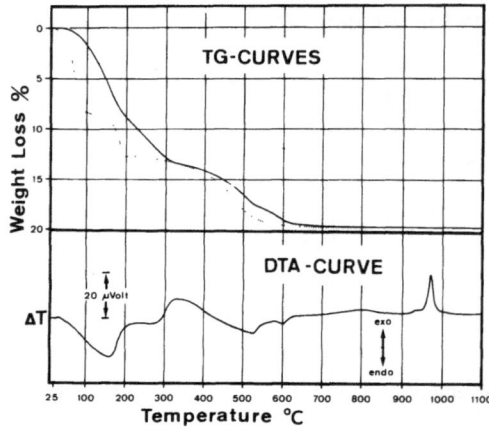

Fig. 31. Effect of heating rate on the weight curve for palygorskite. Solid line: 6 °C/min, dotted line: 0.5 °C/min

a sample consisting of large particles, which means low ratio surface area/weight, will often react more slowly than a sample of equal mass but comprised of very fine particles. Smaller particles also allow closer packing which in turn may effect the shape of a thermal decomposition curve. A loosely packed, coarse material leaves air spaces which reduce thermal conductivity, besides which, such particles may rearrange during heating. Particle size distribution and packing density are parameters which are not so easy to reproduce, therefore they have to be given special attention. As an example Fig. 29 shows the effect of particle size on the decomposition of whewellite, $Ca(COO)_2 \cdot H_2O$, in high vacuum[20]. 6 mg of powdered and of single crystal material are compared. The pronounced effect of the particle size, especially at the higher heating rate is obvious.

Finally, the *atmosphere* surrounding the sample has to be mentioned here, since it may play an active part in certain decomposition reactions and especially in oxidation and reduction processes. Some examples have been given already in the preceding sections. Figure 3 showed the effect of the atmosphere on the decomposition of calcite[21] and Fig. 4 the TG curve of a sample of α-FeOOH during heating up in a reactive atmosphere of hydrogen[22, 23].

Another example is the decomposition of gypsum $CaSO_4 \cdot 2 H_2O$ under various partial pressures of H_2O (Fig. 30). In these experiments cleavage plates of natural gypsum (selenite) of identical size and weight (10 mg) were heated up with 2 °C/min at 20 torr, 120 torr and 600 torr water vapor partial pressure[24-27]. At the lowest partial pressure the dehydration goes directly to γ-anhydrite, during cooling rehydration to the hemihydrate takes place. The intermediate formation of hemihydrate is characteristic for the other two curves. The curve at $p_{H_2O} = 600$ torr shows the overlapping of the dehydration to anhydrite and of the rehydration to the hemihydrate due to back diffusion of water. This leads to a small intermediate maximum at 153 °C. A corresponding reaction in form of a shoulder appears at 184 °C in the TG curve recorded at $p_{H_2O} = 600$ torr. This particular dehydration and rehydration behavior of calcium sulfates is related to structural similarities and to the high mobility of the water molecules between the $CaSO_4$ layers.

3.2. Experimental Parameters

The *rate at which the sample is heated up* is very important, especially in the case of slow or complex reactions. Fast heating rates shift the reactions to higher temperatures and decrease the resolution when several reactions follow each other closely. As a consequence the heating rate must not only be linear but also adapted to the type of reaction which is to be expected. In certain cases stepwise heating may be chosen, since this better approaches the isothermal conditions. Figure 31 shows the effect of the heating rate on the shape of the TG-curve in the case of palygorskite, a complex layer silicate[28]. Only the low *heating rate* (0.5 °C/min) allowed to resolve the individual dehydration steps.

The effects of stepwise heating[29] in shifting the reaction to lower temperatures and in decreasing the interval ΔT as compared to linear heating is shown schematically in Fig. 32. By such special heating techniques it is possible to obtain thermodynamic data also by thermoanalytical methods. This has been demonstrated recently for the

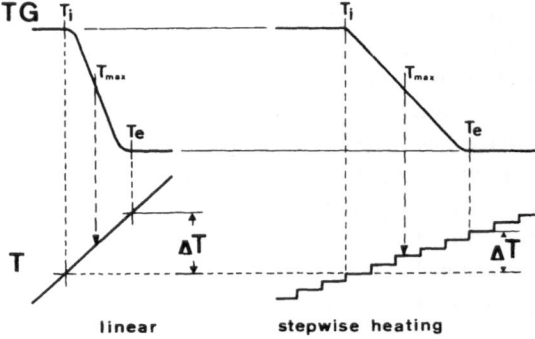

Fig. 32. Schematic presentation of the dependence of dissociation on the heating schedule. The dissociation temperatures used in Fig. 33 correspond to T_{max}

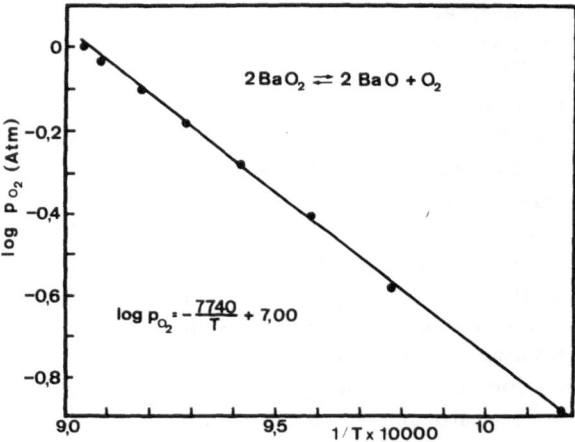

Fig. 33. Dissociation pressure of barium peroxide

dissociation reactions $2 PdO \rightleftharpoons 2 Pd + O_2$ and $2 BaO_2 \rightleftharpoons 2 BaO + O_2$ (Fig. 33). From the linear relation between log of dissociation pressure and $1/T$ the heat of dissociation can be calculated[30-33].

For illustrating the importance of the parameter *time* the application of thermogravimetry for vapor pressure measurements[34-36] will be discussed. The substance is placed into a Knudsen effusion cell which rests on the crucible holder of the thermobalance. The evaporation occurs through the small, calibrated, orifice of the cell at constant temperature and pressure as a function of time. The original curve in Fig. 34 shows the effusion of benzoic acid at 17.25 °C and at a pressure of 1.5×10^{-5} torr outside the cell. The straight slope is in accordance with a constant rate of weight loss, which was in the order of 10^{-7} g sec^{-1}. The vacuum outside the effusion cell should be generally one order of magnitude better than the vapor pressure to be determined. A cold trap near the orifice of the Knudsen cell serves to condense the vaporized species and stabilizes the external pressure. This technique can be applied over the pressure range from 1 to 10^{-6} torr. By simple calculations it gives the vapor

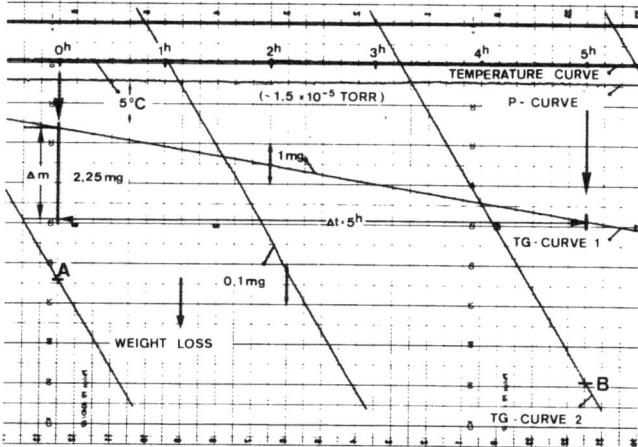

Fig. 34. Vapor pressure of benzoic acid measured on a thermobalance with Knudsen effusion cell

pressure and the heat of vaporization of the substance. However, a number of experimental conditions like sensitivity and reproducibility of the balance, uniform temperature of the sample and of the effusion cell are very critical. In a trial run lasting 10 to 20 hours, the maximum drift of the balance should not exceed ±5 μg, the temperature should be constant within ±0.1 °C. (See Sect. 2.4.)

The possibility of automatic collection and evaluation of thermoanalytical data should be mentioned here in connection with the thermogravimetric vapor pressure measurements. Principally, any desired quantitative evaluation procedure in thermal

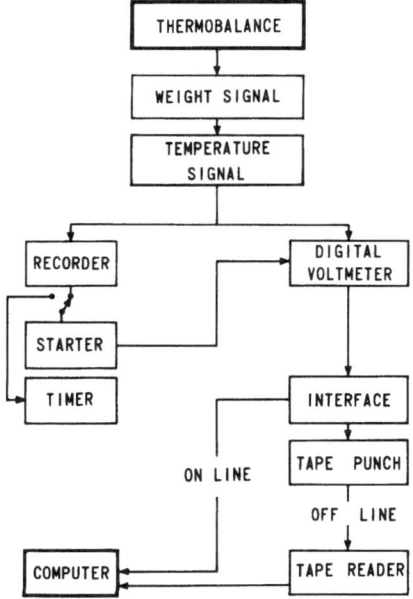

Fig. 35. Data processing for thermogravimetric vapor pressure determination

Table 5. Input of measured TG-data, processing for vapor pressure determination, output of results

---

STEP 1 LISTING OF RATE OF VAPORIZATION (G/SEC) AT CONSTANT TEMPERATURE
CALCULATION OF AVERAGE VALUES AND STANDARD DEVIATION (EXAMPLE FOR 1 HOUR)

NO.	WEIGHT LOSS (G/SEC)	EMF (MV)	NO.	WEIGHT LOSS (G/SEC)	EMF (MV)	NO.	WEIGHT LOSS (G/SEC)	EMF (MV)
1	-0.597E-07	10.80900	2	-0.604E-07	10.80800	3	-0.583E-07	10.80900
4	-0.590E-07	10.80900	5	-0.583E-07	10.80900	6	-0.590E-07	10.80700
7	-0.590E-07	10.81200	8	-0.576E-07	10.81200	9	-0.576E-07	10.81100
10	-0.590E-07	10.80900	11	-0.604E-07	10.80800	12	-0.590E-07	10.80900
13	-0.604E-07	10.80800	14	-0.597E-07	10.81100	15	-0.597E-07	10.80900

RESULTS AFTER FIRST HOUR:

	WEIGHT LOSS (G/SEC)	EMF (MV)	TEMP. (DEG.CEL.)
MEAN VALUES :	-0.592E-07	10.8093	1117.8
STANDARD DEVIATION:	0.886E-09	0.00145	0.12

SEE FIG. 66

STEP 2 DATA OF RATE OF VAPORIZATION AT DIFFERENT CONSTANT
TEMPERATURES FOR CALCULATION OF VAPOR PRESSURES

DATA INPUT   GOLD VAPOR PRESSURE

196.967 ATOMIC OR MOLECULAR WEIGHT

NUMBER OF MEASUREMENT	TEMPERATURE DEG C	RATE MG/ H	ORIFICE SQCM
1	1076.10	0.0918	0.032590
2	1117.80	0.2113	0.032590
3	1099.00	0.0143	0.005685
4	1143.20	0.0266	0.005685
5	1150.00	0.0472	0.005685
6	1203.50	0.1385	0.005685
7	1252.00	0.2956	0.005685
8	1299.00	0.6220	0.005685
...	....	....	....
23	1400.00	4.1880	0.006200

KNUDSEN EQUATION

$$P_s = \frac{\Delta m}{\Delta t} \; \frac{1}{T \cdot r^2} \sqrt{\frac{2\pi \cdot R \cdot T}{M}} \quad \text{dyne cm}^{-2}$$

STEP 3 REGRESSION ANALYSIS OF VAPOR PRESSURE DATA
CALCULATED WITH KNUDSEN EQUATION

REGRESSION ANALYSIS

MEASURED VALUES

CALCULATED VALUES

T(CELS)	T(ABS)	1000/T	LOG10 P	T(ABS)	1000/T	LOG10 P
1076.10	1349.26	0.74115	-4.45529	1349.26	0.74115	-4.60863
1117.80	1390.96	0.71893	-4.08663	1390.96	0.71893	-4.20687
1099.00	1372.16	0.72878	-4.50079	1372.16	0.72878	-4.38498
1143.20	1416.36	0.70604	-4.22436	1416.36	0.70604	-3.97375
1150.00	1423.16	0.70266	-3.97426	1423.16	0.70266	-3.91275
1203.50	1476.66	0.67720	-3.49874	1476.66	0.67720	-3.45243
1252.00	1525.16	0.65567	-3.16247	1525.16	0.65567	-3.06305
1299.00	1572.16	0.63607	-2.83279	1572.16	0.63607	-2.70862
....	....	....	....	....	....	....

STEP 4 DATA OUTPUT OF THERMODYNAMIC PROPERTIES OF GOLD
FOR THE INVESTIGATED TEMPERATURE RANGE 1060 - 1500 K

DATA OUTPUT   GOLD VAPOR PRESSURE

FUNCTION	DEVIATIONS		COEFFICIENTS	
	AVERAGE	MAXIMUM	A	B
LOG(P) = A*1000/T + B	0.098829	0.250609	-18.0816	8.7925

THERMODYNAMIC PROPERTIES   (STANDARD REFERENCE STATE TO 1555.65 DEG K)

DEL H(TO) KCAL/MOL	DEL S(TO,760TORR) CAL/MOL/D
82.6787 +- 1.7065	27.0302 +- 1.1354

110

analysis could be realized by means of a suitable interface and a computer. Such automatic methods for collecting and processing large numbers of individually measured values would be extremely useful; they are available commercially so far only for a small number of thermoanalytical instruments. Figure 35 is a schematic presentation of the components necessary for on-line data processing of vapor pressure data determined by thermogravimetry. The four main steps which are important for the evaluation of the measured values are listed in Table 5 which in some parts has been copied directly from the computer output.

The *atmosphere* and the *pressure* are experimental parameters of general importance since they also determine whether a reaction will take place at a certain temperature or not. One special parameter which connects the atmospheric effect and the balance system is the often neglected buoyancy effect[37-40]. This effect has to be taken into account in all accurate measurements. It is of importance not only for the parts of the balance system and crucible, but also for the sample which changes its weight and volume during the experiment. This is true in many cases also for the surrounding atmosphere. The extent of the buoyancy effects and its correction are shown in TG curves recorded during decomposition of a particularly large sample of $CaCO_3$ in a CO_2 atmosphere (Fig. 36). This experiment has been carried out for testing to accuracy of the thermobalance. Apart from the actually measured, uncorrected weight curve, a second one is shown in which a correction for the buoyancy effect of the crucible and of the curcible holder as a function of temperature was applied. The third curve shown on the TG record takes into account also the buoyancy change of the sample. In addition, the exact sample weights at room temperature, after loss of the adsorbed water at 800 °C, and at the end of the experiment at 1100 °C are shown for CO_2-atmosphere and after correction for vacuum. The extent of necessary corrections can be seen in the figure. It becomes evident therefore that accurate results are obtained only under consideration of the

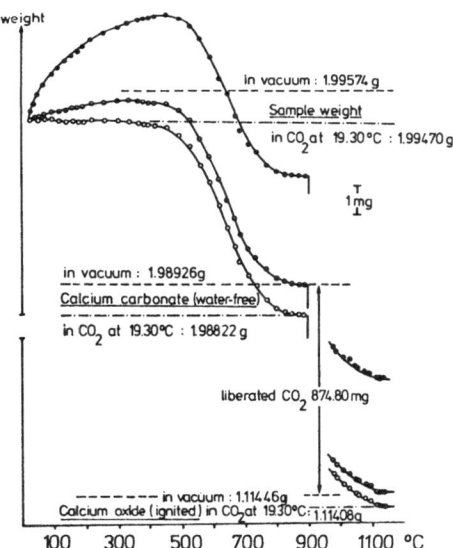

Fig. 36. Evaluation of the buoyancy effect during a decomposition process. —•—•—• uncorrected weight curve, —⊝—⊝—⊝— correction for buoyancy of crucible and holder, —o—o—o— additional correction for sample

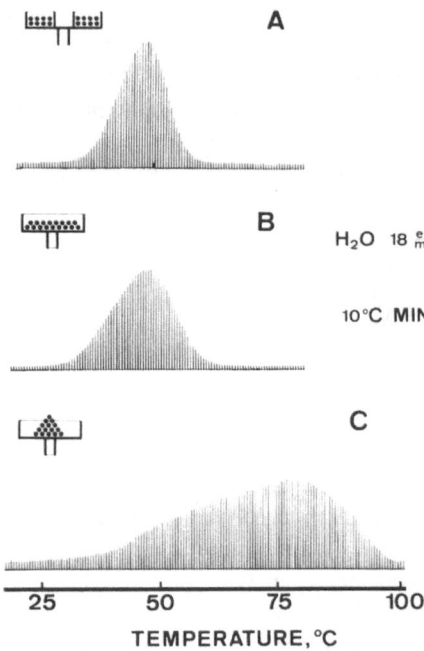

A

B

H_2O 18 $\frac{e}{m}$

10°C MIN

C

25   50   75   100

TEMPERATURE, °C

Fig. 37. Effect of sample packing on the dehydration course of calcium oxalate monohydrate shown by mass spectrometry

buoyancy contributions of the sample, of the crucible and of the crucible holder, all dependent on the temperature. At high temperatures the nearly parabolic buoyancy correction curve may show a pronounced deviation starting from about 700 °C. As could be proven by Wiedemann, this effect is caused by thermal radiation and convection respectively, which lead to a heating of certain parts of the balance system. This problem can be avoided however by placing an intense cooling system between the furnace and the adjacent balance housing.

One important experimental parameter is also the *furnace* which is used in the thermogravimetric apparatus. There always exist pronounced radial and vertical temperature gradients which can be found by calibration runs. As an example, the determination of the vertical temperature distribution by Wiedemann[41] may be referred to. Also the amount and shape of the sample can contribute to a temperature gradient. This fact is of special importance for kinetic studies. Considerable temperature differences — up to several °C — can exist at different locations of the sample holder. Of course this depends to a great extent on the thermal conductivity of the crucible (Pt, Au, Al, Al_2O_3, vitreous silica, graphite) and of the sample holder. Information about the temperature distribution in the furnace may be derived from experiments with specially designed multi-sample holders where equal amounts of the sample are placed into symmetrically arranged small pans.

In all thermoanalytical investigations it is very important how the substance is placed and positioned on the sample holder. This was illustrated in simultaneous TG-DTA measurements in connection with gas analytical investigations by Wiedemann[42]. The three profiles shown in Fig. 37 correspond to the water vapor pressures recorded with a mass spectrometer during thermal dehydration of fine-grained calcium

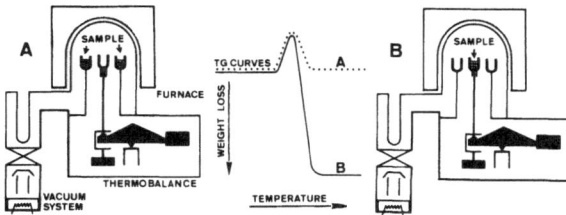

Fig. 38. Explanation for the apparent weight increase at the beginning of decompositions. For curve A the sample was placed outside the balance pan, for curve B the sample was in the usual position on the balance pan

oxalate monohydrate. Arrangements A and B lead to similar, almost symmetrical peaks with rapid dehydration. For arrangement C complete dehydration is delayed, leading to a shift of the peak maximum of almost 40 °C. For thermoanalytical investigations in vacuum, the layer of the substance in the crucible should be as thin as possible, not exceeding 0.5 to 1 mm. In this connection an interesting case of initial weight gain during dehydration of a powdered whewellite sample ($Ca(COO)_2 \cdot H_2O$) in high vacuum[20] (10^{-5} torr) will be discussed briefly. The effect of particle size on the dehydration of this mineral has been already mentioned and is shown in Fig. 29 Part A of this figure proves that the powdered sample shows a strong apparent weight gain when heated at a rate of 8 °C/min, followed by the expected weight decrease due to dehydration. For a better understanding of this effect, another test was carried out whereby equal amounts of the sample were placed into a ring around the empty crucible (Fig. 38, A) and in the other case inside the crucible (Fig. 38, B). The experiment resulted exactly in the same apparent weight increase for both arrangements as can be seen from the corresponding TG-curves for A and B. Simultaneously with the weight increase, a characteristic temporary pressure increase of one order of magnitude occurred in both runs. The test proved that only a fraction smaller than 0.5% of the apparent weight increase can be attributed to a recoil of the molecules leaving the balance pan, practically the entire effect is due to a re-impact phenomenon. It was possible to simulate this effect on a vacuum thermobalance by introducing measured

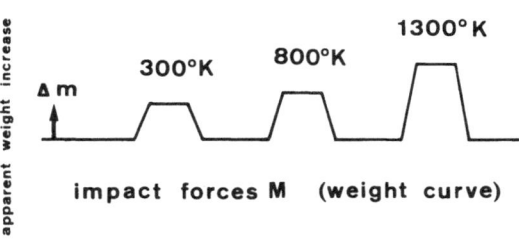

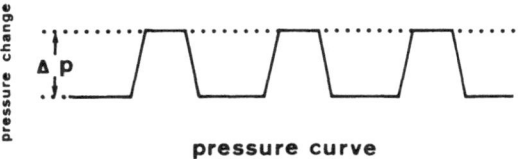

Fig. 39. Apparent weight increase by impact of gas molecules on the balance pan during constant pressure changes

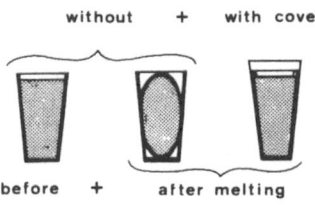

without   +   with cover

before   +   after melting

Fig. 40. Deformation of a metal block sample due to surface tension effects during fusion

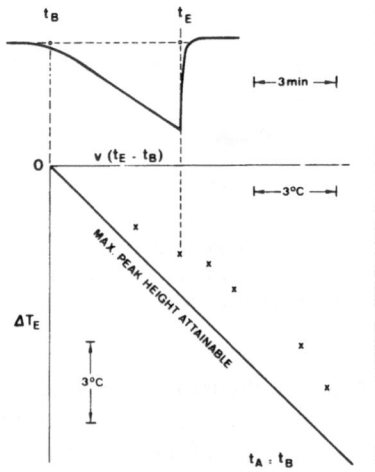

Fig. 41. Melting peak heights of aluminum at various rates of heating

quantities of gases through a separate inlet above the balance pan. As a practical conclusion such dynamic effects always should be taken into consideration when gas producing substances are heated up relatively fast in high vacuum. This helps to avoid wrong interpretation of such measurements. It was found in various experiments that this dynamic effect decreases with smaller heating rates, and that it can also be avoided when much larger pumping cross sections are used. This same phenomenon offers on the other hand an interesting method for calibration of the temperature in the immediate vicinity of the sample container[43]. For this purpose the weight effect of defined pressure impulses Δp on the balance pan was investigated Fig. 39 shows the apparent weight increases caused by impact of gas molecules (CO_2, N_2, O_2, Ar, NH_3) on the balance pan during constant pressure changes from $1 \cdot 10^{-5}$ torr to $5 \cdot 10^{-5}$ torr for different isothermal conditions. In agreement with Maxwell's theory, the measured impact forces are proportional to the square root of the absolute temperature and also to the mean velocity of the gas molecules. Thus it is possible to calculate an unknown temperature from the weight effect measured. The method is limited however to the vacuum region $<10^{-3}$ torr. Standard deviations of the temperature in the order of ± 5 °K and ± 1 °K were found at 1225 °K and at 300 °K respectively. This same method was also used by Wiedemann[44] to follow continuously the evolution of gaseous decomposition products with simultaneous recording of the heating X-ray[45] diffraction pattern (TMBA-X-ray method).

The number of experimental factors which influence the results increases considerably when thermogravimetry is combined with other techniques such as DTA, gas chromatography[46], mass spectrometry, X-ray etc. A systematic discussion of all these additional factors would lead too far, therefore only a representative example will be discussed here. One of the often-applied multiple techniques is the combination TG-DTA. Besides the actual thermal reactions of the sample, the important factors in DTA are the heat capacity and the thermal conductivity of the sample. Optimum heat transfer is required for such thermoanalytical measurements therefore the shape of the sample and its contact with the crucible is of special importance. This problem was investigated by Wiedemann and Van Tets[47] by thermoanalytical

114

characterization of melting and freezing processes and by calorimetric calibration of DTA equipment with metal standards respectively. A series of DTA runs was carried out with metal samples which fitted exactly into the cavity of graphite crucible. After repeated melting and freezing operations a pronounced change in the shape of the DTA-peak was observed. Figure 40 explains this effect, which is caused by the deformation of the metal sample due to surface tension after its first melting. This leads to a change in the contact area between metal sample and crucible wall. This problem can be solved by using a covered crucible whereby the weight of the cover prevents the sample deformation and maintains the contact between sample and crucible wall. Tests were carried out with different metals in pure argon, varying the heating and cooling rates between 0.5 and 10 °C/min. It is very important to keep the gas flow rate constant during such measurements of melting temperatures. Graphical evaluation of the results proved that the maximum temperature deviation ΔT_E is approximately proportional to the square root of the heating rate, in agreement with the theory. The melting time for pure samples under the assumption of an isothermal melting process would be equal to the height of the endothermic peak ΔT_E divided by the heating rate v. Figure 41 shows the actual melting peak heights of aluminum[48] at various heating rates, compared to the theoretical maximum peak height line which was derived from the heating rate v times the melting time $t_E - t_A$. The deviation from the maximum peak height becomes more pronounced at higher heating rates, probably due to surface tension effects.

4. Examples of Application

4.1. Thermal Stability of Platinum Group Metal Oxides

The oxidation behavior of platinum metals is of importance for their applications at high temperatures. These metals form solid oxides and also volatilize as oxides at higher temperature in an oxidizing atmosphere. Most of the gaseous oxides are stable only at high temperatures and usually contain the metal in its highest oxidation states. Palladium is an exception since it dissolves oxygen in the solid state and only forms the solid oxide PdO which dissociates at temperatures above 800 °C.

The platinum metals used in the experiments were catalyst powders with specific surface areas of $10-30$ m^2/g and purity >98%. The oxidation of metals and the decomposition of oxides was investigated with a thermobalance[49] which was combined with an Enraf Nonius High Temperature X-ray Camera. Sample weights were in the order of $5-40$ milligrams. TG and DTA curves were recorded under different atmospheric conditions (air, oxygen, vacuum) and with different heating rates and heating schedules.

Figure 42 and Table 6 show results on the formation and dissociation of the oxides PdO, RuO_2 and IrO_2. The TG-curves were obtained under reduced oxygen pressure (100 torr) and heating rates of 10 °C/min. The assignment of reaction temperatures was carried out according to Fig. 43. The formation and dissociation of the oxides depends strongly on the oxygen pressure, on the heating rate and on the

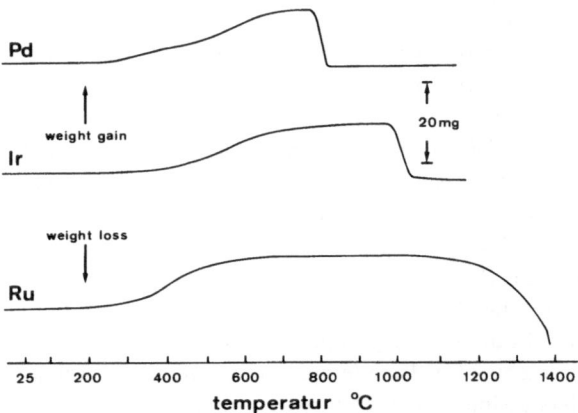

Fig. 42. Formation and dissociation of PdO, IrO_2 and RuO_2. Sample weights 50–100 mg

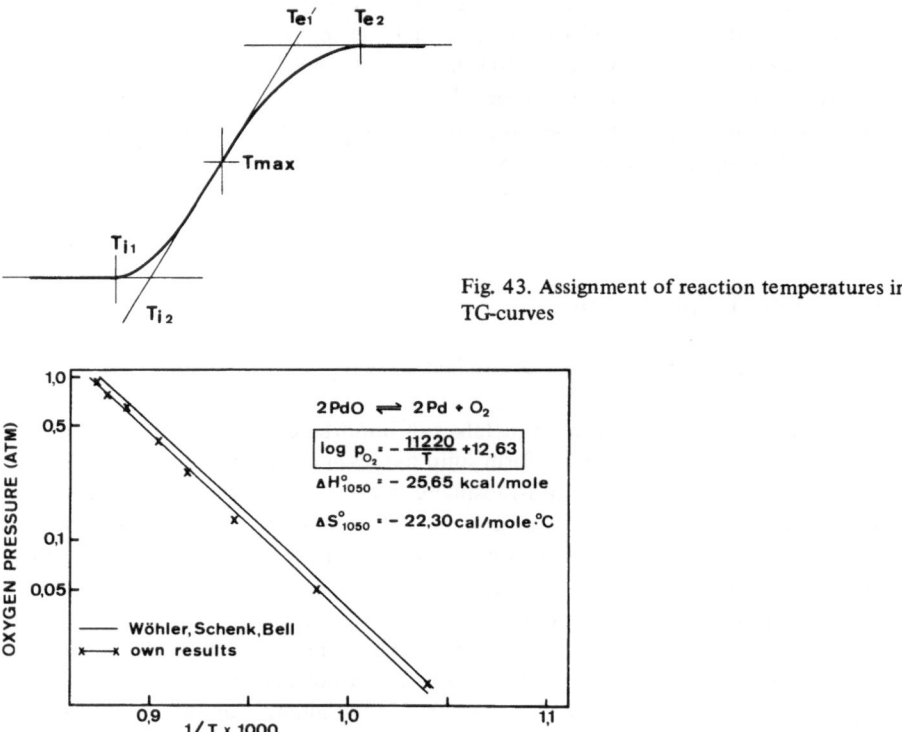

Fig. 43. Assignment of reaction temperatures in TG-curves

Fig. 44. Dissociation pressure of palladium oxide

surface area of the metal powders. Decrease of oxygen pressure favored the complete oxidation of the metal powders.

A series of experimental runs was carried out on the dissociation of PdO at oxygen pressures varying from 717 torr down to 10 torr. The dissociation temperature (T_{max}) of PdO decreases in this range from 872 °C down to 689 °C at 10 torr.

Table 6. Formation and dissociation temperatures of PdO, IrO_2 and RuO_2 (oxygen pressure 100 torr, heating rate 10 °C/min)

Starting material	Oxidation (°C)			Dissociation (°C)		
	T_{i1} T_{i2}	T_{max}	T_{e1} T_{e2}	T_{i1} T_{i2}	T_{max}	T_{e1} T_{e2}
Pd-black, 8.5 m²/g	196 258	567	664 740	762 775	806	819 828
Ir-black, 21.2 m²/g	142 408	560	685 907	945 975	1005	1031[1]) –
Ru-black, 17.0 m²/g	118 312	405	500 698	928 1438	1490	– –

[1]) Temperature influenced by simultaneous vaporization.

Table 7. Dissociation data of PdO, IrO_2 and RuO_2

	ΔH°_{298} (kcal/mole)	Δs°_{298} (cal/mole °C)	$T_{diss.}$ (°C) (p_{O_2} 1 atm.)
PdO	−27.37	−24.00	877
IrO_2	−52.42	−40.52	1124
RuO_2	−71.17	−41.40	1580

A plot of these dissociation temperatures versus the equilibrium oxygen pressures is shown in Fig. 44 in the usual logarithmic form. The oxygen dissociation pressure reaches 1 atm at 877 °C. The heat of dissociation of PdO was calculated from the slope of the line by regression analysis. The medium temperature in the measured range was 777 °C = 1050 °K. Calculated values for $\Delta H^{\circ}_{1050} = -25.65 \pm 0.86$ Kcal/mole and $\Delta S^{\circ}_{1050} = -22.30 \pm 0.77$ cal/mole °C are in agreement with the data reported in the literature. Using the heat capacity equation given by Kelley[50] $\Delta C_p = -6.08 + 12.32 \times 10^{-3} T + 0.20 \times 10^5 T^{-2}$ it is possible to convert these values to the standard state: $\Delta H^{\circ}_{298} = -27.37$ Kcal/mole and $\Delta S^{\circ}_{298} = -24.00$ cal/mole °C for the reaction $PdO_{(s)} \rightleftharpoons Pd_{(s)} + 1/2 O_2$. These values correspond directly to the heat and entropy of formation of PdO from the elements. The relationship between equilibrium oxygen pressure and temperature which is given by the equation in Fig. 44, log $p_{O_2} = -\dfrac{11.220}{T} + 12.63$, applies to $2 PdO_{(s)} \rightleftharpoons 2 Pd_{(s)} + O_2$. Therefore the molar values ΔH and ΔS correspond to half of the values derived from the slope of the line in Fig. 28. A comparison of the dissociation data of PdO to those reported for IrO_2 and RuO_2 is given in Table 7. RuO_2, which has the closest packed structure, also is the most stable oxide.

These measurements on the oxidation of palladium prove that it is possible to obtain thermodynamic data also by dynamic, thermoanalytical methods. The values compare favorably with those obtained by static, calorimetric methods. The slight discrepancies observed at present are due to the effect of the heating modes used in these measurements. Stepwise heating approaches more the isothermal conditions, whereas a linear, even slow heating is often in advance of the actual reaction.

4.2. Formation and Decomposition of Calcium and Cadmium Chromate

These investigations were concerned with the formation of alkaline earth chromates $Me^{2+}Cr^{6+}O_4$ during solid state reactions with Cr_2O_3. Thermogravimetric methods are especially useful for such studies and allow one to follow the oxidation $Cr^{3+} \longrightarrow Cr^{6+}$ in relation to the oxygen pressure and to the thermal behaviour of the alkaline earth compounds used[51]. First of all, the reactions between hydroxides or carbonates and Cr_2O_3 were studied. Figure 45 shows the differences in the case of calcium. Whereas the decomposition of the calcium compound and the oxidation of the Cr_2O_3 are clearly separated in the case of the hydroxide, these two reactions overlap each other when $CaCO_3$ is used. This overlapping was even more pronounced for the mixtures 1 $SrCO_3/0.5\ Cr_2O_3$ and 1 $BaCO_3/0.5\ Cr_2O_3$. All these reactions were complete when carried out in air and lead to formation of the chromates $MeCrO_4 \cdot CdCrO_4$ on the other hand could be prepared by solid state reaction between $CdCO_3$ or $Cd(OH)_2$ and Cr_2O_3 only when pure oxygen was used. Even so, the reaction was not complete, the $CdCrO_4$ decomposed already before all the Cr^{3+} was oxidized to Cr^{6+} (Fig. 45).

The effect of the oxygen pressure on the completeness of the $CdCrO_4$-formation was investigated and is shown in the TG-curves in Fig. 46. Overpressure of about 1 at increases the yield to about 50%. Heating X-ray photographs (Fig. 47) show the various reaction products which are formed during heating up of a mixture 2 $CdCO_3/1\ Cr_2O_3$.

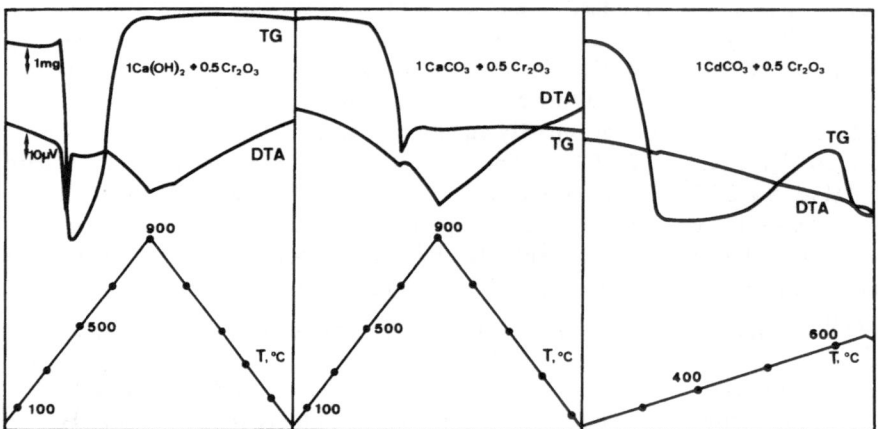

Fig. 45. Formation of $CaCrO_4$ and of $CdCrO_4$ by solid state reaction between different mixtures. A) Sample weight: 96 mg, heating rate: 2 °C/min, atmosphere: air; B) Sample weight: 38 mg, heating rate 2 °C/min, atmosphere: air; C) Sample weight: 43 mg, heating rate 0.5 °C/min, atmosphere: oxygen

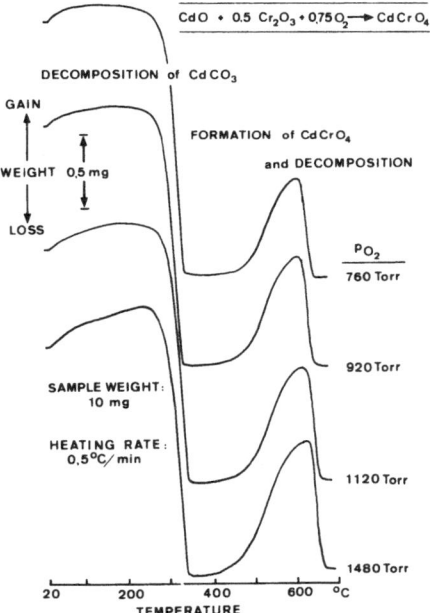

CdO + 0.5 Cr$_2$O$_3$ + 0.75 O$_2$ ⟶ CdCrO$_4$

DECOMPOSITION of CdCO$_3$

GAIN

WEIGHT 0,5 mg

LOSS

FORMATION of CdCrO$_4$

and DECOMPOSITION

P_{O_2}

760 Torr

920 Torr

SAMPLE WEIGHT:
10 mg

HEATING RATE:
0,5°C/min

1120 Torr

1480 Torr

20   200   400   600   °C
TEMPERATURE

Fig. 46. Effect of oxygen pressure on the formation of CdCrO$_4$

This thermal instability of CdCrO$_4$ is probably related to the easy formation of the rather stable spinel CdCr$_2$O$_4$ with 3-valent chromium. The same is true also in the case of MgCrO$_4$, which can be prepared only at high oxygen pressures. These results and the course of the reaction in various systems with Cr$_2$O$_3$ could be confirmed and followed by heating X-ray runs in addition to the TG/DTA investigations.

The chromates of the alkaline earth oxides MeCrO$_4$: MgCrO$_4$, CaCrO$_4$, SrCrO$_4$ and BaCrO$_4$, all have different crystal structures. The coordination number of the

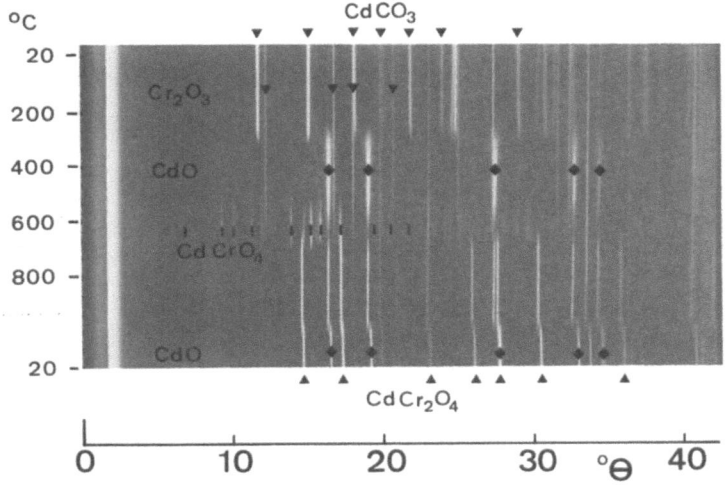

Fig. 47. X-ray heating photograph of the formation and decomposition of CdCrO$_4$ (Heating rate: 1 °C/min, atmosphere: oxygen)

Me-ion increases with the ionic size from 6 in $MgCrO_4$, to 8 in $CaCrO_4$ and ~8 in $SrCrO_4$, to 12 in $BaCrO_4$. On the other hand, the oxygen coordination around Cr^{6+} is the same in all chromates, that is, tetrahedral. The thermal stability of these compounds is primarily related to the size of the Me^{2+}-ion, it increases from <600 °C for $MgCrO_4$ to ~1000 °C ($CaCrO_4$), 1190 °C ($SrCrO_4$) to 1380 °C for $BaCrO_4$. In the case of the heavier chromates complete decomposition according to the reaction $2\, MeCr^{6+}O_4 \rightarrow MeCr_2^{3+}O_4 + MeO + 1.5\, O_2$ is difficult to achieve at these temperatures in an oxygen-containing atmosphere.

4.3. Melting and Freezing Behavior of Metals

These measurements were carried out with a combination thermobalance-DTA, the metal samples were placed in graphite crucibles. A covered crucible was used, the weight of the cover maintained the uniform contact between sample and crucible wall. As an example the melting and freezing curves for pure aluminum are presented in Fig. 48. The DTA-curve of the diagram shows the endothermic melting peak and the exothermic freezing peak during the heating cycle. The TG-curve proves that no weight changes occur, which is possible for certain substances by vaporization already below the melting point. Therefore it is important for quantitative measurements to record not only the DTA-curve but also the TG-curve simultaneously. The lower part of Fig. 48 shows the course of the temperature in the sample during the melting and freezing process, both the DTA-curve and the T-curve were recorded with the same sensitivity to show the principal difference.

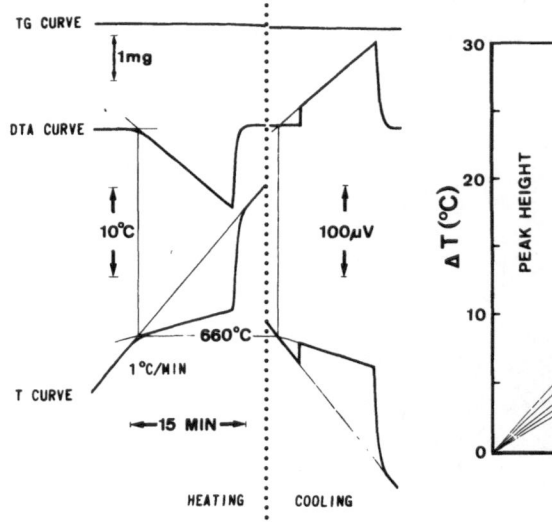

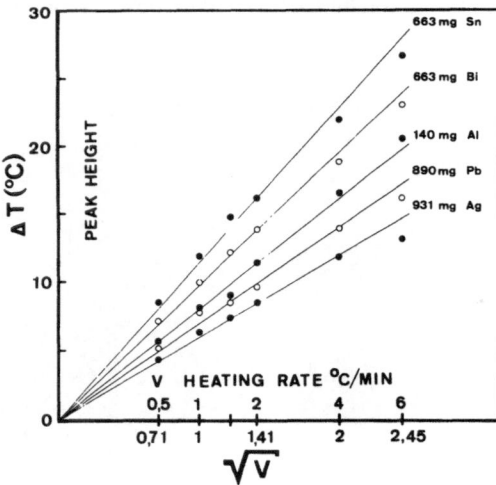

Fig. 48. Melting and freezing curves of aluminum. Sample weight 238 mg, heating rate 1 °C/min, argon atmosphere

Fig. 49. Melting peak heights (ΔT) of different metals at various heating rates

Melting and freezing curves were recorded for various metals[47] at different rates of heating and cooling (0.5, 1, 1.5, 2, 4 and 6 °C/min). The tests were carried out in an atmosphere of purified, dry argon. Figure 49 shows the linear relationship between heating rate and the maximum temperature deviation ΔT in the DTA-curve during the melting process. The temperature deviation ΔT is mainly effected by the thermal conductivity of the metal. One can evaluate the measurement of the temperature T (which lies between sample temperature T^* and furnace temperature T_0) in accordance with the expression

$$T = \frac{R_a T^* + R_k T_0}{R_a + R_k}$$

where R_a is the external thermal resistance between furnace and measuring point and R_k is a contact resistance between sample and differential measuring point.

The exothermic or endothermic heat effect, Q, resulting from chemical reactions or transformations can be determined by means of the equation

$$Q = \frac{1}{R_a} \int_0^\infty \Delta T_b \, dt$$

in which ΔT_b is the transient deviation from the base line during a DTA peak due to the thermal effect. For a constant chart speed and a linear heating rate within the range concerned, the time integral is proportional to the DTA peak area. With the equipment available today, this evaluation of the peak area can be carried automatically by use of an on-line desk calculator. Data like heats of melting, transformations etc. are processed in accordance with the experimental parameters used in the experiments.

4.4. Growth of Alumina-Whiskers

Another, less common application for thermogravimetric methods, is in the area of crystal growth[52–55]. Thereby it is possible to follow not only reactions connected with weight change, but also to control various experimental parameters such as the heating program and the atmosphere. As an example, the growth of alumina whiskers will be discussed. These experiments were carried out by volatilization and oxidation of aluminum in a wet argon/hydrogen atmosphere by use of a thermobalance[54]. The aluminum source was an iron-aluminum alloy Fe_3Al (sample weight lg), the total pressure was kept at 100 torr. Under these conditions alumina whiskers could be grown in the temperature range from 1400 to 1600 °C. The heating rate was 10 °C/min; at the highest temperature the samples were held isothermally for 15 hours. Figure 50 shows a schematic record of such an experiment, the interpretation of the observed weight increase is given in the following diagram (Fig. 51). M_0 is the original sample weight of the Fe_3Al sample, x corresponds to the weight increase resulting from oxidation and deposition of the vaporizing aluminum. The amount of vaporized aluminum (y) was determined by weighing the Fe_3Al sample

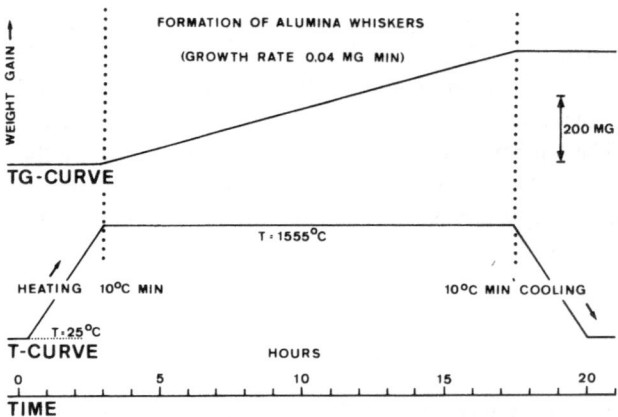

Fig. 50. TG curve showing the formation of alumina whiskers during oxidation of iron-aluminum alloy in wet hydrogen (2%)/argon mixture (isothermal at 1555 °C). Sample weight 1000 mg

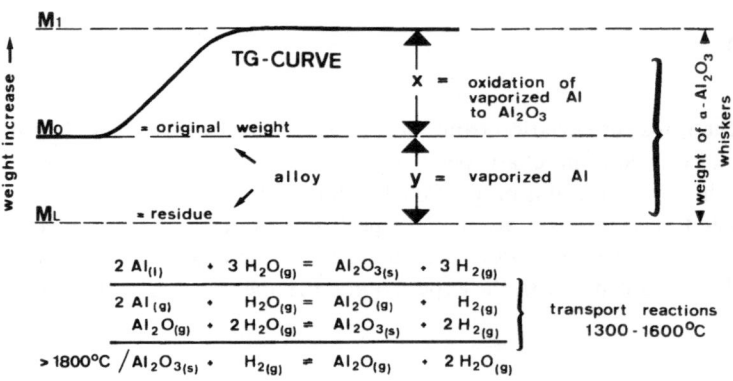

Fig. 51. Schematic presentation of the course of the reaction during formation of alumina whiskers

before and after the run. With covered alumina crucibles, about 90% of the vaporized aluminum was oxidized to alumina and precipitated within the crucible in the form of whiskers. The typical growth of such whiskers in the crucible is shown in Fig. 52. The results proved that the so-called a-whiskers grow in the direction [$11\bar{2}0$] preferentially at lower temperature e.g. between 1450 and 1500 °C. Above this temperature, hexagonal, c-whiskers grow perpendicular to the a-whiskers in the direction [0001]. The orientation of the whiskers to each other can be seen in Fig. 53. Whisker growth did not occur in dry argon/hydrogen atmosphere, besides which the effect of the crucible material is very important. The best results were obtained with closed alumina- or thoria-crucibles, in magnesia crucibles only a polycrystalline reaction layer of the spinel $MgAl_2O_4$ was formed. Graphite crucibles caused carburization of the Fe_3Al alloy. This means that alumina whiskers can be grown in such crucibles or containers as do not react with the metallic source or with the transported aluminum

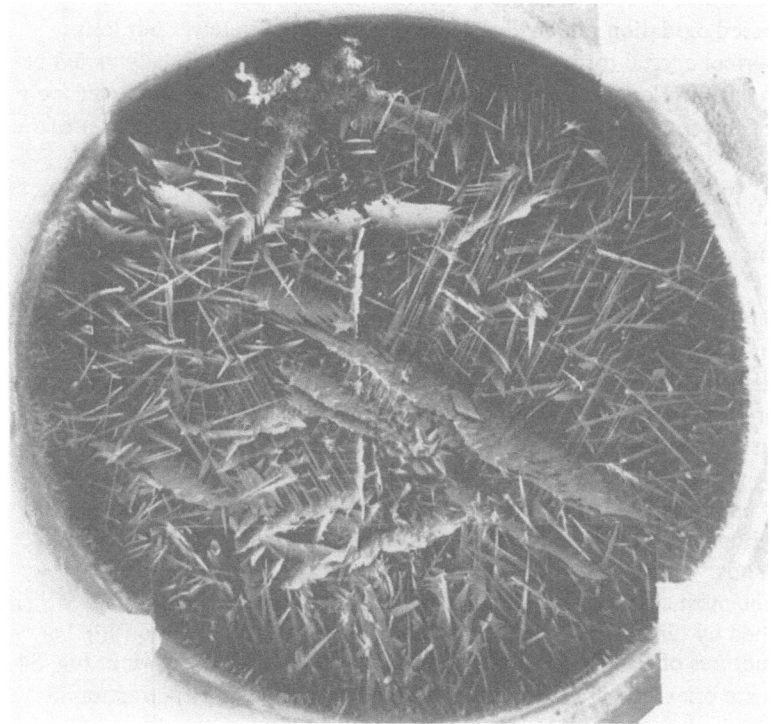

Fig. 52. Typical growth of alumina whiskers in an alumina crucible

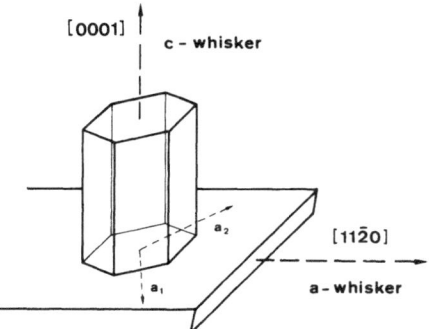

Fig. 53. Orientation of a and c whiskers relative to each other

oxide. When Al-richer alloys were used, strong volatilization of aluminum occurred above 1200 °C without the formation of whiskers. It is important to match the amount of vaporizing aluminum with the H_2O partial pressure available for oxidation of Al to Al-suboxides and finally to Al_2O_3. In using the composition Fe_3Al, it was possible to control the growth rate and the morphology of the alumina whiskers by variation of temperature and time. There is also a difference in whether the whisker growth is carried out in open or in closed crucibles. Considerably more but smaller whiskers were formed in open crucibles, whereas the better temperature homogeneity and the decreased oxidation rate in closed crucibles resulted in fewer but longer whiskers of perfect crystal forms. The thickness of the c-whiskers usually varied between 10–40 μm with lengths up to 500 μm. The longest a-whiskers observed were in the order of the crucible diameter (8 mm) with thickness from 5 μm down to about 200 Å.

4.5. Formation and Dissociation of Barium Peroxide

This is another example of the application of thermogravimetry for determination of equilibrium temperatures in dissociation studies. This also enables one to calculate the heat of dissociation from the linear relation between log of dissociation pressure and l/T. Determination of the specific heat by means of DTA was used afterwards for conversion of the heat of dissociation into the standard values of formation at 298 °K. BaO_2 was chosen for these investigation[56] because it has been investigated in the past by calorimetric methods and therefore gives a possibility for comparing those values obtained from static methods with those obtained from values from dynamic methods.

BaO_2 is the most stable of the alkaline earth peroxides. It is the only one which may be obtained by direct, dry oxidation of the oxide. The close relationship between the structures of BaO (NaCl-type) and BaO_2 (CaC_2-type) is shown in Fig. 54. Peroxide-ions are oriented parallel to the tetragonal c-axis, the [110]-direction in BaO_2 corresponds already to the [100]-direction in BaO. During the dissociation of BaO_2 to BaO the peroxide ions are changed to oxide ions. This results in a shrinkage of the tetragonal c-axis of BaO_2, which now corresponds to the cubic a-axis of BaO.

The preparation of BaO_2 was carried out by decomposition of $BaCO_3$ in high vacuum 10^{-6} mm on the thermobalance (METTLER Thermoanalyzer) at 800 °C to BaO and subsequent oxidation with dry oxygen (Fig. 55). The complete oxidation could be achieved only after cycling in the temperature region 500 to 600 °C. The dissociation of BaO_2 during stepwise heating at 100 mm Hg is shown in Fig. 56. This heating process allows a rather precise determination of the dissociation temperatures. A diagram where the dissociation temperatures thus obtained are plotted versus

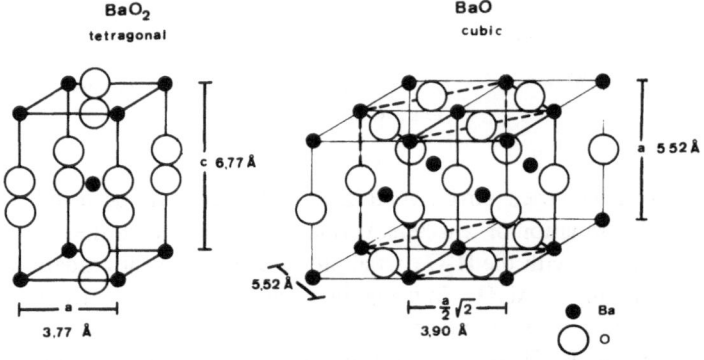

Fig. 54. Structural relation between BaO_2 (CaC_2 structure) and BaO (NaCl structure)

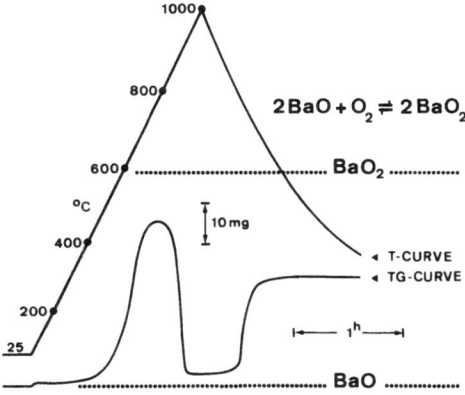

Fig. 55. TG curve showing the oxidation of BaO to BaO_2 followed by decomposition and incomplete reoxidation during cooling. Sample weight: 555.19 mg, heating rate: 10 °C/min, atmosphere: O_2 (720 torr)

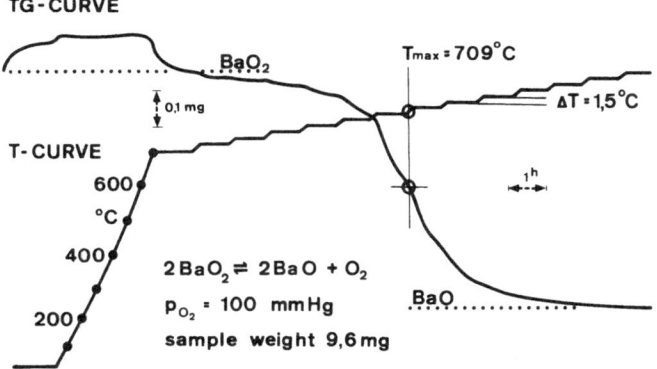

Fig. 56. Dissociation of BaO_2 during stepwise heating, approaching quasi-isothermal conditions

the equilibrium pressures of oxygen, was shown already in Fig. 33. At 1 at oxygen pressure the dissociation temperature is equal to 833 °C. The heat of dissociation of BaO_2 could be derived from this data for a medium temperature of 1040 °K as − 17.8 kcal/mole.

Specific heat measurements on BaO_2 were carried out by DTA for conversion of this value to 25 °C

$$\Delta C_{p BaO_2} = 85.75 - 0.11\ T + 0.14 \cdot 10^{-3}\ T^2.$$

The calculated standard value for the heat of formation

$$2\ BaO_{(s)} + O_{2(g)} \rightleftharpoons 2\ BaO_{2(s)}$$

is $\Delta H^\circ_{298} = +19.3$ kcal/mole.

These data obtained from TG and DTA are in good agreement with the literature values based on calorimetric data.

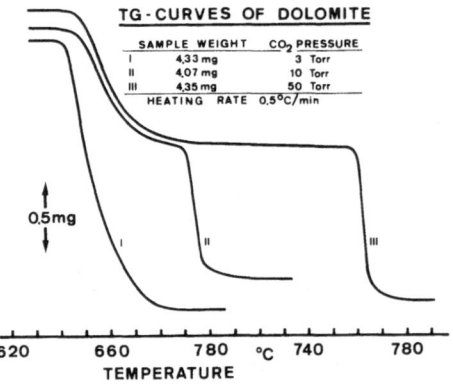

Fig. 57. Decomposition of dolomite at different CO_2-pressures

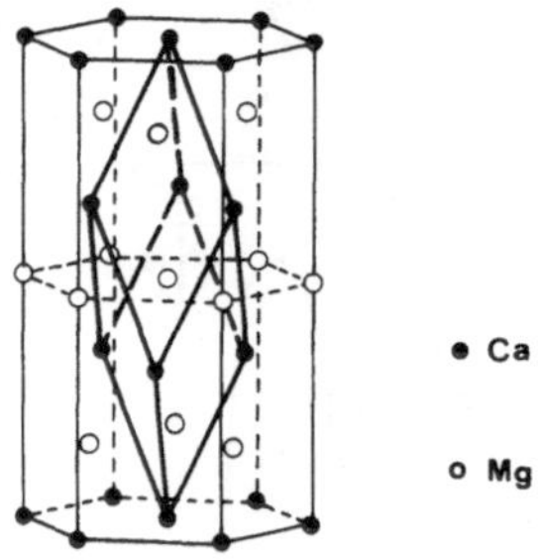

Fig. 58. Dolomite structure

4.6. Thermal Dissociation of Carbonates in Different Atmospheres

During investigations on dissociation equilibria by means of thermogravimetry, it was observed that the detailed mechanism of certain carbonate decompositions[31] is still not completely clarified. The reason for this is that differences in the decomposition mechanism can be introduced by the nature of the substance, by structural variations and especially by even minor changes of the gas atmosphere around and in the sample. Of course, also the effect of the usual experimental conditions has to be taken into account.

The present investigations were carried out on dolomite ($CaMg(CO_3)_2$), smithsonite ($ZnCO_3$) and otavite ($CdCO_3$) by means of TG, DTA, heating X-ray and thermomicroscopy. Different partial pressures of CO_2 and H_2O were used in addition to normal atmosphere and high vacuum. All these carbonates show a very pronounced effect of the atmosphere on their decomposition temperature and on the decomposition products which were formed. The effect of the CO_2 partial pressure on the decomposition of dolomite can be seen from the TG-curves in Fig. 57. The interesting point is that the dolomite structure (Fig. 58) first breaks down completely with the formation of MgO, CaO and evolution of CO_2. If the partial pressure of CO_2 in the atmosphere is higher than 3 torr, the free CaO reacts immediatly and forms calcite which decomposes again at higher temperatures. Therefore, two decomposition steps are present in TG-curves where the CO_2-pressure exceeds 3 torr. These results could be confirmed by heating X-ray. For comparison also mixtures of $CaCO_3$ and $MgCO_3$ in the dolomite ratio were investigated.

The nucleation of these decomposition processes was studied by means of thermomicroscopy on single crystal cleavage plates of calcite, magnesite, dolomite and smithsonite (Fig. 59). The shape of the nuclei was found to be different for these carbonates, which might be also of importance for the decomposition mechanism. The partial pressure of water vapor has a pronounced effect on the decomposition of transition metal carbonates such as $ZnCO_3$ and $CdCO_3$. The evolution of CO_2 is probably catalyzed in the presence of water vapor and shifted to considerably

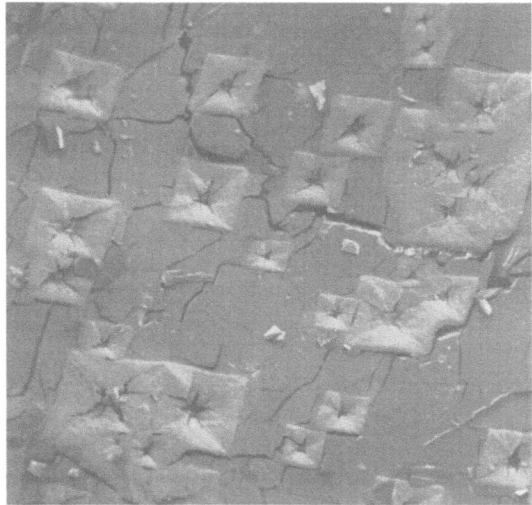

Fig. 59. Nuclei formation on a cleavage plate of calcite

lower decomposition temperatures. The reaction mechanism here is different from dolomite or calcite due to formation of basic carbonates.

The effect of the partial pressure of oxygen and of CO_2 is also very important for the decomposition behaviour of siderite and rhodochrosite. The formation of the iron oxides was followed by TG and by high temperature X-ray diffraction. Below 10^{-6} mm Hg oxygen pressure only Fe_3O_4 was formed.

4.7. Thermomolecular Beam Analysis of $CaCO_3$-Dissociation

Experiments on the thermal decomposition of calcite single crystals were carried out in air, in vacuum and in CO_2[57]. Electron micrographs show that the decomposition nuclei are different in shape and in morphology depending on the atmosphere (Fig. 59). These differences will also effect the kinetics of calcite decomposition. The decomposition starts preferentially at dislocations and also along twin boundaries, preferred growth of CaO may occur topotactically on the rhombohedral cleavage planes (100) and on ($1\bar{1}1$) and (111).

The simultaneously recorded heating X-ray pattern of calcite in vacuum, the TMBA-curve and the mass spectrometric curve for CO_2 are shown in Fig. 60 and in Fig. 61. It can be seen that the decomposition of calcite in vacuum (10^{-4} torr) starts already at 420 °C and that it is complete at 660 °C. The equipment and experimental procedure for thermomolecular beam analysis has been discussed in detail in Section 2.4.

The simultaneous presence of the $CaCO_3$-reflections and of the CaO-reflections over a wide temperature range, and the persistence of certain calcite reflections point to topotaxy. This behaviour was generally observed during decomposition and recrystallization processes. The growth of the CaO crystallites with increasing temperature can be also seen in Fig. 61 (decrease of line broadening in the diffraction pattern).

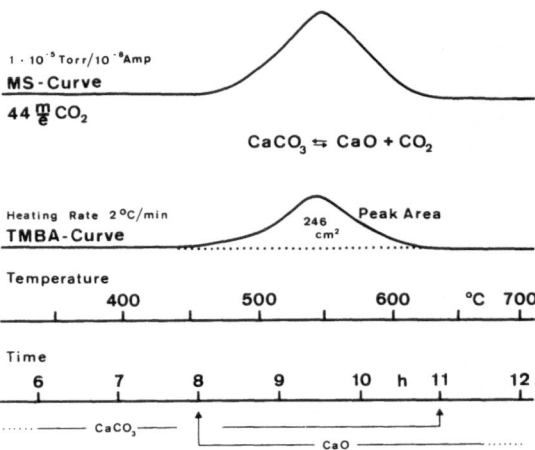

Fig. 60. TMBA- and MS-curves of the decomposition of calcium carbonate. Heating rate 0.5 °C/min

Due to the strong expansion anisotropy of calcite, the reflections move into different directions and even cross each other in the case of high index reflections (*e.g.* (300) and (0.0.12) during heating (Fig. 61). Linear thermal expansion coefficients which can be calculated from this X-ray pattern ($\alpha_a = -5.10^{-6}/°C$, $\alpha_c = 27.10^{-6}/°C$ between 20 and 600 °C) agree with the data reported in the literature. Therefore this method can give information on the thermal expansion as well, provided that the temperature range is large enough for causing measurable shifts of the reflections. respectively (Fig. 62). This means that the thermal stability of such blue pigments increases with increasing ionic radius or atomic weight of the alkaline earth ion.

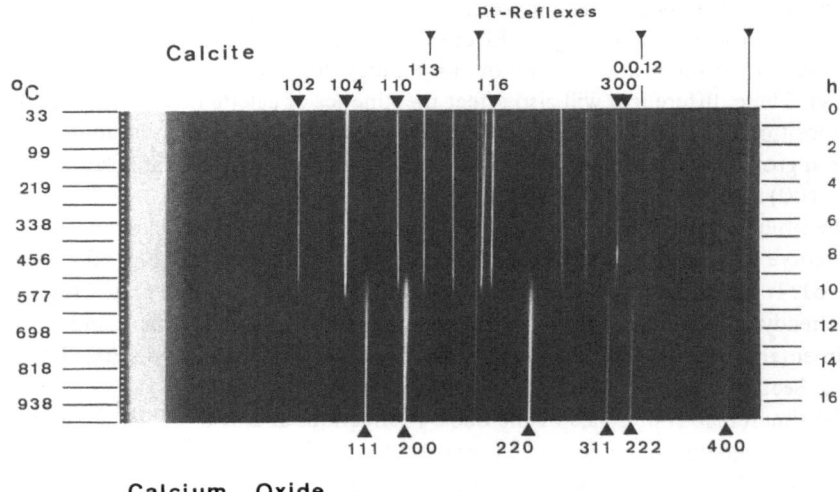

Fig. 61. Simultaneously recorded X-ray heating pattern of the $CaCO_3$-decomposition

4.8. Formation and Stability of Egyptian Blue

Egyptian Blue, $CaCu[Si_4O_{10}]$, and the isostructural sheet silicates $SrCu[Si_4O_{10}]$ and $BaCu[Si_4O_{10}]$ can be synthesized by solid state reaction methods. Information on the reaction steps may be obtained by use of a thermobalance[58, 59]. It was possible thereby to establish the effect of various Ca-, and Cu-containing compounds on the formation of the Ca–Cu-silicate and also the importance of adding fluxing agents such as sodium carbonate, borax, PbO. Without such additions the reaction proceeds very slowly leading to fine-crystallized, impure products which do not show the striking and intense blue color of these alkaline earth copper silicates. One of the most effective agents for accelerating the reaction was borax. The compound $CaCu[Si_4O_{10}]$ forms at about 900 °C. With heating rates of 2 °C/min it remains stable in oxidizing atmosphere up to about 1080 °C. Above this temperature it decomposes due to the reduction $Cu^{2+} \longrightarrow Cu^{1+}$ which is reversible. However $CaCu[Si_4O_{10}]$ does not form again during cooling down in spite of the reoxidation of $Cu^{1+} \longrightarrow Cu^{2+}$. Thermoanalytical runs proved that the stability of the corresponding compounds $SrCu[Si_4O_{10}]$ and $BaCu[Si_4O_{10}]$ is higher, e.g. 1155 °C and 1170 °C respectively (Fig. 62). This means that the thermal stability of such blue pigments increases with increasing ionic radius or atomic weight of the alkaline earth ion.

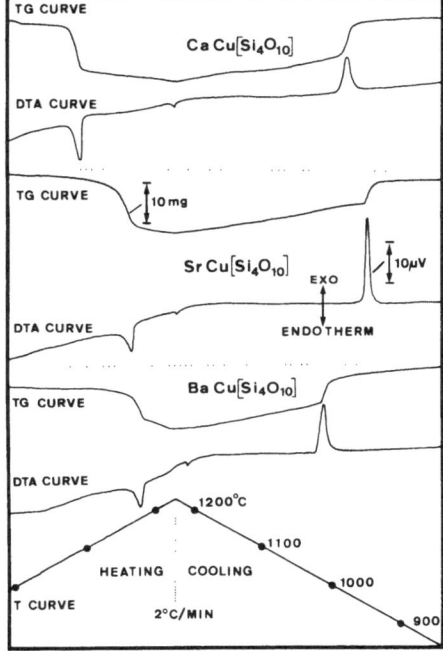

Fig. 62. Decomposition and reoxidation of Egyptian Blue and of the isostructural Sr and Ba compound in air, heating rate 2 °C/min

It is interesting to note that the thermal decomposition is completely reversible in the case of $BaCu[Si_4O_{10}]$ which reforms during cooling down. Single crystals of Ca-, Sr- and Ba-copper-silicate could be grown using borax, PbO or Na_2SO_4 as fluxes

Fig. 63. Scanning electron micrograph of Egyptian Blue crystal (200x)

and heating times of 30 hours at about 900 °C. These quadratic platelets with their characteristic intergrowth are shown in Fig. 63. Some of original, Egyptian samples which were also studied by these methods were exceptionally pure and well-crystallized and showed identical properties as the synthetic samples.

Thermal stability tests were carried out in inert and reducing atmospheres such as N_2 and CO. In both cases the decomposition started at lower temperatures, at about 950 and 800 °C respectively, with formation of red-brown colored reaction products due to the presence of Cu_2O.

4.9. Thermal Synthesis of Pentazinc Hexahydroxide Dicarbonate

Investigations on the reaction of solid hydroxides with gaseous carbon dioxide and water vapor have shown the formation of compounds similar to pentazinc hexahydroxide dicarbonate. In a series of investigations, tests were made to establish how far the zinc hydroxide carbonate (Merck), which was completely decomposed to zinc oxide in an atmosphere of CO_2-H_2O vapor, could be synthesized again during the linear programmed cooling[60]. A special furnace was developed for this kind of investigation (see Section 2.4.).

The zinc hydroxide carbonate sample (Merck) was weighed on the thermobalance and linearly heated or cooled in the water vapor furnace. The H_2O-CO_2 atmosphere was generated by a flow of CO_2 through the water vaporizer into the sample chamber. The condensed water flows back into the flask, the CO_2 leaves through the gas outlet. An additional flow of CO_2 through the balance prevents any water condensation in the balance chamber or on the sample holder. The tests could only be started after both CO_2 gas flows were adjusted to a constant rate and the vaporizer showed a constant return flow of condensed water. Flowmeters were used to adjust and control the gas flow rates.

Figure 64 shows several curves of these measurements. The clearly visible weight gain at the beginning of the test is the result of water vapor adsorption on the original sample at room temperature. After constant flow conditions of the vapor mixture were reached, heating was started at 10 °C min^{-1}. Due to heating an even larger amount of water adsorption is noted. In the range of 50–100 °C the adsorbed water is released again and the actual decomposition of the sample starts only then. Compared to air as atmosphere, the decomposition temperature is increased in a CO_2–H_2O atmosphere.

The decomposition is nearly completed from 300 °C on. Continuous heating to 600 °C yields only little additional weight loss, until pure zinc oxide finally remains. During each test, the sample was held isothermally for a certain time after reaching the corresponding maximum test temperature, to enable it to establish decomposition equilibrium. During the following cooling, a noticeable weight gain was detected between 200–150 °C. The results of these qualitative investigations confirmed that a substance similar to pentazinc hexahydroxide dicarbonate is formed in a gas environment of CO_2-water vapor and within a temperature range from room temperature to 200 °C.

As the test showed, the formation of the hydroxide carbonate and the speed at which this takes place are dependent on the composition of the test atmosphere, especially on the water pressure. The zinc oxide did not show any weight gain in a CO_2 atmosphere with a moisture ratio (p/p_o) less than 0.1; i.e., no formation of hydroxide carbonate takes place. Only above a moisture ratio of 0.35 did the zinc oxide take up more water in a relatively short period of time than was necessary for the formation of basic zinc carbonate. In the moisture ratio range of 0.1–0.3, the formation was slower and did not lead to complete formation of hydroxide carbonate.

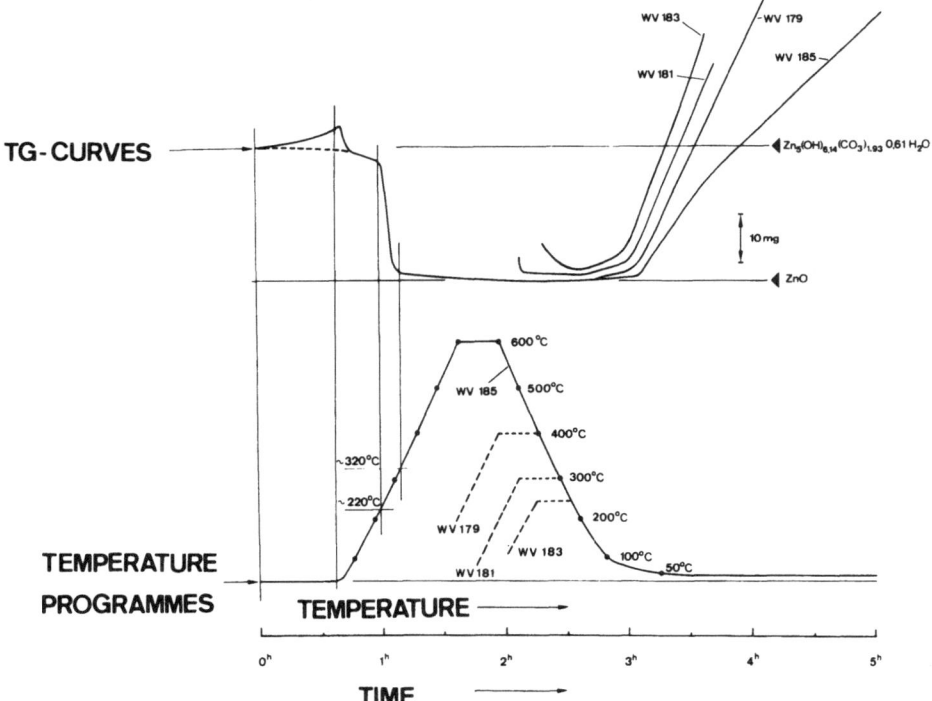

Fig. 64. TG curves of the decomposition and formation processes of zinc-hydroxide-carbonate as a function of different heating programs

All samples for the tests discussed here were formed during more or less linear cooling. If, however, a stepwise isothermal cooling process was used and stopped above room temperature, a better crystalline substance was obtained.

4.10. Vapor Pressure Determinations by TG and DTA Measurements

The thermogravimetric analysis for vapor pressure is based on the effusion method. The Knudsen effusion method uses measurements of the flow rate of vaporized molecules leaving the effusion cell by an orifice (see Section 2.2.). For the calculation of vapor pressure, it is necessary to record the following data: Loss in weight as a function of time simultaneously with sample temperature and pressure measured outside the effusion cell. Measurement of vapor pressure of organic and inorganic compounds by use of the Knudsen effusion method are possible in the range from 1 to 10^{-5} torr. A second method for measuring vapor pressure data is simultaneous thermogravimetric and differential thermal analysis. Thereby the heat of vaporization and vapor pressure data can be calculated. This simultaneous method allows measurements in the range from 1 to 760 torr, under isobaric conditions assuming that the measured pressure is identical to that of the vaporized substance. The apparatus used for this work was schematically illustrated in Section 2.4. The reaction chamber used for the Knudsen effusion method is positioned above the balance. For vapor pressure measurements, the effusion cell is partly filled with the sample and placed in the reaction chamber. The measurements are done under high vacuum conditions. For the simultaneous differential thermal and thermogravimetric analysis sample and reference were measured in special closed aluminum crucibles[61]. A steel ball was put on each

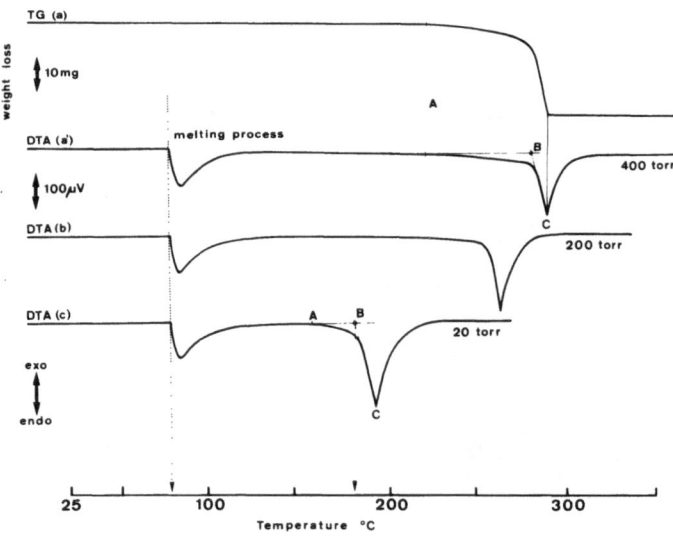

Fig. 65. Simultaneous TG-DTA curves of o-acetophenetitide measured at various pressures.
A: start of vaporization, B: boiling point, C: end of vaporization

crucible acting as a valve which opens if the inside pressure is higher than the outside pressure plus the weight of the ball.

As an application of the simultaneous TG- and DTA-method, the characteristic thermograms of o-acetophenetidide which were recorded at various pressures are shown in Fig. 65. The experimental details were as follows: sample weight 25 mg, heating rate 4 °C/min, pressure range 10—500 torr.

4.11. Vapor Pressure Measurements and Data Processing by Thermogravimetry/ Computer Combination

The present work demonstrates the application of thermogravimetry for vapor pressure determination[62, 63]. The optimum use of the apparatus by automation of the measuring program, collection of the data and possible sources for errors will be discussed with some examples.

For measuring low vapor pressures within the range from 1 to 10^{-6} torr, the well-known Knudsen effusion method was applied (see Section 2.2.). The alumina Knudsen cell was used for vapor pressure measurements of gold and of sodium chloride. The orifice is located in the center of the lid.

A schematic TG-curve of the vaporization of gold under isothermal conditions is shown in Fig. 66. The temperature and the rate of vaporization are recorded in usual form and simultaneously punched on a tape by means of a data transfer system (see Fig. 35 and Table 5).

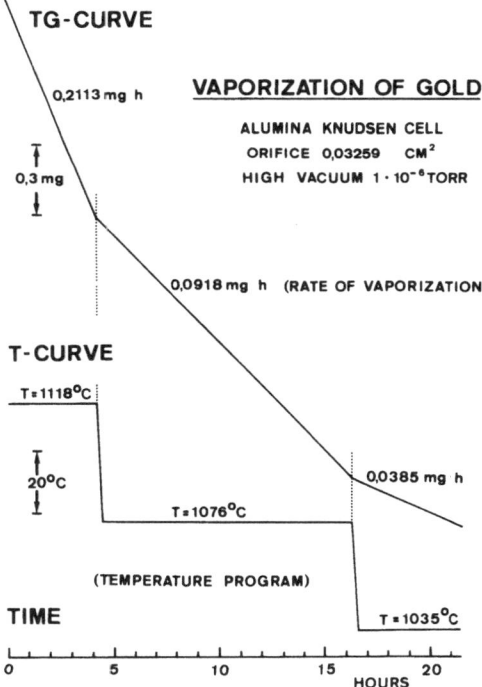

Fig. 66. Stepwise, isothermal vaporization of gold in a thermobalance with Knudsen cell

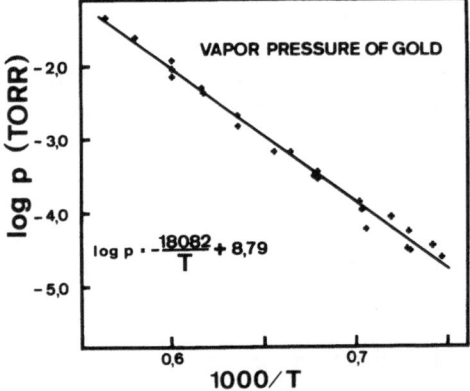

Fig. 67. Vapor pressure of gold measured between 1338 and 1773 °K

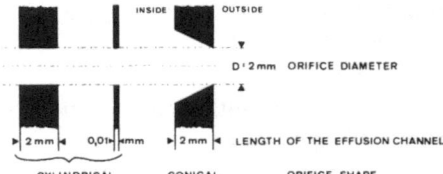

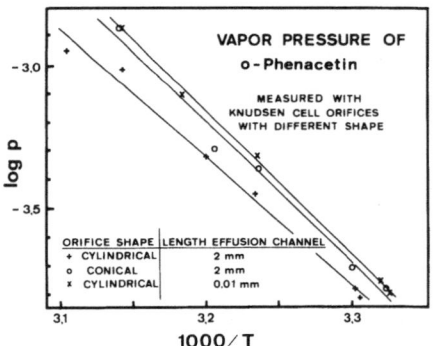

Fig. 68. Effect of shape and length of orifice on measured vapor pressure data

The rate of vaporization (g/sec), the EMF voltage and the temperature are listed in Table 5.

The optimum data are selected in *step 1* by using a precalculated standard deviation, which is directly related to the constancy of temperature. These selected sets of data are used in *step 2* for calculation of equilibrium vapor pressures by means of the Knudsen equation. A regression analysis of these vapor pressure data is then carried out in *step 3*.

The printed data output *step 4* lists the values for the constants A and B of the general vapor pressure equation and the enthalpy and entropy of vaporization or sublimation.

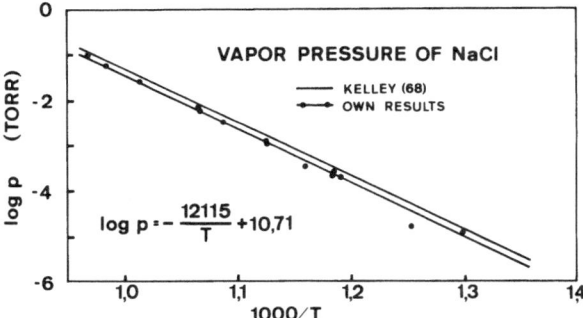

Fig. 69. Vapor pressure of sodium chloride

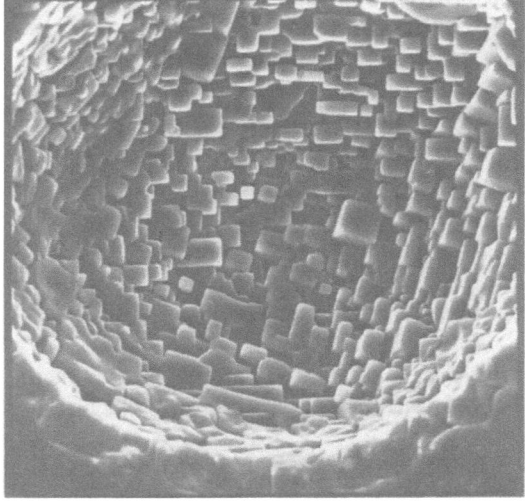

Fig. 70. Thermal etch figures on sodium chloride

The calculated vapor pressure of gold is shown as a function of the temperature in the usual form in Fig. 67. It is compared to data which were measured by the National Bureau of Standards and evaluated by analogous procedures[64].

An important requirement for such vapor pressure measurements with Knudsen cells is the temperature distribution in and around the cell. If the temperature is not homogeneous within the cell, condensation and crystallization of the vaporized species may occur in the colder regions. This was observed *e.g.* with gold. In other cases the orifice was reduced in diameter up to complete blockage when vaporized metals like aluminium where oxidized due to the partial pressure of oxygen. To prevent these deposits, ultra high vacuum is necessary or the use of graphite cells instead of alumina cells.

Ideally, the Knudsen effusion method can be applied for determination of vapor pressures if molecular flow is verified ($p < 10^{-3}$ torr). Precise measurement of vapor pressures is possible in the intermediate region between molecular flow and laminar flow ($p > 10^{-3}$ torr-1 torr) if the dimension of the cell orifice is compatible with the mean free path of the gas molecules. According to Winterbottom[65], the cell orifice should be in the order of 1/10 of the mean free path. Therefore larger orifice diameters are required for relatively low vapor pressures, since the effusion rate is directly proportional to the vapor pressure. The influence of the orifice shape and diameter on the vapor pressures calculated from the effusion rate will be shown in one example.

For this purpose a special Knudsen cell with exchangeable orifice plates has been designed. When the equilibrium pressure, which corresponds to the saturation pressure p_s is reached, effusion of a part of the vaporizing substance takes place through the orifice. The amount depends strongly on the shape and length of the orifice (Fig. 68). From the substance which vaporizes in all directions, only a partial beam can pass freely through the orifice.

The effusion rate $\Delta m/\Delta t$ at identical orifice diameter and temperatures was slowest for the cylindrical hole with long effusion channel and fastest for a cylindrical orifice with very short effusion channel. This means that the back scattering of the vapor molecules is strongly reduced if the ratio L/D between length L of the effusion channel and diameter D of the orifice is small, which is in agreement with theoretical calculations by Clausing[66, 67]. There are three contributions to the Knudsen cell current which originate from the cell interior, from the channel wall, and the cell lid.

To prove that this method can be applied also to solids with strong ionic bonding, NaCl was investigated as an example. Also in this case an aluminia Knudsen cell was used, the orifice diameter was calibrated with gold. Figure 69 shows a graphical presentation of the vapor pressure data obtained, compared with data from the literature (Kelly)[68].

The slight shift results from the differences in dissociation in such measurements. This could be proved by simultaneous mass spectrometric measurement which showed the presence of various vaporized species.

$$^{23}Na^+, \ ^{35}Cl^+, \ ^{37}Cl^+, \ ^{23}Na^{35}Cl^+, \ ^{23}Na^{37}Cl^+, \ ^{23}Na_2^{35}Cl^+, \ ^{23}Na_2^{37}Cl^+ \ etc.$$

The main dissociation occurs at a temperature of 650 °C in high vacuum. For such substances which show dissociation, special equations have been derived by Stranski and Hirschwald[69]. Therefore it is very important in such vapor pressure measurements to control the process by simultaneous mass spectrometry in order to avoid wrong interpretation of the data obtained.

The crystalline NaCl-deposits formed outside of the Knudsen cells showed different morphology. Besides bulk crystals, mainly dendrites but also whiskers were formed.

Thermal etching in high vacuum proved that growth faces were mainly (100). The depth of the quadratic, terrace-like etch figures depends strongly on the saturation of the vapor atmosphere. It increases with increasing undersaturation which means a high rate of vaporization. Figure 70 shows such typical etch figures in the case of high rates of vaporization.

4.12. Dehydration of Copper Sulfate Pentahydrate

In the decomposition of $CuSO_4 \cdot 5 \cdot H_2O^{57)}$ in a flowing air atmosphere, nucleation begins only after a substantial loss of water (approx. 1 1/2 mole H_2O), similar to the case of gypsum. The nucleation starts from the surface and proceeds rapidly throughout the crystal. The first step is the formation of bubble-like craters on the crystal surface which is followed immediately by the peripheric crystallization of $CuSO_4 \cdot 3 H_2O$. Figure 71 a, b shows some features of such nuclei after crystallization. They all have a spherical habitus and a geode-like cavity in the center. In addition to such nuclei, fissures in the direction of the c-axis are observed when the crystals are heated rapidly. Individual nuclei may separate from the surface, due

a(260x)   b(740x)   c(180x)

Fig. 71. Surface nucleation during thermal decomposition of $CuSO_4 \cdot 5 H_2O$

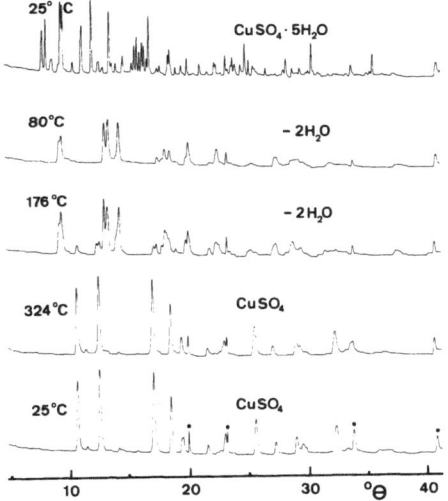

Fig. 72. Photometer curves of X-ray photographs taken during the decomposition process of copper sulfate pentahydrate

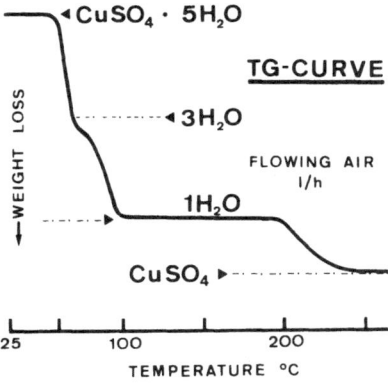

Fig. 73. TG curve of the thermal decomposition of $CuSO_4 \cdot 5 H_2O$ to $CuSO_4$

probably to the different expansion behavior of both phases. Figure 71c shows the surface after complete loss of 2 H_2O; there are no intermediate spaces left between the nuclei. Further dehydration causes only small changes of the surface until anhydrous $CuSO_4$ is formed. Photometer curves (Fig. 72) of a heating X-ray photograph, which was taken under identical conditions as the microphotographs, showed that all known intermediate phases – $CuSO_4 \cdot 3\ H_2O$, $CuSO_4 \cdot 1\ H_2O$ and $CuSO_4$ – are formed during the dehydration reaction. The TG-curve of the decomposition of $CuSO_4 \cdot 5\ H_2O$ to $CuSO_4$ is presented in Fig. 73.

5. Conclusions

One of the main points of this review on modern thermogravimetry was to show the versatility and some of the areas where TG-methods may be used. The main trends of the rather rapid developments in this field are the following:

Automatic data collection and processing is becoming more and more common in modern instrumentation. This not only shortens the time for accurate evaluation of the results, but also makes it possible to take into account various experimental factors and to make necessary corrections.

As far as the equipment is concerned, the miniaturization of many components has lead to construction of smaller apparatus without sacrificing the accuracy, efficiency or versatility. Of importance for this development was also the availability of better and new construction materials, e.g. for measuring heads, thermocouples, furnaces etc.

The theoretical background of various thermoanalytical methods is now well established. An important contribution to the quantitative evaluation of the results was made by the use of standard reference materials. These allow not only the calibration of the instrument but also the comparison and correlation of results which were obtained with different instruments.

The possibilities for combining thermoanalytical methods like TG with other techniques and to carry out such measurements in certain cases simultaneously are increasing. This is of great significance for the easier and sometimes unambiguous interpretation of results.

Thermoanalytical methods, including thermogravimetry, have been used originally in the earth sciences and in chemistry. With todays multifunctional and highly sensitive instruments, these methods are becoming important for many other, new fields of science and technology.

6. References

[1] Wagner, C., Grünwald, K.: Z. Physik. Chem., Abt. *B 40*, 455 (1938)
[2] Honda, K.: Sci. Repts. Tôhoku Imp. Univ., Ser. *IV*, 97 (1915)
[3] Guichard, M.: Bull. Soc. Chim. France, Ser. *37*, 62, 251, 381 (1925)
[4] Chevenard, P., Waché, X., de la Tullaye, R.: Bull. Soc. Chim. France, Mém. Ser. (*5*) 11, 41 (1944)

5) Duval, C.: Inorganic thermogravimetric analysis. Amsterdam-London-New York: Elsevier Publishing Co. 1963

6) Wendlandt, W. W.: Thermal methods of analysis. 2nd edit. New York-London-Sydney: John Wiley & Sons 1974

7) Garn, P. D.: Thermoanalytical methods of investigation. London-New York: Academic Press 1965

8) Eyraud, Ch.: J. Physique Radium *14*, 638 (1953)

9) Berg, L. G.: Introduction to thermal analysis. Moscow: Izd. Akad. Nauk SSSR 1969

10) Saito, H.: Proc. Imp. Acad (Tokyo) *2*, 58 (1926)

11) Paulik, F., Paulik, J.: Thermoanalysis. Budapest: Akad. Kiadó 1963

12) Redfern, J. P.: Proceedings of First International Thermoanalysis Conference 1965. London: Macmillan 1966
Schwenker, R. F., Garn, P. D.: Proceedings of Second International Thermoanalysis Conference 1968. London-New York: Academic Press 1969
Wiedemann, H. G.: Proceedings of the Third International Conference on Thermal Analysis. Basel-Stuttgart: Birkhäuser Verlag 1972
Buzás, I.: Proceedings of the Fourth International Conference on Thermal Analysis. Budapest: Akad. Kiadó 1975

13) Wiedemann, H. G.: Chem. Ing. Technik *36*, 1105 (1964)

14) Saito, H.: Thermobalance analysis. Tokyo: Gijitsu Shoin 1962

15) Keattch, C. J., Dollimore, D.: An Introduction to Thermogravimetry, Heyden, London-New York-Rheine, 2nd edition, 1975

16) a) McAdie, H. G., Garn, P. D., Menis, O.: Selection of Differential Thermal Analysis Temperature Standards through a Cooperative Study, GM-758, 759, 760; U.S. Natl. Bur. Stand. Spec. Publ. 260–40, Washington DC, 1972
b) Garn, P. D., Menis, O.: ICTA Certified Reference Materials for Differential Thermal Analysis from 180 K to 330 K, GM-757; U.S. Natl. Bur. Stand. Certificate, Washington DC, in press
c) Garn, P. D., Menis, O.: ICTA Certified Reference Material Polystyrene for Glass Transition Measurements, GM-754; U.S. Natl. Bur. Stand. Certificate, Washington DC, in press
d) U.S. Natl. Bur. Stand. Certificate: Standard Reference Material 745, Gold Vapor Pressure; Washington DC, May 14, 1969
e) U.S. Natl. Bur. Stand. Certificate: Standard Reference Material 720, Synthetic Sapphire (Al_2O_3); Washington DC, August 26, 1970

17) Oswald, H. R., Dubler, E., in: Reviews on analytical chemistry. Euroanalysis II, Budapest 1975. Fresenius, W. (ed.). Budapest: Akadémiai Kiadó 1977, p. 191

18) Wiedemann, H. G.: Ph. D. Thesis, Univ. of Bern, Switzerl., 1971

19) Wiedemann, H. G.: Thermochim. Acta *7*, 131 (1973)

20) Wiedemann, H. G., in: Vacuum microbalance techniques. Vol. 7, Massen, C. H., Van Beckum, H. J. (ed.). New York: Plenum Press 1970, p. 217

21) Jucker, H., Wiedemann, H. G., Vaughan, H. P.: Preprint of a Paper Presented at the Pittsburgh Conference on Analytical Chemistry and Applied Spectroscopy, March 2, 1965

22) Peters, H., Möbius, H. H.: Z. Phys. Chem. *209*, 298 (1958)

23) Mann, G.: Dipl. Thesis, Univ. Rostock 1957

24) Kuntze, R. A.: Canad. J. Chem. *43*, 2522 (1965)

25) Fowler, A., Howell, H. G., Schiller, K. K.: J. Appl. Chem. *18*, 366 (1968)

26) Wiedemann, H. G., Bayer, G.: Z. Anal. Chem. *276*, 21 (1975)

27) Günter, J. R., Oswald, H. R.: Bull. Inst. Chem. Res. Kyoto Univ. *53*, 249 (1975)

28) Hirsiger, W., Müller-Vonmoos, M., Wiedemann, H. G.: Thermochim. Acta *13*, 223 (1975)

29) Bayer, G., Wiedemann, H. G., in: Thermal Analysis, Proc. 4th Int. Conf. on Thermal Anal., Budapest 1974, Buzás, I. (ed.). Budapest: Akadémiai Kiadó 1975, Vol. 1, p. 763

30) Wiedemann, H. G., Bayer, G.: Chimia *30*, 351 (1976)

31) Bayer, G., Wiedemann, H. G., in: Proc. First Europ. Symp. on Thermal Anal., Univ. of Salford 1976, Dollimore, D. (ed.). London: Heyden 1976, p. 256

32) Paulik, F., Paulik, J.: J. Thermal Anal. *5*, 253 (1973); ibid. *8*, 557 (1975)

33) Paulik, J., Paulik, F.: J. Thermal Anal. *8*, 567 (1975)

34) Motzfeldt, K.: J. Phys. Chem. *59,* 139 (1955)

35) Cordes, J. F., Schreiner, S.: Z. Anorg. Allg. Chem. *299,* 87 (1959)

36) Wiedemann, H. G.: Thermochim. Acta *3,* 355 (1972)

37) Simons, E. L., Newkirk, A. E., Aliferis, I.: Anal. Chem. *29,* 48 (1957)

38) Peters, H., Wiedemann, H. G.: Z. Anorg. Allg. Chem. *298,* 202 (1959); ibid. *300,* 142 (1959)

39) Wiedemann, H. G.: Z. Anorg. Allg. Chem. *306,* 84 (1960)

40) Duval, C.: Mikrochim. Acta [Wien], 705 (1958)

41) Wiedemann, H. G.: Chem. Ing. Techn. *36,* 1105 (1964)

42) Wiedemann, H. G., in: Thermal Analysis, Proc. 2nd Int. Conf. on Thermal Anal., Worcester, Mass. 1968, Schwenker, R. F., Garn, P. D. (ed.). New York: Academic Press 1969, Vol. 1, p. 229

43) Wiedemann, H. G., Bayer, G., in: Progress in Vacuum Microbalance Techniques, Vol. 3, Proc. 12th Conf. on Vac. Microbal. Techn., Lyon 1974, Eyraud, C., Escoubes, M. (ed.). Heyden, London, 1975, p. 103

44) Wiedemann, H. G., in: Thermal Analysis, Proc. 3rd Int. Conf. on Thermal Anal., Davos 1971, Wiedemann, H. G. (ed.), Birkhäuser Basel 1972, Vol. 1, p. 171

45) Gerard, N.: Journ. of Physics E: Scientific Instruments *7,* 509 (1974)

46) Wiedemann, H. G.: Abstr. 4th Scandinav. Symp. on Thermal Anal., Arrhenius Lab., Univ. of Stockholm, Aug. 20–22, 1975, p. 18

47) Wiedemann, H. G., Van Tets, A.: Thermochim. Acta *1,* 159 (1970)

48) Van Tets, A., Wiedemann, H. G., in: Thermal Analysis, Proc. 2nd Int. Conf. on Thermal Anal., Worcester, Mass. 1968, Schwenker, R. F., Garn, P. D. (ed.). New York: Academic Press 1969, Vol. 1, p. 121

49) Bayer, G., Wiedemann, H. G.: Thermochim. Acta *11,* 79 (1975)

50) Kelley, K. K.: Bur. Mines Bull. 584, Washington D. C. 1960

51) Bayer, G., Wiedemann, H. G.: Thermal Analysis Proceedings 5th ICTA, Kyoto, Japan, August 1–6, 1977, p. 437, edited by Chihara, H., Kagaku Gijutsu-Sha

52) Minagawa, S., Gejyo, T.: J. Am. Ceram. Soc. *52,* 282 (1969)

53) Schmitt, J.: Bull. Soc. Fr. Ceram. *73,* 3 (1966)

54) Wiedemann, H. G., Sturzenegger, E., Bayer, G., Wessicken, R.: Naturwissenschaften *61,* 65–69 (1974)

55) Sears, G. W., Devries, R. C.: J. Chem. Phys. *39,* 28, 37 (1963)

56) Wiedemann, H. G., Bayer, G.: Chimia *30,* 351 (1976)

57) Wiedemann, H. G., Bayer, G.: Z. Anal. Chem., Band *276,* Heft 1, 21 (1975)

58) Bayer, G., Wiedemann, H. G.: Naturwissenschaften *62,* 181 (1975)

59) Bayer, G., Wiedemann, H. G.: Sandoz Bulletin *40,* 19 (1976)

60) Wiedemann, H. G.: Thermochimica Acta *7,* 131–149 (1973)

61) Pfefferkorn, J., Wiedemann, H. G.: Progress in Vacuum Microbalance Techniques. London: Heyden and Son 1972, Vol. II, p. 221

62) Wiedemann, H. G.: Thermochim. Acta *3,* 355 (1972)

63) Wiedemann, H. G., Bayer, G.: Chemtech. 381 (1977)

64) Paule, R. C., Mandel, J.: Nat. Bur. Stand. (U.S.) Spec. Publ. *260,* 19 (1970)

65) Winterbottom, W. L., Hirth, J. P.: J. Chem. Phys. *37,* 384 (1962)

66) Clausing, P.: Ann. Phys. *12,* 961 (1932)

67) Clausing, P.: Z. Phys. *66,* 471 (1930)

68) Kelley, K. K.: Bur. Mines. Bull. 383, Washington 1935

69) Hirschwald, W., Stolze, F., Stranski, I. N.: Z. Phys. Chem., Neue Folge *42,* 96 (1964)

Received April 24, 1978

Determination of Molecular Weights by Light Scattering

Malcolm B. Huglin

Department of Chemistry and Applied Chemistry, University of Salford, Salford M5 4WT,
England

Table of Contents

M. B. Huglin

I. Introduction

The scattering of X-rays, neutrons and visible light is different in origin, since it occurs respectively on free electrons, neutrons and bound electrons. Consequently, what has been termed by Serdyuk and co-workers[1] the scattering capactiy per unit volume of the substance is dictated, in the same order by the electronic density, the scattering length and the refractive index. The joint use of all three types has recently been deployed to quantify the distribution of RNA and protein within the 50S sub-particle of *E. coli* ribosomes[2] as well as to elucidate the structure of synthetic co-polymers in solution[3]. There is, of course, a large difference in the wavelengths appropriate to small angle X-ray scattering and light scattering (LS), but the essential complimentarity of both seems conveniently to extend the range accessible to either method alone, and the relevant theoretical expressions can be deduced from the same basic approach[4]. It is possible, although not yet immediately feasible, that the wave-length range in scattering studies may be extended yet further, when the Synchrotron is fully developed.

LS from solution is an invaluable tool for yielding the dimensions[5, 6] of the scattering particles and thermodynamic information[7–13] such as virial coefficients, chemical potentials and excess free energy of mixing. Its main use is to provide the molecular weight[4, 6, 14–16] of the scattering substance, *i.e.* the solute. If the difference in refractive index between solute and solvent is large enough and with the availability of sufficiently sensitive apparatus there is, in principle, virtually no limit to the range of molecular weight measurable *via* LS. Thus, the molecular weights of diethyl ether[17] (74) and bacteria[18] (ca. 1×10^{11}) have been determined by this means. Nonetheless, for molecular weights below ca. 2×10^4 it is probably more convenient to utilise vapour pressure osmometry, although this technique does yield a differently averaged value.

1. Molecular Weight Averages

If the solute is monodisperse with respect to molecular weight, as in the situation for purified globular proteins for example, then the same value for the molecular weight will be yielded irrespective of the experimental method and its underlying theory. Many synthetic polymers as well as such naturally occurring ones as cellulose, rubber and bacterial dextrans are polydisperse and the experimentally determined molecular weight is an average value. Two of the most important averages are the number average molecular weight $\overline{M}_n$ and the weight average molecular weight $\overline{M}_w$, which are defined as:

$$\overline{M}_n = \sum_i (n_i M_i)/\sum_i n_i \tag{1}$$

$$\overline{M}_w = \sum_i (n_i M_i^2)/\sum_i (n_i M_i) \tag{2}$$

Here the whole solute is considered to comprise n_i moles of species i having molecular weight M_i (the definitions are unaltered if n_i is considered as the number of molecules of species i). A typical colligative property such as the osmotic pressure Π exerted by the solution is related under ideal conditions to the overall concentration of the solution c (g/ml) and to the molecular weight M of the solute by an expression of the form:

$$\Pi \propto c/M \tag{3}$$

The total osmotic pressure is the sum of the osmotic pressure Π_i exerted by each species, viz

$$\Pi = \sum_i \Pi_i \propto \sum_i (c_i/M_i) \tag{4}$$

Denoting mass of species i by g_i and using the relation $n_i = g_i/M_i$ reduces Eq. (1) to

$$\overline{M}_n = \Sigma g_i / \Sigma(g_i/M_i) \tag{5}$$

Moreover, the mass in both numerator and denominator of Eq. (5) may be replaced by mass per unit volume, *i.e.* concentration, thus

$$\overline{M}_n = \Sigma c_i / \Sigma(c_i/M_i), \text{ whence}$$

$$\Sigma(c_i/M_i) = \Sigma c_i / \overline{M}_n = c/\overline{M}_n \tag{6}$$

Combination of Eqs. (4) and (6) shows that the molecular weight yielded by a measurement of osmotic pressure is a number average value:

$$\Pi \propto c/\overline{M}_n \tag{7}$$

As will be demonstrated later, LS can be quantified by an experimental quantity termed the excess Rayleigh ratio between solution and solvent, R, which is related to the concentration and molecular weight of solute by the following expression valid under ideal conditions:

$$R \propto cM \tag{8}$$

As before, we may write

$$R = \Sigma R_i \propto \Sigma(c_i M_i) \tag{9}$$

and Eq. (2) as

$$\overline{M}_w = \Sigma(g_i M_i / \Sigma g_i = \Sigma(c_i M_i)/\Sigma c_i \tag{10}$$

$$\therefore \Sigma(c_i M_i) = \overline{M}_w \Sigma c_i = c\overline{M}_w \tag{11}$$

Combination of Eqs. (9) and (11) shows that the molecular weight yielded by LS is a weight average quantity, viz

$$R \propto c\overline{M}_w \tag{12}$$

The conventional colligative properties, all of which yield $\overline{M}_n$, are depression of freezing point, elevation of boiling point, osmotic pressure and relative lowering of vapour pressure (on which vapour pressure osmometry depends). We have noted that LS yields a different type of molecular weight average and it is commonly considered not to belong to the group of colligative properties. Eisenberg[4] has pointed out that LS is, in fact, colligative (i.e. depends on the number of mols of scattering solute per unit volume) and only appears otherwise since the practical equations are expressed in terms of concentration in mass/volume. Although this is perhaps a semantic point, it is nonetheless correct. Verification can be accomplished by recasting the concentration (mass/volume) in Eqs. (3) and (8) [or equally in Eqs. (7) and (12)] in terms of the number of mols/volume, n, where n = c/M. The following results show that Π and R are both proportional to n, that is, they are both colligative in nature. LS has the additional characteristic of being also strongly dependent on M:

$$\Pi \propto n$$
$$R \propto nM^2$$

For ease of notation, the weight average molecular weight $\overline{M}_w$ will henceforth be written simply as M. Relative molar mass of a substance is defined[19] as the ratio of the average mass per molecule of specified isotopic composition of the substance to one twelfth of the mass of an atom of the nuclide ^{12}C. It is dimensionless. When it is obtained from an experiment involving a known mass of substance or a known concentration (mass/volume) of it in solution, the value obtained is a molar mass or molecular weight expressed in practical units of mass/mol. The customary c.g.s. system employing concentration in gram/vol will yield M in units of g/mol, whereas the value of M in units of kg/mol will be a thousand times smaller if the practical concentration are the S.I. one of kg/volume. In this review molecular weights are taken to be in the familiar units of gram/mol.

2. Types of Light Scattering

The basics of LS were formulated a considerable time ago by Mie, Zernicke, Prins, Smoluchowski, Rayleigh, Gans and Debye. Scattering by independent particles is usually classified into three types, although there is not complete unanimity about which name should be attributed to each. Here, we shall follow Oster[20] and refer to them as (a) Rayleigh, (b) Debye and (c) Mie scattering. The class is dictated by two parameters namely the relative refractive index $\widetilde{n}_r$ and the relative size parameter D/λ. The former is the refractive index of the particle relative to that of the surrounding medium, whilst in the latter D denotes the major dimension of the particle and λ is the wavelength of the light in the scattering medium. Approximate criteria for the classification of LS are given in Table 1.

When unpolarised light of intensity I_0 is incident on a particle, a small fraction of it, of intensity i_θ, is scattered at an angle θ between the directions of incident

Table 1. Criteria for classification of LS

Type	Refractive Index requirement	Relative size requirement
Rayleigh	$\lvert(\tilde{\eta}_r - 1)\rvert \ll 1$	$(D/\lambda) < 1/20$
Debye	$\lvert(\tilde{\eta}_r - 1)\rvert \ll 1$	$1 > (D/\lambda) > 1/20$
Mie	$\lvert(\tilde{\eta}_r - 1)\rvert$ not $\ll 1$	$(D/\lambda) > 1$

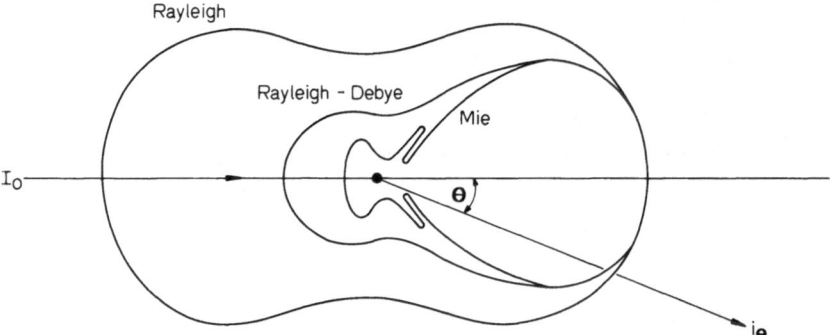

Fig. 1. Polar scattering diagram[21]: Scattering envelopes according to the three LS theories for particles of increasing size but the same scattering power

beam and observation. The resultant polar scattering diagram[21] in Fig. 1 illustrates the effect of increasing particle size but the same relative scattering power according to types (a), (b) and (c). For type (a), scattering is symmetrical about 90° (or 270°). For type (b) wherein the particles are larger, interference effects (to be considered more fully later) between parts of the same particle are destructive in the backward direction with the result that scattering is dissymmetrical about 90° (or 270°). Type (c) is much more complex and diffraction patterns begin to appear in the polar scattering diagram. The rigorous Mie theory has been developed for spheres and poses formidable computational problems in order to calculate the intensity of scattered light. We shall be concerned here only with Rayleigh and Debye scattering.

II. Theory

Many standard texts include sufficient LS theory to render a detailed account unnecessary here. Reference is particularly recommended to treatments by Flory[22], Morawetz[23], Tanford[24], Tsvetkov, Eskin and Frenkel[25], Kerker[26], and Oster[20]. Other books[5, 6, 27] are devoted exclusively to LS from polymer solutions. It will be useful, nonetheless, to summarise the approach[24] in order to emphasise some significant features of the final result.

1. Scattering by Dilute Gases

If the electric field strength of the incident light is E_0, then the magnitude of the dipole induced in the irradiated material will be directly proportional to E_0 *via* a proportionality constant termed the polarisability, α. This dipole acts as a source of new scattered radiation of a different electric field strength E_s, the magnitude of which depends on the second derivative with respect to time of the dipole moment. The useful experimental quantities are the intensities of scattered (i_s) and incident (I_0) light. They, in turn, are proportional to E_s^2 and E_0^2 respectively averaged over a vibrational period, that is, from time = 0 to time = $\lambda_0/\tilde{c}$, where λ_0 is the wavelength of light in vacuo and $\tilde{c}$ is the velocity of light. One obtains thereby:

$$i_s = 16\,\pi^4\alpha^2 I_0 \sin^2\theta_1/\lambda_0^4 r^2 \tag{13}$$

In Eq. (13), θ_1 is the angle between the dipole axis and the line joining the dipole to the observer. The distance from observer to dipole is r and, because it is typically ca. 10^5-10^6 times greater than the size of the dipole itself, r is effectively the distance between observer and all parts of the dipole. As might be expected *a priori,* the scattering is enhanced by a large incident light intensity and a highly polarisable material. The observation that scattering of sunlight by gas molecules or dust particles imparts a blue (*i.e.* short λ_0) coloration also leads one to anticipate some form of inverse dependence of scattering intensity on λ_0. The fact that light scattering photometers do not all necessarily exploit this dependence by using very low wavelength sources will be discussed later.

The number of particles per unit volume, n, as well as α govern the value of the dielectric constant ϵ:

$$\epsilon = 1 + 4\,\pi n\alpha \tag{14}$$

Moreover, Maxwell's relationship, $\epsilon = \tilde{n}^2$ where $\tilde{n}$ is the refractive index affords

$$\tilde{n}^2 = 1 + 4\,\pi n\alpha \tag{15}$$

Changing from number concentration to weight concentration and initial Taylor series expansion of $\tilde{n}$ give α as

$$\alpha = M(d\tilde{n}/dc)/2\,\pi\mathcal{N} \tag{16}$$

where $\mathcal{N}$ is the Avogadro number, c (mass/volume) is the concentration of particles of molecular weight M (mass/mol) in the gas, and $d\tilde{n}/dc$ (volume/mass) is the variation of the refractive index of the dilute gaseous solution with concentration. Hence the mass of a particle is $M/\mathcal{N} = c/n$ and the intensity of LS from a single particle is, *via* Eqs. (13) and (16):

$$i_s = 4\,\pi^2 M^2 I_0 \sin^2\theta_1 (d\tilde{n}/dc)^2/\mathcal{N}^2\lambda_0^4 r^2 \tag{17}$$

A noteworthy feature of this result is the dependence of the scattered light intensity from a single particle on the molecular weight of the particle raised to a power greater than unity. As pointed out by Flory [22], if this were not so, the method would not be amenable to the measurement of molecular weight. Experimentally, one determines the scattering per unit volume. Hence there are n $(= \mathcal{N} c/M)$ particles per unit volume and multiplying i_s in Eq. (17) by this number shows that the scattering per unit volume depends directly on M, viz:

$$i_s = 4 \pi^2 M c I_0 \sin^2 \theta_1 (d\widetilde{n}/dc)^2 / \mathcal{N} \lambda_0^4 r^2 \qquad (18)$$

If we were considering a different property such as heat capacity for which the particle makes a contribution directly proportional to its mass raised to the power of unity, the observed value would depend solely on the total quantity of particles in the given volume and would be independent of the number of molecules into which the particles are divided. Measurements of this type of property cannot yield molecular weights. Similarly, for example, an aqueous solution containing 72 g of polyacrylic acid is neutralised by 1000 ml of 1.0 M NaOH and no information regarding the molecular weight of the polymer is afforded, since the titre depends only on the total quantity of acid groups present. Exactly the same titre would be needed to neutralise a solution containing 72 g acrylic acid or even 72 g of a mixture in any proportion of acrylic acid and polyacrylic acid.

With vertically polarised light, i_s will be independent of angle and $\sin^2\theta_1$ will vanish from Eq. (18), because it has a value of unity. For more commonly used unpolarised light it is readily shown that Eq. (18) reduces to:

$$i_\theta = 2 \pi^2 (d\widetilde{n}/dc)^2 M c (1 + \cos^2\theta) / \mathcal{N} \lambda_0^4 r^2 \qquad (19)$$

where the angle of observation θ means the angle between the direction of the incident beam and the observed scattered beam of intensity i_θ in this direction.

2. Scattering from Polymer Solutions

Since the scattering from a single particle is proportional to the square of its molecular weight, macromolecules are especially suitable for study by LS. Moreover, the presence of any smaller molecules makes a relatively minor contribution unlike the situation prevailing in osmotic pressure measurements. In a binary solution of polymer and solvent, scattering from the latter comprises a finite, but small, part of the total and we shall term the difference between scattering from solution and solvent the excess scattering and use the same symbol for it, i_θ, as in the preceding Section II. 1.

Assuming in the first instance that the polymer molecules are quite independent, then the situation is analogous to that in the preceding section, except that the scattering molecules (polymer) are now surrounded by solvent molecules of refractive index $\widetilde{n}_0$ instead of by free space of refractive index 1.0. The analogue of Eq. (15) becomes

$$\widetilde{n}^2 - \widetilde{n}_0^2 = 4\,\pi n\alpha \tag{20}$$

and the expression replacing Eq. (19) is

$$i_\theta = 2\,\pi^2 \widetilde{n}_0^2 (d\widetilde{n}/dc)^2 (1 + \cos^2\theta) McI_0 / \mathscr{N} \lambda_0^4 r^2 \tag{21}$$

It is convenient to recast this as

$$R_\theta = KMc \tag{22}$$

where

$$R_\theta = i_\theta r^2 (1 + \cos^2\theta)/I_0 \tag{23}$$

and

$$K = 2\,\pi^2 \widetilde{n}_0^2 (d\widetilde{n}/dc)^2 / \mathscr{N} \lambda_0^4 \tag{24}$$

$$\equiv K'(d\widetilde{n}/dc)^2 \tag{25}$$

The total polarisability per unit volume of solute plus solvent has been invoked in Eq. (22) so that the dimensions of the Rayleigh ratio R_θ in Eq. (23) are not cm^2 as it appears, but actually cm^{-1}. For pure liquids and polymers of small-moderate molecular weight, scattering at an angle of 90° is used and the Rayleigh ratio in such instances is given simply as

$$R_{90} = i_{90} r^2 / I_0 \tag{26}$$

With unpolarised light in the visible region and at room temperature, the value of R_{90} for common liquids is usually[26, 28, 29] in the range $5 \times 10^{-6} - 50 \times 10^{-6}\ cm^{-1}$, whilst for polymer solutions in these liquids R_{90} is much greater, to an extent dependent on the values of c, M and $d\widetilde{n}/dc$. When dealing with a solution, it is strictly more exact to write the Rayleigh ratio in Eq. (22) as ΔR_θ instead of R_θ, since the difference between solution and pure solvent is implied (and is also determined experimentally). Similar imprecise usage of the word "scattering" intended synonymously as "Rayleigh ratio" is also to be noted and accepted. Hence Eq. (22) states that for a solute of a certain molecular weight M, the scattering increases directly as the weight concentration; for a solute of exceedingly high molecular weight it is only necessary to use very dilute solution to obtain appreciable scattering. Also, for a series of solutes of differing molecular weight but all at the same concentration, the scattering is proportional to M. In general, the ratio c/R_θ for solutions of different concentration should always have the same value, constant $\times M^{-1}$, just as for osmotic pressure measurements the ratio Π/c should also always have the same value, constant $\times \overline{M}_n^{-1}$. The nature of the constants differs; in the former the constant is K^{-1}, whilst in the latter it is $\mathscr{R}T$, where $\mathscr{R}$ is the Universal Gas Constant.

3. Allowance for Non-Ideality

To obtain a correct form of Eq. (22) allowing for thermodynamic non-ideality of the solution, fluctuation theory originally developed by Einstein, Zernicke, Smoluchowski and Debye has been adapted to polymer solutions.

The solution is regarded as comprising arbitarily chosen volume elements δV which

(a) are very small relative to the wavelength of light in the medium λ ($\lambda = \lambda_0/\widetilde{n} \approx \lambda_0/\widetilde{n}_0$ for dilute solution) and

(b) contain a small number of polymer molecules and a large number of solvent molecules.

Fluctuations in refractive index occur within a volume element and arise from variations in density and concentration. The former are responsible for scattering by the solvent and may be ignored in the present context since solution and solvent scatterings are subtracted. Regarding the dissolved polymer, it is only necessary, therefore, to consider scattering caused by local fluctuations in concentration.

The concentration c of the solution may be written as

$$c = c' + \delta c \tag{27}$$

where c' is the fluctuating concentration over the whole solution and δc is the (positive or negative) magnitude of the small concentration fluctuation.

Correspondingly, the actual polarisability α of a volume element at any instant of time is

$$\alpha = \alpha' + \delta\alpha \tag{28}$$

where α' is the average polarisability of a volume element and $\delta\alpha$ is the contribution due to the concentration fluctuation δc. Since δV is the volume of one volume element, there are $1/\delta V$ volume elements per unit volume and Eq. (13) yields eventually for the scattered intensity per unit volume:

$$i_s = 16\,\pi^4 \sin^2\theta_1\, I_0\, \langle(\delta\alpha)^2\rangle/\lambda_0^4 r^2 \delta V \tag{29}$$

where $\langle(\delta\alpha)^2\rangle$ is the time average of $(\delta\alpha)^2$ for a single volume element or, equivalently, the average value at any instant for a large number of volume elements. At constant pressure p and temperature T:

$$\delta\alpha = \left(\frac{\partial\alpha}{\partial c}\right)_{T,p} \cdot \delta c \tag{30}$$

The quantity $(\partial\alpha/\partial c)_{T,p}$ is obtained from Eq. (20) using $n = 1/\delta V$, whence Eq. (29) reduces to the following expression after conversion to conditions of unpolarised light:

$$i_\theta = 2\,\pi^2\widetilde{n}^2(d\widetilde{n}/dc)^2(1 + \cos^2\theta)I_0\, \langle(\delta c)^2\rangle/\lambda_0^4 r^2 \tag{31}$$

Fluctuation theory gives Eq. (32), which is converted to Eq. (33) *via* thermodynamic transpositions and approximation of partial molar volume of solvent to its molar volume V_1:

$$\langle(\delta c)^2\rangle = k\ T/\left(\frac{\partial^2 G}{\partial c^2}\right)_{T,p} \tag{32}$$

$$= -k\ T/\left(\frac{\delta V}{cV_1}\right)\left(\frac{\partial\mu_1}{\partial c}\right)_{T,p} \tag{33}$$

Here, k is the Boltzmann constant ($=\mathscr{R}/\mathscr{N}$); is the Gibbs free energy of the whole solution and μ_1 is the chemical potential of the solvent in solution. Moreover,

$$\left(\frac{\partial\mu_1}{\partial c}\right)_{T,p} = -V_1\left(\frac{\partial\Pi}{\partial c}\right)_T \tag{34}$$

and the concentration dependence of osmotic pressure Π for a non-ideal solution is a virial expansion involving coefficients A_2 and A_3. This expansion enables $\partial\Pi/\partial c$ to be formulated so that Eqs. (31), (33) and (34) yield

$$i_\theta = \frac{2\pi^2\tilde{n}_0^2(d\tilde{n}/dc)^2 I_0(1+\cos^2\theta)c}{\mathscr{N}\lambda_0^4 r^2(1+2\ A_2 c+3\ A_3 c^2+\ldots)} \tag{35}$$

For most practical purposes the influence of the third virial coefficient A_3 is slight in dilute solution so that the following form of Eq. (35) is adequate

$$Kc/R_\theta = 1/M + 2\ A_2 c \tag{36}$$

As the second virial coefficient of non-ideality, A_2, is generally finite, the molecular weight is given by

$$M = 1/\underset{c\to 0}{\mathrm{Lim}}(Kc/R_\theta) \tag{37}$$

4. Allowance for Destructive Interference

In the Rayleigh scattering just outlined the relative size parameter was small. Table 1 indicates that a rough criterion of this is that the major dimension of the scattering particle should not exceed ca. $\lambda/20$. For example, for a dilute aqueous solution of the particles irradiated with green mercury light the upper limit of the dimension would be given as

$$\lambda/20 = (\lambda_0/\tilde{n})/20 \approx (\lambda_0/\tilde{n}_0)/20 = (546/1.33)/20 \approx 200\ \text{Å}$$

Debye scattering deals with the situation wherein the relative size parameter is large. Since the root mean square radius of gyration of the particle, $\langle s^2\rangle^{1/2}$, is a

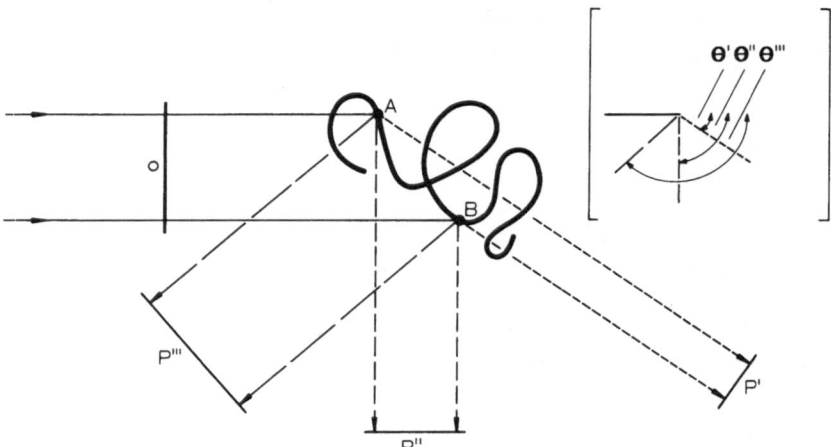

Fig. 2. Destructive interference for a particle having a large relative size parameter

dimension independent of the precise shape, the size parameter is conveniently expressed as $\langle s^2 \rangle^{1/2}/\lambda$. If the dimensions of the particle begin to approach λ in magnitude, light scattered from different regions of the large particle reaches the observer with different phases. Consequently, the value of i_θ is reduced due to interference and the measured scattered light intensity is less than it should be.

Figure 2 illustrates the incident light reaching a reference plane in phase and being scattered subsequently from two different points A and B of the large particle. Perpendicular planes P′, P″ and P‴ are drawn in the scattered beam for three scattering angles $\theta' < \theta'' < \theta'''$. Considering the last of these, for example, the path length between 0 and P‴ *via* A is smaller than the corresponding length *via* B so that the scattered light from A reaches the observer out of phase with the light scattered from B. A similar situation obtains for scattering at the two other angles, but the effect is less pronounced the smaller the angle. At a viewing angle of zero the scattered light intensity is not weakened at all by interference. The particle scattering factor $P(\theta)$, which is a function of both relative size parameter and angle of observation, describes the angular dependence of LS from a large particle. It is defined as[30]

$$P(\theta) = \frac{\text{Scattered intensity for the large particle}}{\text{Scattered intensity in absence of interference}} = \frac{R_\theta}{R_0} \tag{38}$$

where R_0 signifies the Rayleigh ratio for a viewing angle of zero and $P(\theta)$ varies in magnitude according to the angle. $P(\theta) = 1$ only when $\theta = 0$. It may be shown that with small particles $P(\theta) = 1$, because there is no reduction in LS intensity at $\theta > 0$. For large particles considered here, $P(\theta)$ serves as a correction factor in Eq. (36), the amended form of which is now

$$Kc/R_\theta = (1/M)[1/P(\theta)] + 2\,A_2 c \tag{39}$$

For low values of θ the truncated expansion of $P(\theta)$ yields

$$Kc/R_\theta = (1/M)[1 + (\widetilde{\mu}^2/3) \langle s^2 \rangle] + 2 A_2 c \qquad (40)$$

where

$$\widetilde{\mu} \equiv (4 \pi/\lambda) \sin(\theta/2) \qquad (41)$$

The two types of correction thus give the molecular weight as

$$M = 1/\mathop{\mathrm{Lim}}_{\substack{c \to 0 \\ \theta \to 0}}(Kc/R_\theta) \qquad (42)$$

A full treatment of particle scattering functions has been written by Kratochvíl[30] and the modes of effecting the limits indicated in Eqs. (37) and (42) will be described in a later Section (III.4).

An alternative[31], but less frequently used, formulation is to regard the scattering factor as a function of the whole parameter $\widetilde{\mu}$ instead of merely θ. Appropriate changes of subscript then render the relevant variables as $P(\widetilde{\mu})$, $R_{\widetilde{\mu}}$ and $K_{\widetilde{\mu}}$.

5. Subsidiary Aspects: Turbidity, Depolarisation, Absorption, Refractive Index Increment

By analogy with absorption spectroscopy we may identify an extinction coefficient due to scattering with the turbidity τ which is defined as:

$$\tau = (1/\ell)\ln(I_0/I_t) \qquad (43)$$

I_t is the intensity of light transmitted by a sample of path length ℓ. It can be shown that $\tau = (16 \pi/3)R_\theta$ so that Eqs. (22) and (40) may be expressed with τ in place of R_θ and an optical constant H ($= 16 \pi K/3$) in place of K. Although τ is usually too small to be measured as such directly, some experimental results are often reported in the form of Hc/τ even though Kc/R_θ is the measured quantity.

So far the scattering particle has been assumed to be optically isotropic. For anisotropic particles the direction of the electric field associated with the incident light may not coincide with the induced movement of the electron cloud. Consequently, the light scattered at 90° is not plane polarised perpendicular to the plane of the incident beam and the direction of observation; it exhibits a weak component in the horizontal direction. The ratio of the horizontally to vertically polarised components of the unpolarised incident beam (known as the depolarisation ratio ρ_u, although polarisation ratio might be a better term) will, therefore, not be zero. The general expression for ρ_u involves the three principal polarisabilities of the scattering molecules and, from the form of this expression, ρ_u attains a maximum value of 0.5 for long rod-shaped particles in which one of the polarisabilities is much larger than the other two. Values of ρ_u range from 10^{-3} for the rare gases to 0.10 for gaseous carbon disulphide and from 0.01 to 0.05 for low molecular weight organic molecules in the gaseous state[23]. A depolarisation ratio can refer to any specified scattering angle θ (see Casassa and Berry[32]), but most commonly relates to $90°$. The intensity of light scattered at $90°$ from anisotropic molecules is increased over the value predicted on the basis of isotropy by the Cabannes factor. For unpolarised (u) incident light at an angle $90°$ the Cabannes factor f_c^u is given by

$$f_c^u = (6 + 6\,\rho_u)/(6 - 7\,\rho_u) \tag{44}$$

where:

$$\rho_u = \frac{H_u(\text{solution}) - H_u(\text{solvent})}{V_u(\text{solution}) - V_u(\text{solvent})}$$

$H_u(\text{solution})$ and $H_u(\text{solvent})$ are the readings using a horizontal analyser, whilst $V_u(\text{solution})$ and $V_u(\text{solvent}$ are the corresponding data with a vertical analyser. When using vertically (v) polarised incident light the Cabannes factor f_c^v is given by

$$f_c^v = (3 + 3\,\rho_v)/(3 - 4\,\rho_v) \tag{45}$$

where the depolarisation ratio ρ_v is obtained as

$$\rho_v = \frac{H_v(\text{solution}) - H_v(\text{solvent})}{V_v(\text{solution}) - V_v(\text{solvent})}$$

$H_v(\text{solution})$ and $H_v(\text{solvent})$ are the readings using a horizontal component analyser whilst $V_v(\text{solution})$ and $V_v(\text{solvent})$ are the corresponding data with a vertical analyser.

The expressions for scattered light intensity (and Rayleigh ratio) must be corrected by dividing by the appropriate Cabannes factor. Effectively this is equivalent to replacing the optical constant K as defined in Eq. (24) by Kf_c^u and by $2\,Kf_c^v$ for unpolarised and vertically polarised incident light respectively.

For solutions of high polymers the Cabannes correction is close to unity, although the occurrence of stereoregularity can present a notable exception. The necessity to apply the factor to oligomers will be apparent from Fig. 3 where a value of $f_c^u = 1$ is only attained for polyethylene oxides[33] of $M > \text{ca. } 10^4$. In a later section (IV.1) it is shown how LS can be used to measure M for a low molecular weight simple solute. In such circumstances it is imperative to employ the isotropic Rayleigh ratio. Thus, the measured R_{90} for pure benzene ($\lambda_0 = 546$ nm and $T = 25\,°C$) is 16.3×10^{-6} cm^{-1} and the measured value of ρ_u is 0.42, whence from Eq. (44), $f_c^u = 2.78$. Hence the isotropic Rayleigh ratio is $16.3 \times 10^{-6}/2.78 = 5.85 \times 10^{-6}$ cm^{-1}. A marked concentration dependence of f^c is often observed for simple compounds or oligomers[34, 35] and in such circumstances the limiting value at infinite dilution is employed in the LS equation by some workers. In common with Maron and Lou[36] the author is inclined to believe it more theoretically sound to use the separate different f_c values for each concentration.

Stabilisers introduced to suppress degradation of polymers at elevated temperature can introduce errors due to their absorption or fluorescence. Absorbing groups are frequently active in the UV region. In general, neither absorption nor fluorescence poses a serious problem within the range λ_0 = 400–600 nm used for most LS experiments. Since it is the excess scattering which is measured, fluorescence from solvent need not be corrected for, unless the solute has a quenching action on it. Brice et al.[37] have described a procedure for correcting for fluorescence from solute which has proved adequate in many instances. Alternatively, special filters may be inserted into the nosepiece of the LS photometer, which transmit a negligible amount of fluorescence.

When a polymer absorbs very strongly in the visible region, near IR incident radiation is used. In a very coloured solution the scattered intensity is reduced by a factor $\exp(-\epsilon\ell)$ where ϵ is the absorption coefficient of the solvent. Hence i_θ must be multiplied by $\exp(+\epsilon\ell)$ in order to obtain the true scattered intensity undiminished by absorption effects. For small values of $\epsilon\ell$, the quantity $\exp(\epsilon\ell)$ approximates well to $(1 + \epsilon\ell)$ so that Eq. (42) becomes[38].

$$M = 1/\lim_{\substack{c \to 0 \\ \theta = 0}} [Kc/R_\theta (1 + \epsilon\ell)] \tag{46}$$

An example is provided by the work of Valtasaari[39] on solutions of cellulose in the coloured complex solvent FeTNa, which necessitates this correction even though the transmittance maximum of this solvent lies conveniently at λ_0 = 546 nm. In the absence of an absorption correction the molecular weight obtained will be lower than its true value.

An important characteristic of a solution with regard to its LS is the specific refractive index increment $\widetilde{dn}/dc$ (frequently denoted also by the symbol ν). As will

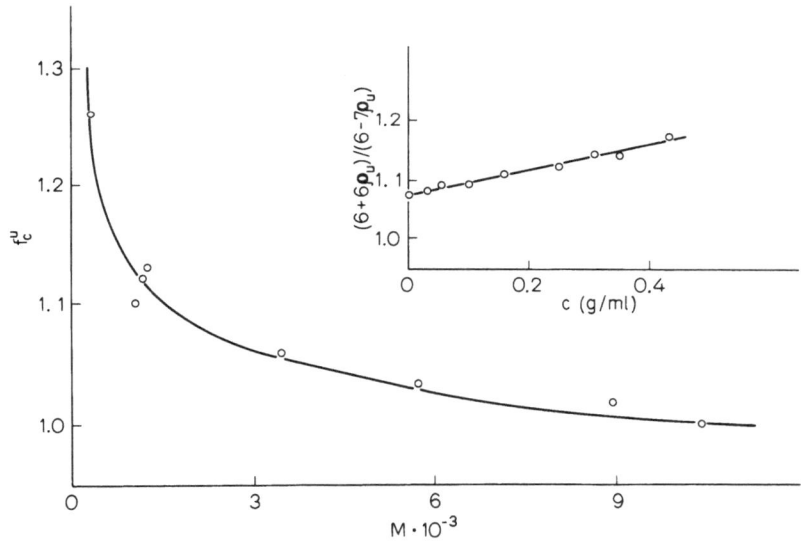

Fig. 3. Cabannes factor as a function of molecular weight for solutions of polyethylene glycol in methanol[33] at T = 25 °C and λ_0 = 546 nm. [Inset – refer to Section IV.2: Dependence of Cabannes factor on concentration for solutions indicated in Fig. 59]

be apparent from Eq. (19), the scattered intensity is proportional to the square of this quantity. The difference between the refractive indices of solution and solvent is measured by either differential refractometry or interferometry[40]. Separate absolute measurement of each refractive index is insufficiently accurate. The quantity $(\tilde{n} - \tilde{n}_0)/c$ sometimes exhibits a concentration dependence and extrapolation yields the value appropriate to the expressions in which it is used:

$$\nu \equiv d\tilde{n}/dc = \lim_{c \to 0} (\tilde{n} - \tilde{n}_0)/c \qquad (47)$$

For a given solute, ν depends on $\tilde{n}_0$, λ_0 and T, the last two of these being maintained constant. For a molecular weight determination one attempts to select a solvent affording the largest value of $| d\tilde{n}/dc |$ (since the value can be positive or negative). The value of $d\tilde{n}/dc$ is constant for a given polymer – solvent pair and independent of M except for oligomers, in which case $d\tilde{n}/dc$ only attains constancy above a certain molecular weight[35], as illustrated in Fig. 4.

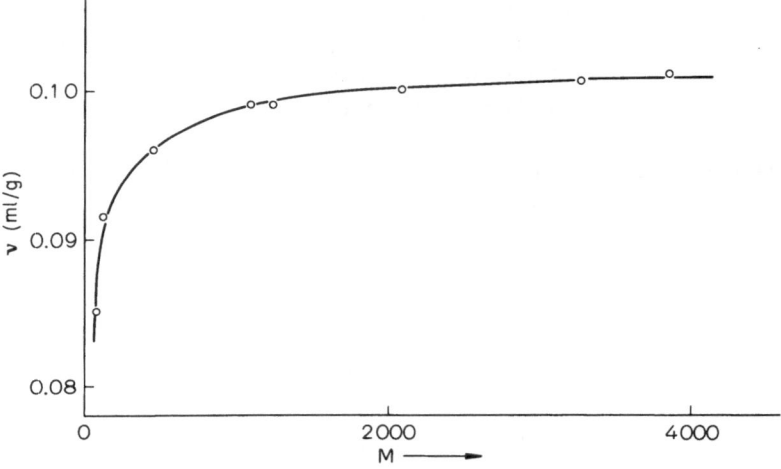

Fig. 4. Dependence of ν on molecular weight for polypropylene glycol[35] in acetone at 25 °C

6. Brillouin Scattering

An alternative to the common device of determining relative intensities is a study of the fine structure of the scattered beam. This entails resolving the spectrum of scattered light into its three peaks, viz. a central peak and two side ones. The need is thus obviated to refer to I_0 or, according to the apparatus, the scattering power of a standard calibration material. The method is used mainly for determining diffusion constants and thermodynamic properties of liquids.

In the Brillouin scattering method the side peaks serve as references and a measurement of the ratio of the intensity of the central peak to the sum of intensities of the side peaks, J, is related to M in a manner analogous to R_θ for the total iso-tropic LS. Under thermodynamically ideal conditions, the expression (48) is the analogue of Eq. (22), viz.[41]

$$K_1 K_2 c/(J - J_0) = 1/M \qquad (48)$$

whilst allowance for non-ideality [cf. Eq. (36)] gives:

$$K_1 K_2 c/(J - J_0) = 1/M + 2 A_2 c + \dots \qquad (49)$$

Here J_0 is the value of J for the pure solvent. The constant K_1 (cm^3 gm^{-2} mol) is defined as

$$K_1 = (C_p/\mathscr{R} T^2 [d\widetilde{n}/dc)/(d\widetilde{n}/dT)]^2 \qquad (50)$$

where the heat capacity at constant pressure C_p relates to 1 cm^3 of solvent. Because of the presence of $d\widetilde{n}/dT$, the value of K_1 is affected by both solute and solvent. On the other hand, the constant K_2 depends only on the solvent. For a relaxing solvent, $K_2 = J_0$, but for a non-relaxing one, $K_2 < J_0$ viz

$$K_2 \quad = (J_0 + 1)[(C_p/C_v) - 1]/(C_p/C_v)(1 + f) \qquad (51)$$

$$\text{where, } f = [2b/(1 - b)] + (C_p/C_v)b^2/(1 - b)^2 \qquad (52)$$

$$b \quad = 1 + \beta_T(d\widetilde{n}/dT)/\widetilde{\alpha}(d\widetilde{n}/dp) \qquad (53)$$

Here, C_v is the heat capacity of solvent at constant volume; $\widetilde{\alpha}$ (deg^{-1}) is its coefficient of thermal expansion; β_T (cm^2 dyne^{-1}) is the coefficient of isothermal compressibility. From Eq. (49) it is seen that the molecular weight of solute is simply:

$$M = 1/\underset{c \to 0}{\text{Lim}} [K_1 K_2 c/(J - J_0)] \qquad (54)$$

III. Instrumentation

In recent years much information has accrued from the study of LS under the influence of external factors[6, 42] such as applied electric field, applied pressure, hydrodynamic flow etc. Pulse induced critical scattering has also been developed to investigate in an elegant manner phase equilibria and changes in polymer solutions. These refinements are not, however, germane to the objective of molecular weight determination, for which the fundamental instrumental requisites are[32, 43]

    (a) stable light source to provide a collimated beam on the sample,
    (b) optical components for detecting scattered light,
    (c) a cell assembly for solution and solvent, which allows for passage of incident and scattered light — preferably with temperature control,

(d) electrical and electronic components, comprising a circuit for operation of light source, photomultipliers in the monitor and scattering detector and some form of instrumentation (*e.g.* galvanometer) to provide a read out.

As well as requirements (a) – (d), most (but not all) photometers include the facility for measuring polarisation and altering the wavelength by means of filters. Neutral transmittance filters can be inserted to reduce I_0 by a known amount so as to render the reading comparable in magnitude with that of the much smaller quantity, i_θ. The scattered intensity may be decreased similarly under warranted conditions *e.g.* phase separation studies at temperatures approaching the critical separation point. The range of λ_0 available from a single source is not exceptionally wide. The following sources and wavelengths have been used: Hg lamp 365, 436, 546 and 578 nm; Ar ion laser 476, 480, 514 and 533 nm; Kr ion laser 531, 547 and 568 nm; He-Cd laser 442 nm; He-Ne laser 633 and 1080 nm; Nd-YAG laser 1064 nm. A quantum doubling substance in conjunction with a laser halves the wavelength; *e.g.* barium sodium niobate with an IR nyodinium laser gives λ_0 = 532 nm.

Only light from within the scattering volume should impinge on the receiving phototube, and the incident beam should be correspondingly large enough. Concomitantly, the beam must be sufficiently narrow that measurements at small angle can be made without picking up the transmitted beam (which eventually falls on the light trap). A useful criterion of good resolution is that the volume viewed by the receiver should vary directly as $1/\sin\theta$, which condition is generally found to hold to within better than 1% in well alligned instruments. Effectively, the product of $\sin\theta$ and the galvanometer or recorder deflection G_θ should remain constant at all angles. An example[44] is provided in Fig. 5.

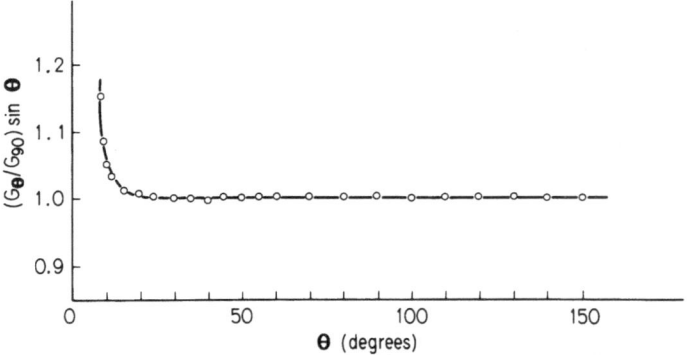

Fig. 5. Angular variation of scattering from benzene (T = 30 °C, λ_0 = 436 nm) measured on the LS photometer LS-2 of Utiyama and Tsunashima[44]. G_θ and G_{90} denote the recorder deflections at angles $\theta°$ and $90°$ respectively

1. Light Scattering Photometers

Critical reviews of LS photometers, especially those which are commercially available, have been made recently enough to render all but the main points and recent developments unnecessary here. The main instruments are (a) Brice-Phoenix (U.S.A.), (b) Shimadzu-Seisakasho (Japan), (c) Sofica (France), (d) Aminco (U.S.A.), (e) Polymer Consultants (U.K.). Instrument (e), known also as the Peaker photometer, is manufactured now only to special order by A. D. Whitehead, Ardleigh, Colchester, U.K. and must be regarded as obsolescent. It has been reported recently[16] that instrument (d) is unlikely to remain long in production.

If the geometry is precisely defined, as in photometer (d), there is no need for a calibration standard. R_{90} is proportional to G_{90}/G_0 where G_{90} and G_0 are the photocurrent readings at 90° and 0°, the known constant of proportionality involving the size of the slits and the photocell nosepiece and their distance apart. No temperature control is provided. Correction factors and calibration procedures for instrument (a) have been covered over the years by Kratohvil and co-workers[45, 46]. In particular, reflection effects constitute important corrections with this instrument. When they are made, the results agree to within 3% of those obtained on an apparatus [photometer (c)] which is virtually free from such effects. A modification of photometer (a) by Roche and Tanner[47] enables the intensity of scattered light to be measured free from reflection effects, over an extended angular range of 30°−150° and with a higher resolution than is possible with the unmodified model. A resolution of 0.02° has been achieved by Aughey and Baum[48] in their modified low-angle Brice-Phoenix photometer, which is now available from the same manufacturers who supply the unmodified instrument. Details have been described by Livesey and Billmeyer[49], and more recently, by Levine et al.[50]. The latter report on its use within the range 2°−35°.

The Shimadzu is a basically similar instrument to photometer (a), but has much superior facilities for temperature control. The low angle PG-21 instrument, superseded subsequently by an improved version LS-2 are both modifications of the Shimadzu by Utiyama et al.[44, 51]. In the later version the temperature stability within 5°−100 °C is probably the best available in any instrument, and this photometer is capable of allowing routine measurements at angles down to 9°.

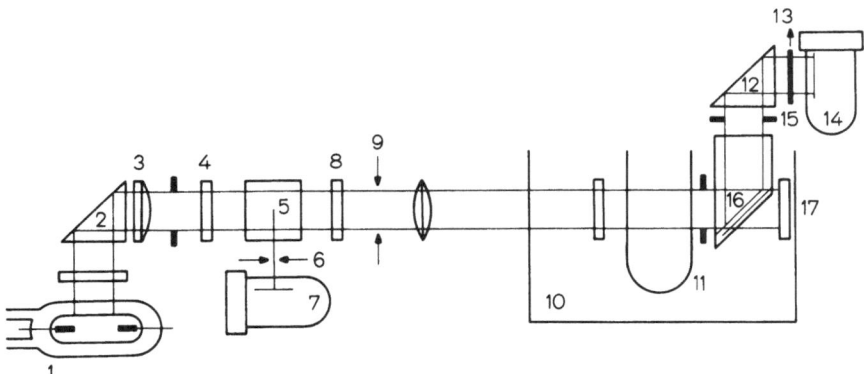

Fig. 6. The Sofica LS photometer: 1 − Hg Pump cooled by high pressure water, 2 − total reflection prism, 3 − achromatic condensing lens, 4 − wavelength filters, 5 − glass diffuser, 6 − iris diaphragm, 7-reference photomultiplier, 8 − polariser, 9 − variable slit, 10 − constant temperature vat, 11 − entrance slit, 12 − total reflection prism, 13 − manual shutter, 14 − photomultiplier, 15 − exit slit, 16 − air blade total reflection prism, 17 − light trap

The Sofica (Fig. 6), which is of a more radical design than instruments (a) and (e), is unique in that the sample cell and observation prism are immersed in a bath (normally containing toluene for $T \not> $ ca. 70 °C; mesitylene can be used for temperatures up to ca. 120 °C), that not only serves as a thermostat, but also eliminates unwanted refraction and reflection effects. For all the photometers, the detector is a photomultiplier tube rotatable to the desired angle either manually or, in instrument (c), by a motor. The newer, so-called automatic, version of instrument (c) (Fica 50) probably offers the widest versatility of commercial non-absolute photometers. The temperature range is from −10 °C to 300 °C with a stability of 0.1 °C. The lowest angle of measurement is claimed to be 15° (instead of 30°), although this is still insufficient for the instrument to qualify as a low angle LS photometer. A high degree of sophistication is provided, particularly with regard to extended range of wavelengths (λ_0 = 365, 436, 546 and 578 nm), dilution, subtraction of solution and solvent readings, repetitive calculations and print-out.

Serious errors are inevitable, if extrapolations to $\theta = 0°$ are made from data obtained for insufficiently low angles. Commenting on the feasibility of accurate measurements in the 6°−10° range, Levine et al.[50] quote the views of other workers as "unlikely"[52], "unfeasible"[53], "impractical"[54] or "impossible"[55]. Nevertheless, these quoted workers and others[56, 57] stressed the need for very low angle data to supplement those obtained fairly readily at $\theta = 30°-150°$; a need especially critical for measuring the molecular weights of biopolymers of high M. Some low angle instruments have been constructed, but two recent commercial ones merit particular mention.

The Bleeker small angle LS photometer[58] comprises a primary beam section, a sample chamber and a rotatable detection system, the first and last parts being provided with diaphragms and filter holders. The scattered light can be detected with an accuracy of better than 0°2′ within the range 0°30′−130°. The detector is a photomultiplier tube intercepting a very small angle for higher angular resolution. Modulation of the primary beam by means of a built-in chopper can be effected for very low LS intensities. The source is a 100 W high pressure Hg lamp with neutral density filters. For its primary intended use a planparallel solid sample is contained between glass slides in a holder. For solutions and gels, special cuvettes are available which have planparallel walls, the separation of which is adjustable so as to alter the thickness of sample, that is, volume of solution. The Rayleigh ratio is obtained without a secondary scattering standard, because the geometry is such that energies rather than intensities are measured. If Eq. (23) is written so that the relevant symbols imply intensities rather than intensities per unit volume, then

$$R_\theta = i_\theta r^2/I_0 V \tag{55}$$

where V is the scattering volume. In the focal plane of lens L2 (Fig. 7) before the photomultiplier, two pinholes are mounted. Using the one of large diameter at $\theta = 0°$, all the primary beam is collected onto the photocathode. Hence the incident intensity is:

$$I_0 = \epsilon_0/A \tag{56}$$

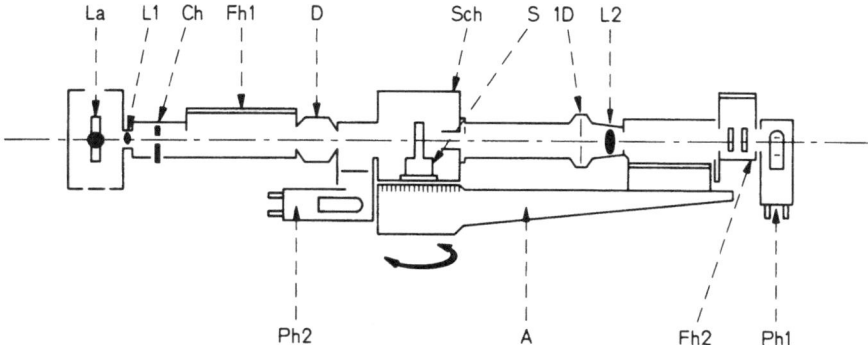

Fig. 7. The Bleeker small angle LS photometer[58]: La − light source, LI condenser lens, Ch − chopper disc, Fhl − filter holder, D − iris diaphragm and interchangeable pinholes of various sizes, Sch − sample chamber, S − sample holder, ID − iris diaphragm, L2 − receiver lens, Ph2 − detector photomultiplier, A − measuring arm, Fh 2 − filter holder, Phl − detector photomultiplier

where ϵ_0 is the energy measured at $\theta = 0°$ and A is the cross-sectional area of the primary beam. For scattering at $\theta \neq 0°$, the pinhole of smaller diameter is used to increase the angular resolution. In this case, therefore, the scattered energy is measured over an area $\pi r^2 \tan^2 \gamma$ (where γ = radius of smaller pinhole used/focal length of lens L2). Hence if ϵ_θ is the scattered light energy, the scattered intensity is

$$i_\theta = \epsilon_\theta / \pi r^2 \tan^2 \gamma \qquad (57)$$

Since the scattering volume is given by $V = A\ell$, where ℓ is the thickness of sample, Eq. (56) and (57) in conjunction with Eq. (55) yield

$$R_\theta = \frac{\epsilon_\theta r^2 A}{\epsilon_0 \pi r^2 A\ell \tan^2 \gamma} = \left(\frac{\epsilon_\theta}{\epsilon_0}\right)\left(\frac{1}{\pi \ell \tan^2 \gamma}\right) = \left(\frac{G_\theta}{G_0}\right)\left(\frac{1}{\pi \ell \tan^2 \gamma}\right) \qquad (58)$$

where the ratio $\epsilon_\theta / \epsilon_0$ has been replaced by the measured ratio G_θ / G_0 of photocurrents. The cuvette is not ideally suitable for dilution *in situ* to obtain solutions of different concentration, but the angular range and the realisation of absolute Rayleigh ratios render this a potentially very attractive photometer meriting wider use than is so far apparent from the literature.

The newest addition is the expensive Chromatix KMX-6 low angle LS photometer (Fig. 8). This also yields absolute values of R_θ *via* two measurements, viz. (a) a reading of the photomultiplier signal G_θ from the sample at an angle θ and (b) a reading of the signal G_0 caused by the illuminating beam transmitted through the sample. Since the former is much weaker than the latter, attenuators are inserted when measuring G_0 so that G_0 may be obtained as a signal reading that is ca. 25% of the full-scale reading of G_θ for the same setting of amplifier gain and photomultiplier voltage. The final expression for the Rayleigh ratio is

$$R_\theta = (G_\theta / G_0)/(q/\sigma\ell) \qquad (59)$$

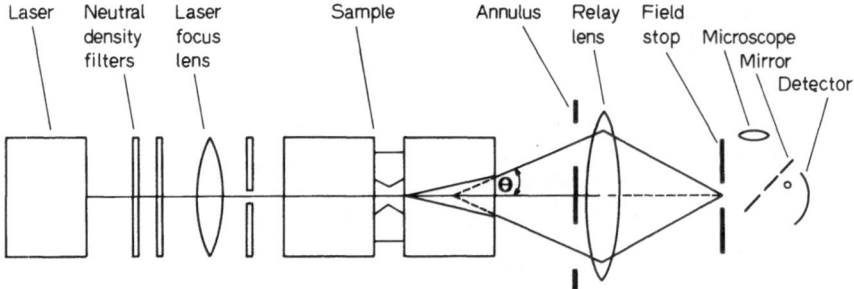

Fig. 8. Schematic of the Chromatix KMX-6 low angle laser LS photometer[59]

The factor $(q/\sigma l)$ may be obtained from independent optical and geometrical measurements, since q is the transmittance of the attenuator used for the incident beam, σ is the solid angle of detection of scattered light and l is the thickness of sample.

The philosophy of design of this instrument is founded on the limitations imposed by conventional light sources such as a Hg lamp, which requires good collimation to allow resolution of the scattering signal as a function of angle. This obstacle is normally met by a collimated beam of large cross-sectional area, which in turn demands consideration as two further problems viz (a) limitation of angular resolution by the change in scattering angle across the scattering volume as seen by the detection aperture and (b) high probability of dust or other contaminant being in the relatively large scattering volume. A laser beam permits a very large power output to be transmitted through a very small cross-sectional area, while maintaining a small angle of divergence. However, early applications of lasers to LS concentrated on their high power output without due recognition of the significance of spatial aspects. Such sources were thus incorporated with optical systems that were geared for use with Hg lamps, without regard to optics more appropriate to a laser.

The radically different approach in the Chromatix instrument is to implement a low power (3 mW) He-Ne laser (λ_0 = 633 nm) so that its spatial characteristics enable scattering to be measured at angles below 2° and with an illuminated sample diameter of only ca. 100 microns; this alleviates the power angular resolution. Furthermore, there is only a very slight probability of adventitious dust particles being present within the small illuminated area. The small sample volume (ca. 0.15 ml), also, is of importance when the quantity of available polymer is strictly limited on economic or other grounds. Fluorescence absorption is unlikely for most substances at the wavelength in question and, despite the dependence of i_θ on λ_0^{-4}, a signal to noise ratio of better than 100 is claimed even from such a weakly scattering liquid as water. The high sensitivity affords sufficient scattering even for very dilute solutions. In general, as intimated by Utiyama, a LS photometer should possess a sensitivity such that the Rayleigh ratio of the most dilute solution used is at least twice that of the pure solvent alone, that is, the excess scattering should not be less than the scattering of the solvent. A nomogram[59] (Fig. 9), relating to the Chromatix instrument, shows the lowest concentration c_{min} (g/ml) at which the excess R_θ equals the Rayleigh ratio of water, as a function of M for aqueous solutions of several biologically important macromolecules. Because this is a low angle photometer, (θ from 7° down to 2°), obtaining and

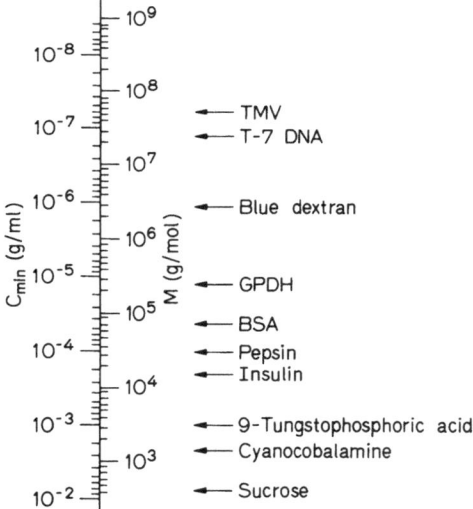

Fig. 9. Nomogram for Chromatix photometer[59] showing the minimum concentration at which the excess Rayleigh ratio of aqueous solutions of polymers of different molecular weight equals the Rayleigh ratio of pure water

manipulating data for molecular weight determination is greatly simplified. Readings of R_θ at a low angle of, say, 4° may be taken effectively to be the same as R_θ at $\theta = 0°$. Hence extrapolation to zero angle is unnecessary and Eq. (39) reduces to Eq. (36), requiring only extrapolation to infinite dilution.

For solutes of exceptionally high molecular weight there has been a discernible move towards conducting LS experiments in the near IR region ($\lambda_0 \approx 1000$ nm), which effectively reduces the relative size parameter $\langle s^2 \rangle^{1/2}/\lambda$, even though $\langle s^2 \rangle^{1/2}$ itself is very large for such materials. The associated problems of absorption and

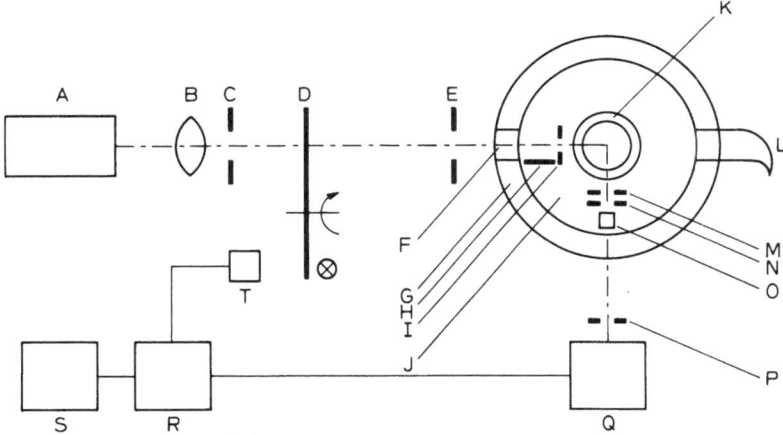

Fig. 10. LS photometer[60] for use at $\lambda_0 = 1086$ nm: A – laser, B – lens, C – shutter, D – rotating chopper, E – shutter, F – entrance window, G – thermostat vessel, H – metal shield, I – shutter, J – thermostat liquid, K – cylindrical LS cuvette, L – light trap, M – shutter, N – shutter, O – prism, P – shutter, Q – photomultiplier, R – amplifier, S – x-t recorder

calibration have been discussed by Meyerhoff[60] and Jennings[18], the instrument constructed by Meyerhoff and Burmeister[60] being illustrated schematically in Fig. 10. Modification[18] of an existing commercial instrument for use with a Nd YAG laser (λ_0 = 1060 nm) is shown in Fig. 11. It is clear from Eqs. (40) and (41) that, for a given sample, data extrapolated to infinite dilution will require a much smaller subsequent extrapolation to $\widetilde{\mu}^2 = 0$ (that is, to a value of 1/M) for, say, λ_0 = 1086 nm than for λ_0 = 436 nm. Examples will be provided later and it will be demonstrated also that in certain circumstances the LS behaviour falls outside the Rayleigh-Debye region unless the relative size parameter is reduced by using light of large wavelength.

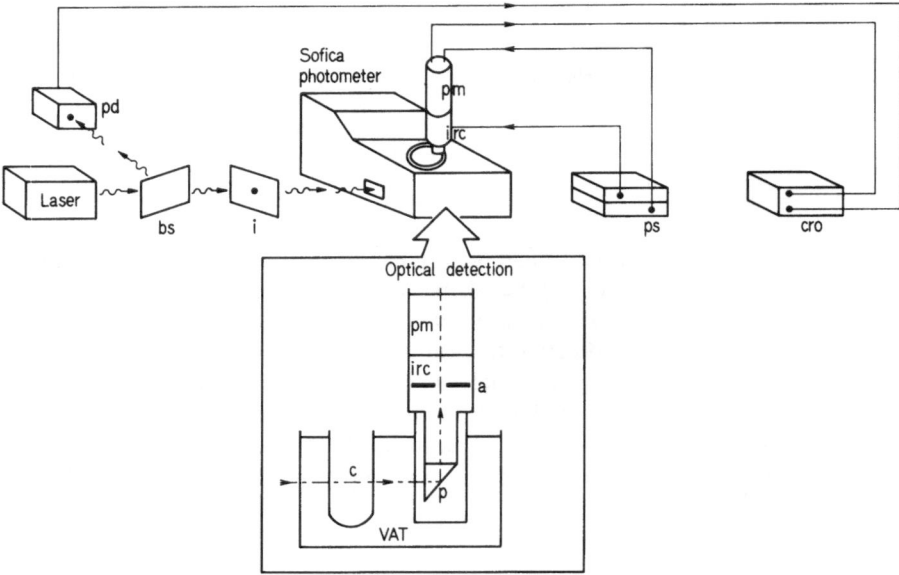

Fig. 11. Modification[18] of Sofica photometer for use at λ_0 = 1060 nm: pd – monitoring photodiode, bs – beam splitter, i – iris, pm – photomultiplier, irc – infrared converter, ps – power supplies, cro – cathode ray oscilloscope, a – aperture, p – reflecting prism, c – cell

2. Subsidiary Aspects: Clarification, Refractive Index Increment and Calibration

An indispensable requirement for reliable LS data is the freedom of the medium from extraneous matter such as dust particles. Clarification procedures have been reviewed[61] and consist of (a) centrifugation followed by careful removal of the upper clarified portion and/or (b) filtration; through ultrafine filters. Contamination during transference in (a) can be overcome by using the LS cell as centrifuge tube; the dust therein remains held at the bottom for the duration of the LS readings. For aqueous solutions, which are notoriously difficult to clarify, the addition of salt is often found effective in bringing down the dust. Electrostatic attraction of dust particles by organic droplets appears to be the underlying cause of the effectiveness of a rather specific expedient due to Bernadi[62]. Aqueous nucleic acid solution is emulsified by shaking with a binary organic mixture which, after gravity separation and centrifugation, yields a highly clarified aqueous solution. Filters are selected on the basis of their chemical resistance to the solvent used as well as their efficiency, viz. pore size 0.1 μm Polycarbonate filters (Nuclepore Corporation, U.S.A.) have been reported[50] to produce markedly clearer solutions at faster flow rates than the more commonly used cellulose ester filters (Millipore Corporation) of the same nominal pore size. The special problems associated with solutions of native DNA have been discussed. A useful microfiltration apparatus, which prevents re-introduction of dust into filtered solution through contact with laboratory air, has been described by Levine et al.[50], and is shown in Fig. 12. The liquid is placed in the upper chamber and the lid closed. Filtration into the LS cell and recycling back into the upper chamber are carried out and repeated several times by manipulation of the stopcocks. Finally the filled cell is sealed and placed in the LS photometer. A rather similar device has been reported by Casassa and Berry[32].

As indicated, the specific refractive index increment is best measured by differential refractometry or interferometry. Experimental procedures as well as tabulated values of $d\widetilde{n}/dc$ for many systems have been presented elsewhere[40, 63]. The relevant wavelength and temperature are those used for LS. The value of λ_0 is invariably 436 or 546 nm, but with the advent of laser LS, values of $d\widetilde{n}/dc$ at other wavelengths are required. These can be estimated with good reliability using a Cauchy type of dispersion ($d\widetilde{n}/dc \propto 1/\lambda_0^2$). For example the values of $d\widetilde{n}/dc$ for aqueous solutions of the bacterium T-ferrioxidans at 18 °C are 0.159, 0.141 and 0.125 ml/gm at $\lambda_0 = 488$, 633 and 1060 nm respectively[64].

Modifications of existing differential refractometers have been made for precise evaluation of $d\widetilde{n}/dc$ at high λ_0, which has been particularly necessary where values at other wavelengths are not known. The apparatus of Jennings et al.[64] is based on the immersed prism technique of Debye (also used in the commercial Shimadzu-

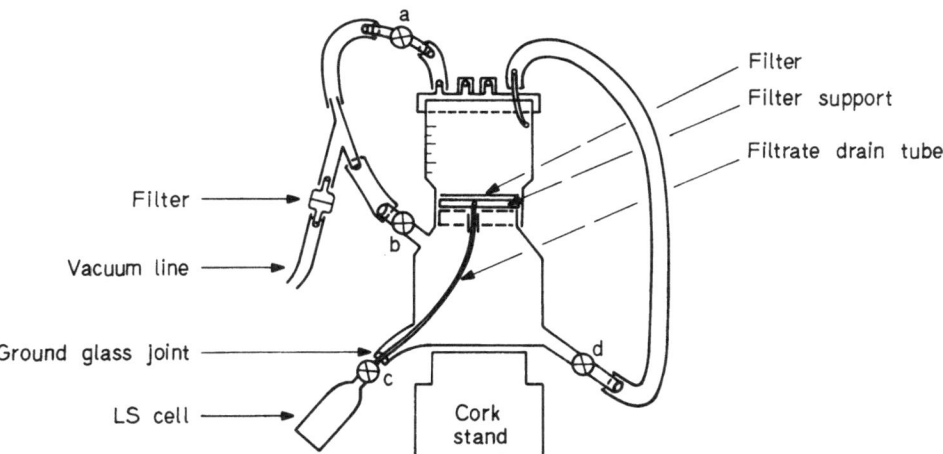

Fig. 12. Microfiltration apparatus[50]: a, b, c and d denote respectively the upper vacuum stopcock, the lower vacuum stopcock, the cell stopcock and the recycling stopcock

Seisakusho differential refractometer). For a solution of known concentration, the difference in refractive index, $(\tilde{n} - \tilde{n}_0)$, between the liquid in the prism and that in the solvent bath causes refraction of the light beam, which is observed as a displacement δ of the image at a distance d from the prism centre. The overall sensitivity depends on the precision of measuring δ and the refractive index difference is given by

$$\tilde{n} - \tilde{n}_0 = \delta/2\,d\tan(\gamma/2) \tag{60}$$

where γ is the angle of the prism, that is, the angle subtended by the sides through which the light beam is transmitted. Advantages of the laser beam used include (a) high intensity, which is especially critical for systems such as bacterial suspensions that strongly attenuate the beam because of their intense scattering power. This is probably the reason for the dearth of experimental $d\tilde{n}/dc$ values for such systems at visible wavelengths, (b) the low divergence and high penetrability of the beam enable emergent light to be projected over a large distance d, thereby yielding a higher value of δ.

The instrument of Meyerhoff[60] (Fig. 13) is based on the differential refractometer of Bodmann. It measures an image displacement which is proportional to $(\tilde{n} - \tilde{n}_0)$ via an instrumental calibration constant the value of which must be established by means of standard solutions of known $(\tilde{n} - \tilde{n}_0)$. Values of the latter quantity have been collated for several aqueous salt solutions, but relate mainly to $\lambda_0 = 436$, 546 and 589 nm. Calibration at $\lambda_0 = 1086$ nm and T = 20 °C may be effected with aqueous KCl for which $d\tilde{n}/dc$ (at c = 0) has a reported[65] value of 0.1351 ml/g. This datum and other useful data relative to the increasingly used (laser source) wavelengths are assembled in Table 2.

As will be seen later (Section V.1), meaningful molecular weights in multicomponent systems can be determined, if the specific refractive index increment appertains to conditions of constant chemical potential of low molecular weight solvents (instead of at constant composition). Practically, this can be realised by dialysing the solution against the mixed solvent and then measuring the specific refractive index increment of the dialysed solution. The theory and practice have been reviewed[4, 14, 15, 72].

There is little doubt that the disagreement among many reported molecular weights can be attributed in some measure to uncertainty in calibration procedures and/or failure to effect calibration in conjunction with associated corrections. Reference is recommended to full discussions by Utiyama[29], Kratohvil[73] and Casassa,

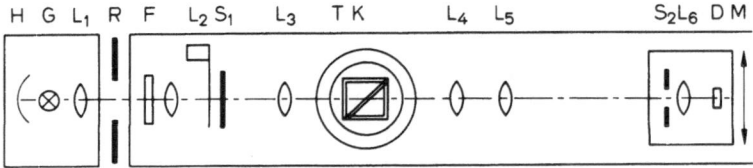

Fig. 13. Differential refractometer[60] for use at $\lambda_0 = 1086$ nm: H – hollow mirror, G – incandescent halogen lamp, $L_1 - L_6$ – lenses, R – heat reflector, F – interference filter, $S_1 - S_2$ – slits, T – heating mantle, K – differential cell, D – silicon photocell, M – micrometer thread

Table 2. Refractometric and LS data at less common wavelengths λ_0

System	Quantity	Value	λ_0(nm)	Temp (°C)	Ref.
Acetone	$\tilde{n}_0$	1.3574	633	23	66)
Benzene	$\tilde{n}_0$	1.4971	633	23	66)
Carbon disulphide	$\tilde{n}_0$	1.6207	633	23	66)
Carbon tetrachloride	$\tilde{n}_0$	1.4582	633	23	66)
Chloroform	$\tilde{n}_0$	1.4444	633	23	66)
Cyclohexane	$\tilde{n}_0$	1.4254	633	23	66)
Methanol	$\tilde{n}_0$	1.3274	633	23	66)
Methyl ethyl ketone	$\tilde{n}_0$	1.3773	633	23	66)
Toluene	$\tilde{n}_0$	1.4921	633	23	66)
Water	$\tilde{n}_0$	1.3324	· 633	23	66)
Water	$\tilde{n}_0$	1.324	1060	25	67)
Bacterium S. marcescens in water	$d\tilde{n}/dc$(ml/g)	0.175	488	18	64)
Bacterium S. marcescens in water	$d\tilde{n}/dc$(ml/g)	0.158	1060	18	64)
Bacterium E. coli in water	$d\tilde{n}/dc$(ml/g)	0.152	1060	18	64)
Bacterium T. ferrioxidans in water	$d\tilde{n}/dc$(ml/g)	0.125	1060	18	64)
Bacterium T. ferrioxidans in water	$d\tilde{n}/dc$(ml/g)	0.159	488	18	64)
Bovine plasma albumen in water	$d\tilde{n}/dc$(ml/g)	0.192	488	18	64)
Bovine plasma albumen in water	$d\tilde{n}/dc$(ml/g)	0.181	1060	18	64)
β-Lactoglobulin in water	$d\tilde{n}/dc$(ml/g)	0.171	1060	18	64)
β-Lactoglobulin in water	$d\tilde{n}/dc$(ml/g)	0.184	488	18	64)
Tobacco mosaic virus in water	$d\tilde{n}/dc$(ml/g)	0.181	488	18	64)
DNA in water	$d\tilde{n}/dc$(ml/g)	0.183	488	18	64)
Sucrose in water	$d\tilde{n}/dc$(ml/g)	0.144	488	18	64)
Cellulose trinitrate in acetone	$d\tilde{n}/dc$(ml/g)	0.0982	1086	25	60)
Polystyrene in benzene	$d\tilde{n}/dc$(ml/g)	0.102	633	25	41)
Polystyrene in benzene	$d\tilde{n}/dc$(ml/g)	0.1021	644	20	68)
Polystyrene in toluene	$d\tilde{n}/dc$(ml/g)	0.1060	644	20	68)
Polystyrene in toluene	$d\tilde{n}/dc$(ml/g)	0.1065	633	25	41)
Polystyrene in toluene	$d\tilde{n}/dc$(ml/g)	0.1051	1086	25	60)
Polystyrene in toluene	$d\tilde{n}/dc$(ml/g)	0.1034	1086	20	60)
Polystyrene in dimethyl formamide	$d\tilde{n}/dc$(ml/g)	0.1064	644	20	68)
Polystyrene in methyl ethyl ketone	$d\tilde{n}/dc$(ml/g)	0.2110	644	20	68)
Polybenzyl acrylate in methyl ethyl ketone	$d\tilde{n}/dc$(ml/g)	0.1623	644	20	68)
Polyethyl acrylate in methyl ethyl ketone	$d\tilde{n}/dc$(ml/g)	0.0852	644	25	68)
Polymethyl acrylate in methyl ethyl ketone	$d\tilde{n}/dc$(ml/g)	0.0915	644	20	68)
Polymethyl methacrylate in methyl ethyl ketone	$d\tilde{n}/dc$(ml/g)	0.1102	644	20	68)
Polymethyl methacrylate in acetone	$d\tilde{n}/dc$/ml/g)	0.1276	644	20	68)
Polyvinyl bromide in tetrahydro-furan	$d\tilde{n}/dc$(ml/g)	0.112	644	–	69)

Table 2. (continued)

System	Quantity	Value	λ_0(nm)	Temp (°C)	Ref.
KCl in water	$d\tilde{n}/dc$(ml/g)	0.1351	1086	20	65)
NH$_4$NO$_3$ in water	$d\tilde{n}/dc$(g/g)	0.1239	633	–	66)
NaCl in water	$d\tilde{n}/dc$(g/g)	0.1766	633	–	66)
KCl in water	$d\tilde{n}/dc$(g/g)	0.1371	633	–	66)
Carbon disulphide	ρ_u	0.653	633	23	66)
Carbon disulphide	ρ_u	0.632–0.669	633	23	70)
Toluene	ρ_u	0.491	633	23	66)
Toluene	ρ_u	0.506–0.528	633	23	70)
Benzene	ρ_u	0.419	633	23	66)
Benzene	ρ_u	0.438–0.453	633	23	70)
Chloroform	ρ_u	0.211	633	23	66)
Chloroform	ρ_u	0.204	633	23	70)
Carbon tetrachloride	ρ_u	0.036	633	23	66)
Carbon tetrachloride	ρ_u	0.031–0.099	633	23	70)
Cyclohexane	ρ_u	0.046	633	23	66)
Cyclohexane	ρ_u	0.059	633	23	70)
Acetone	ρ_u	0.155	633	23	66)
Acetone	ρ_u	0.226	633	23	70)
Methanol	ρ_u	0.049	633	23	66)
Methanol	ρ_u	0.050	633	23	70)
Carbon disulphide	ρ_v	0.485	633	23	66)
Carbon tetrachloride	ρ_v	0.0184	633	23	66)
Chloroform	ρ_v	0.118	633	23	66)
Cyclohexane	ρ_v	0.0235	633	23	66)
Benzene	ρ_v	0.265	633	23	66)
Toluene	ρ_v	0.325	633	23	66)
Acetone	ρ_v	0.084	633	23	66)
Methyl ethyl ketone	ρ_v	0.077	633	23	66)
Methanol	ρ_v	0.025	633	23	66)
Water	ρ_v	0.025	633	23	66)
Benzene	$10^6 \times R_{90}$(cm^{-1})	8.765	633	23	71)
Toluene	$10^6 \times R_{90}$(cm^{-1})	10.31	633	23	71)
Methanol	$10^6 \times R_{90}$(cm^{-1})	1.349	633	23	71)
Water	$10^6 \times R_{90}$(cm^{-1})	0.490	633	23	71)

and Berry[32]; illustrations involving several modes of calibration have been provided by Levine et al.[50] and by Jennings and Plummer[74]. The latter workers appear to have been the first to have conducted a completely reliable calibration of a LS photometer on the basis of Mie theory for solutions of large monodisperse latices. Calibration involving calculation from the geometry of the apparatus is generally complex and is rarely used now. In general for non-absolute photometers the purpose of the necessary calibration is to relate in instrument response (such as a galvonometer reading G_{90} at an angle 90°) to the actual Rayleigh ratio. The three major modes are (a) comparison of G_{90} for a standard substance with the known value of R_{90} for that substance (b) comparison of G_{90} from a colloidal suspension of small spheres with the turbidity τ determined by subsidiary photometric transmission measurements, (c) use of a polymer of accurately known M, and comparing the known value of 1/M with the result of extrapolating Kc/G_θ to infinite dilution and zero concentration.

Provided the standard is stable, free from fluorescence and exhibits sufficient scattering, method (a) is the simplest and most frequently used. Benzene has long been employed for this purpose, although there was a controversy for many years regarding its true R_{90} values at λ_0 = 436 and 546 nm; two schools of thought subscribed to values which differed between them by about 40%. Today, the situation has been resolved in favour of the higher values and, indeed, when adopting them for calibration, one obtains the same value of M as that found by an absolute technique[41] not requiring any calibration. The most recent values[75] of R_{90} (cm^{-1}) and ρ_u for benzene as a function of temperature T(°C) and at λ_0 = 436 nm are:

$$R_{90} = 10^{-6}\,(45.4 + 0.109\,T)$$

$$\rho_u = 0.486 - 0.00148\,T$$

It is convenient to employ a secondary standard of known Rayleigh ratio relative to that of the primary standard. Thus a solid cylindrical rod of flint glass (manufactured by Schott and Gen, Mainz, W. Germany) is recommended for use with the Sofica photometer and the value of the ratio of the relative scatterings r_{bg} of benzene to glass is quoted; r_{bg} has a value of the order of unity. The relation between the true Rayleigh ratio R_{90} and the observed galvanometer reading G_{90} for the solution in question is

$$R_{90} = [G_{90}/r_{bg}\,G_{90}\,(\text{glass})]\,R_{90}\,(\text{benzene})\,[\tilde{n}/\tilde{n}\,(\text{benzene})]^2 \tag{61}$$

In Eq. (61), G_{90} (glass) is the galvo reading for the glass standard and the product $r_{bg}\,G_{90}$ (glass) denotes the reading which would have been obtained, if benzene itself had been used instead of glass. The final factor in Eq. (61) is the "refractive index squared" correction of Hermans and Levinson[76] which is unity only if the refractive index of the substance studied, $\tilde{n}$, is the same as that of the primary calibrant, $\tilde{n}$ (benzene). Although this correction for the different refracting powers is very frequently used, results of Jennings and Plummer[74] in Table 3 suggest that the true form of the correction might be some intermediate function between "refractive index" and "refractive index squared". In fact, within experimental error, the factor seems to be simply $\tilde{n}/\tilde{n}$ (water) for these data involving calibration with aqueous solutions.

Since the scattering due to dust is negligible relative to the very large scattering from colloidal suspensions, procedure (b) is inherently attractive. Using the definition of turbidity τ and the interrelation between τ and R_{90}, the calibration constant Φ of the instrument is given by

$$(c/G_{90})_{c=0}/(c/\tau)_{c=0} = \Phi(16\pi/3)\exp(\tau\ell) \tag{62}$$

Hence c(g/ml) is the concentration of colloidal suspension; G_{90} is the reading on the LS photometer at θ = 90°, ℓ is the path length, which equals the cell diameter when using a cylindrical cell; τ is the turbidity obtained from measurements of optical density. Table 4 gives the results of calibrating a Sofica instrument with colloidal

Table 3. Molecular weight M of a monodisperse polystyrene fraction in solvents of different refractive index $\tilde{n}_0$; role of refractive index correction factor[74]

Solvent	$\tilde{n}_0$	Apparent M[1]	M[2] using correction in $\tilde{n}_0^2$	Correction in $\tilde{n}_0$
Benzene	1.50	51300	65300	57800
Dioxan	1.42	54200	61900	57900
Methyl ethyl ketone	1.38	55400	60100	57500

[1]) Obtained from LS readings and average value of calibration constant Φ for the wavelength used (546 nm).

[2]) Obtained from LS readings, average value of Φ and invoking correction factor.

Table 4. Calibration constants obtained with colloidal silicas and with Mie particles[74]

	$\Phi \times 10^5$ (cm^{-1}) at $\lambda_0 = 436$ nm	at $\lambda_0 = 546$ nm
Ludox HS (colloidal silica)	48.0	14.6
Syton 2X (colloidal silica)	47.0	14.7
Polystyrene latex (Mie particles)	47.5	15.2

silica as well as with a larger Mie scatterer. Excellent accord is evident among the values of Φ obtained. For subsequent use in an actual molecular weight determination, the value of Φ must be multiplied (as indicated in Table 3) by the appropriate refractive index correction. Tomato bushy stunt virus, in aqueous solution is such an intense scatterer that its optical density (and hence τ) is measurable. It is rarely used, since it is a far less accessible material than Ludox (E. I. du Pont de Nemours) or Syton 2X (Monsanto Chemicals Ltd).

Just as analysis of pure urea gives two amino groups and one carbonyl group per molecule and a calculated molecular weight of 60, so also is the exact number and identity of amino acid residues in a protein molecule often known. Thus, irrespective of the accuracy of a molecular weight measurement, the molecular weight of intact un-ionised human haemoglobin must of necessity be exactly 64458 (including four haeme groups)[32]. Appealing as the use of such a material for procedure (c) is, the increasing current availability of highly monodisperse synthetic polymer makes these latter substances the preferred choice today. In comparison with method (a), more work is entailed when using polymer solutions as calibrants, since effectively one has to conduct an entire LS experiment at different angles and different concentrations of standard polymer in order to obtain the calibration constant. Similar considerations apply to silicotungstic acid which is of lower molecular weight but which has been recommended and shown suitable as calibration standard for several types of LS photometer. In solution it is a Rayleigh scatterer and allows measurements free from the complication of dissymmetry. The formula is $SiO_2 \cdot 12\ WO_3 \cdot 26\ H_2O$ and $M = 3311$. In aqueous salt solution, the dissolved molecule is $H_4SiW_{12}O_{40}$ and $M = 2879$. These solutions develop colour on contact with metal, and only glass and plastic cells and filtration apparatus must be used[50].

Some Rayleigh ratios at less common wavelengths have been assembled with cognate refractometric data in Table 2.

3. Brillouin Scattering

The whole field of Brillouin scattering spectroscopy is a rapidly developing one with regard to instrumentation. For the present purpose it will suffice to reproduce schematically a device which has been used to measure molecular weights[41, 77]. In Fig. 14 the source is a low power (10 mW) He-Ne laser and the sample is contained in a standard Brice-Phoenix square turbidity cell. The scattered light is analysed with a piezoelectrically driven scanning Fabry-Perot interferometer (Fig. 15). Typical spectra[77] for pure solvent and solution of known concentration are shown in Fig. 16. Occasional spikes in the traces are due to the presence of dust particles passing through the incident light beam.

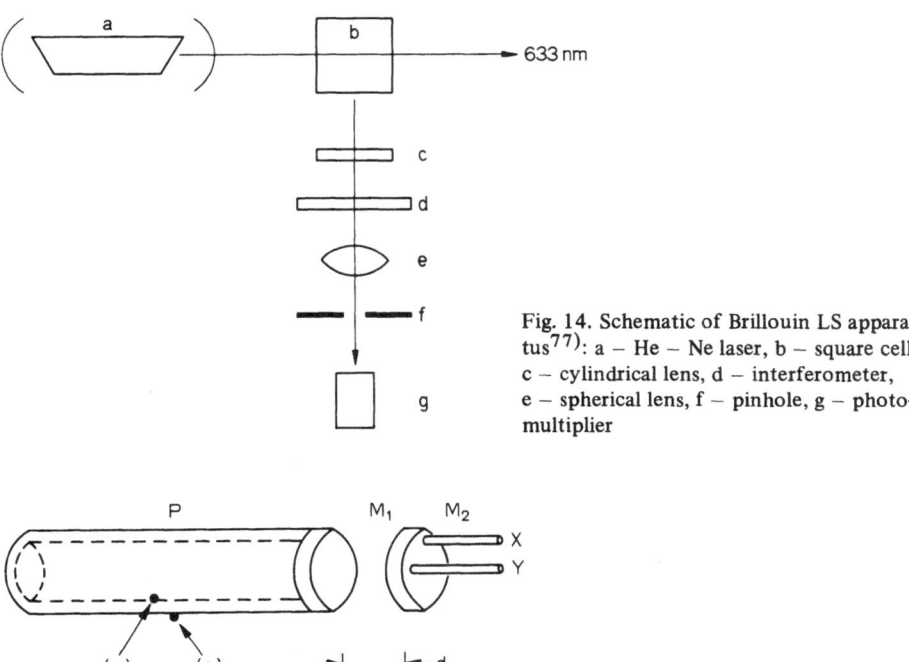

Fig. 14. Schematic of Brillouin LS apparatus[77]: a – He – Ne laser, b – square cell, c – cylindrical lens, d – interferometer, e – spherical lens, f – pinhole, g – photomultiplier

Fig. 15. Simplified diagram of scanning Fabry – Perot interferometer[77] showing piezoelectric ceramic tube (P), multilayer dielectric mirrors (M_1, M_2), micrometer adjustment for parallelism (X, Y). Mirror spacing (d) is adjustable

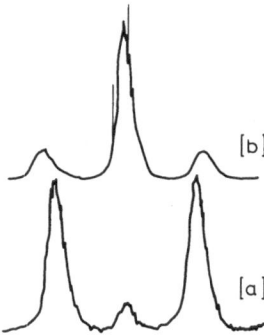

Fig. 16. Brillouin LS spectra[77] at 25 °C for (a) pure water and (b) solution of 300 g of glycerol/kg of water

4. Treatment of Experimental Data

LS data necessitate extrapolation to infinite dilution. The clarification procedure may well introduce a small change (usually, an increase) in concentration from the initially known one. The values of c should therefore be determined on the clarified solution by an appropriate analytical procedure such as spectrophotometry for solutions of DNA using the known phosphorous extinction coefficient, acid-base titration for solution of polyacrylic acid or by evaporating an aliquot to dryness. When high vacuum conditions are obligatory not only for preparing the polymer, but also for dissolving and diluting it and making LS measurements, there is less freedom of choice and spectrophotometry is normally used. Extrapolation to c = 0 is facilitated if A_2 [cf. Eq. (36)] is small as obtains in a poor solvent or is zero under conditions of thermodynamic ideality. The latter state is often realised at a temperature different from ambient. If this should be an elevated temperature, filtration rather than centrifugation is the normal mode of clarification. The temperature in some polymer-solvent systems is below ambient as illustrated in Fig. 17 where a large value of A_2 obtains in the good solvent[78], methyl ethyl ketone at 30 °C, whereas the extrapolation to c = 0 presents no uncertainty at all under the ideal condition[78] prevailing in ethyl lactate at 11.7 °C. In principle, only a single LS measurement on one solution suffices to yield M in such a system provided that the data at different angles have been extrapolated to $\theta = 0$.

For large particles, Eq. (40) may be treated *via* different methods the most popular of which is the Zimm plot. Essentially, all the data are displayed on a grid which yields on appropriate extrapolations, two lines representing (a) the dependence of $(Kc/R_\theta)_{\theta=0}$ on concentration and (b) the dependence of $(Kc/R_\theta)_{c=0}$ on angle. The common intercept of these lines is 1/M. The abscissa in the Zimm plot is $k_1 \sin^2(\theta/2) + k_2 c$ where a value of unity is normally (but not necessarily) assigned to the arbitrary constant k_1 and the other constant k_2 is also selected rather arbitrarily so as to afford the most convenient spread of data points on the graph. For the two extrapolated lines (a) and (b), the abscissa become $k_2 c$ and $k_1 \sin^2(\theta/2)$ respectively. Methods of obtaining an optimum value of k_2 have been proposed[79]. Frequently, an involuted Zimm plot can be unravelled[80] by assigning a negative value to k_2 (Fig. 18).

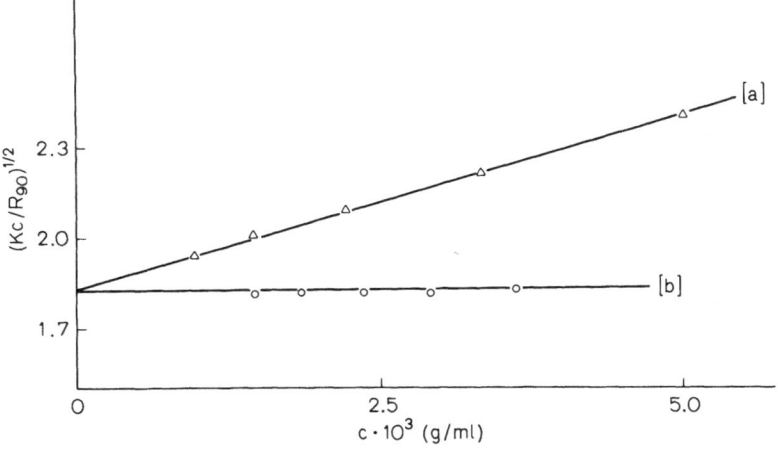

Fig. 17. Square root plot of 90° LS for a solution of low molecular weight polyphenyl acrylate[78] in (a) methyl ethyl ketone at 30 °C (non ideal conditions) and (b) ethyl lactate at 11 · 7 °C (thermodynamically ideal conditions). The value of K is different for plots (a) and (b)

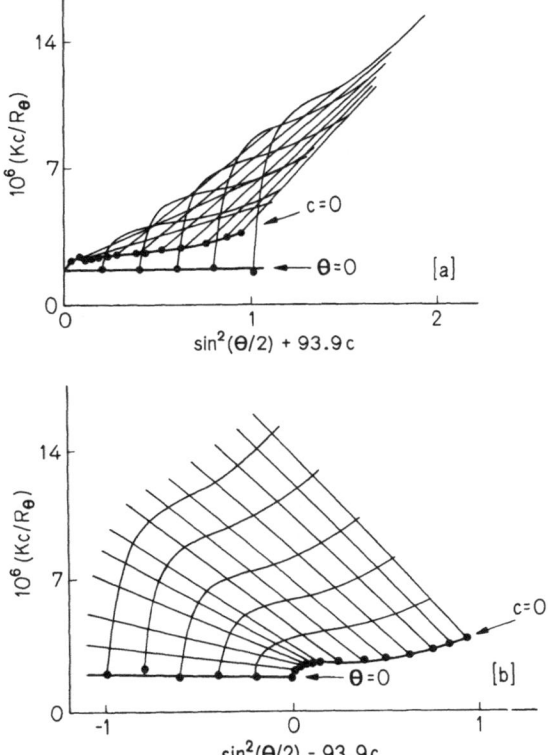

Fig. 18. Use of a negative value for the plotting constant k_2 for solutions of amylose[80] in di-methyl sulphoxide: (a) $k_2 = +93.9$, (b) $k_2 = -93.9$. The primary data points have been omitted for clarity

An exactly equivalent and equally acceptable alternative to the variable $\sin^2(\theta/2)$ is the parameter $\tilde{\mu}^2$, which is identical to [constant x $\sin^2(\theta/2)$] for an experiment involving light of one fixed wavelength[31]. It is convenient to omit the optical constant K from the plots. Hence the ordinate becomes simply c/R_θ and the molecular weight is given by M = 1/K (intercept). For reasons outlined elsewhere[32], the author is inclined to the belief that the Zimm grid offers few advantages in real terms, and that it is preferable to construct separate plots for the concentration dependence of c/R_θ at each fixed angle and for the angular dependence of c/R_θ at each fixed concentration. One important factor militating against the indiscriminate use of the Zimm plot is that the angle and concentration scales cannot be selected simultaneously so as to give the optimum separation of data points for accurate extrapolations of both dependences.

Square root plots proposed by Berry[81] are often more linear than the conventional ones and hence allow an easier extrapolation. Thus, at zero angle

$$(c/R_\theta)_{\theta=0}^{1/2} = (1/MK)^{1/2}(1 + MA_2c) \tag{63}$$

and the resultant intercept at c = 0, when the left-hand-side of Eq. (63) is plotted against c, yields:

$$M = 1/K\,(\text{intercept})^2$$

At zero concentration:

$$(c/R_\theta)_{c=0}^{1/2} = (1/MK)^{1/2}\left[1 + (8\pi^2/3\lambda^2)\langle s^2\rangle \sin^2(\theta/2)\right] \tag{64}$$

and the resultant intercept at θ = 0, similarly yields the molecular weight, when the left-hand-side of Eq. (64) is plotted versus $\sin^2(\theta/2)$:

$$M = 1/K\,(\text{intercept})^2$$

The procedure is outlined schematically in Fig. 19. Such plots can usually be constructed without difficulty, but there are theoretical reasons why they need not necessarily be parallel when the heterogeneity of the sample is high.

The dissymmetry method is useful especially if the instrument does not afford facilities for a wide angular scan of scattered intensities. For large particles the scattering envelope is not symmetrical and, as already indicated in Fig. 1, the forward scatter is larger than that in the backward direction. Hence the dissymmetry Z_d is greater than unity, where

$$Z_d = R_{45}/R_{135} = P(45)/P(135) = G_{45}/G_{135} \tag{65}$$

Here the symbols R, P and G denote respectively the Rayleigh ratio, particle scattering function and instrument scattering reading. It is possible to take other angles such as 60° and 120°, which are also symmetrical about 90°. However, the angles 45° and 135° are most frequently selected, and the widely used Brice-Phoenix photo-

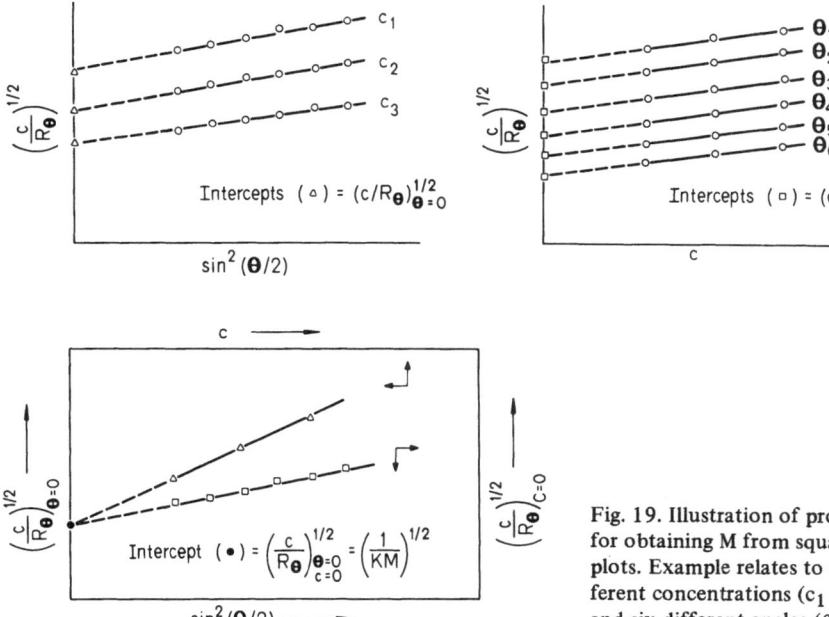

Fig. 19. Illustration of procedure for obtaining M from square root plots. Example relates to three different concentrations ($c_1 - c_3$) and six different angles ($\theta_1 - \theta_6$)

meter is equipped with a special semi-octagonal cell for measuring specifically at these angles. The appropriate dissymmetry is the one relating to infinite dilution, which is termed the intrinsic dissymmetry, $[Z_d]$. The dependence of Z_d on c is not linear, but the variation of $1/(Z_d - 1)$ on c is normally sufficiently linear to allow extrapolation to c = 0 and thereby obtain the value of $[Z_d]$ from the intercept in such a plot, thus[30]:

$$[Z_d] = 1 + 1/(\text{intercept}) \tag{66}$$

For readings at 90° only:

$$Kc/R_{90} = (1/M)[1/P(90)] + 2 A_2 c \tag{67}$$

and neglect of the particle scattering function, that is assumption that P(90) is unity, gives only an apparent molecular weight M*

$$\lim_{c \to 0} (Kc/R_{90}) = 1/M^* \tag{68}$$

The correction is effected by assigning the appropriate value of $1/P(90)$, which may be found from tables[30] wherein it is correlated with $[Z_d]$.

When this is done, the molecular weight corrected for dissymmetry, M_d, is given by

$$M_d = 1/P(90)/\lim_{c \to 0} (Kc/R_{90}) \tag{69}$$

Table 5. Approximate values of $[Z_d]$ and $1/P(90)$ for different particle shapes

Coil		Rod		Sphere		Disc	
$[Z_d]$	$P^{-1}(90)$	$[Z_d]$	$P^{-1}(90)$	$[Z_d]$	$P^{-1}(90)$	$[Z_d]$	$P^{-1}(90)$
1.02	1.02	1.02	1.01	1.02	1.01	1.02	1.01
1.12	1.09	1.13	1.09	1.12	1.08	1.13	1.09
1.21	1.15	1.22	1.15	1.20	1.14	1.20	1.14
1.31	1.22	1.33	1.24	1.32	1.21	1.31	1.21
1.62	1.46	1.62	1.48	1.62	1.39	1.62	1.41
2.03	1.81	2.04	2.01	2.00	1.60	2.07	1.69
3.05	2.99	–	–	3.02	2.09	2.99	2.25
4.01	4.99	–	–	3.91	2.44	4.00	2.93

Some selected and approximate values of $[Z_d]$ and $1/P(90)$ are quoted as illustration in Table 5. A wider range of exact values is available in tabulated and graphical form in the literature[30]. Inspection of Table 5 shows that the corrected molecular weight is ca. 8% greater than the uncorrected one, when $[Z_d] \approx 1.1$, and this applies irrespective of particle shape. In fact it is unnecessary to know the shape for intrinsic dissymmetries up to ca. 1.3, but thereafter such a knowledge is required in order to select $1/P(90)$ from the measured $[Z_d]$.

Tables of this sort are valid for Gaussian coils only. In thermodynamically good solvents the Gaussian behaviour of chain molecules is perturbed by what is called the "excluded volume effect"[30]. The $P/(\theta)$ function depends on the distribution of mass within the particle and this, in turn, is changed if the volume effect is operative. A useful parameter for quantifying the effect is $\widetilde{\epsilon}$, which is defined as

$$\widetilde{\epsilon} = (2\widetilde{\nu} - 1)/3 \tag{70}$$

where $\widetilde{\nu}$ is the exponent in the Mark-Houwink equation relating the intrinsic viscosity of the polymer in the particular solvent (the same solvent as used in LS) to the molecular weight. In an ideal solvent $\widetilde{\nu} = 0.50$ and $\widetilde{\epsilon} = 0$. The magnitude of $\widetilde{\epsilon}$ increases with the goodness of the solvent, since under these conditions $\widetilde{\nu} > 0.50$ and hence $\widetilde{\epsilon} > 0$.

The methods are illustrated in Fig. 20 where the data[78] refer to a sample of poly(phenyl acrylate) in methyl ethyl ketone at $\lambda_0 = 436$ nm and T = 30 °C. Plots (a) and (b) utilise a full angular scan; their common intercept in conjunction with the optical constant K yield a molecular weight of 1.09×10^6. A lower molecular weight of $M^* = 0.746 \times 10^6$ is given by the intercept of plot (c) which invokes only data for 90° scattering. The intercept of plot (d) yields an intrinsic dissymmetry of 1.62. Reference to Table 5 for this value of $[Z_d]$ gives a corresponding value of 1.46 for $1/P(90)$ assuming a random coil configuration in solution; hence the molecular weight, M_d, corrected for dissymmetry is $0.746 \times 10^6 \times 1.46 = 1.09 \times 10^6$. The value of $\widetilde{\nu}$ for this system has been found[78] to be 0.72 and hence $\widetilde{\epsilon} = 0.15$. Tables[30], not reproduced here, relate $[Z_d]$ to a relative size parameter $\langle h^2 \rangle_d^{1/2}/\lambda$, where $\langle h^2 \rangle$ is the mean square end-to-end distance of a coil and equals $6\langle s^2 \rangle$; the subscript d indicates

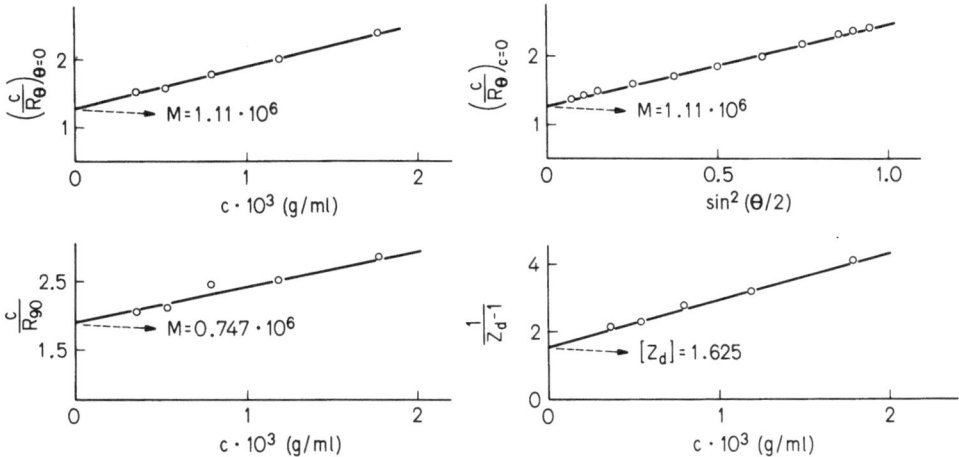

Fig. 20. Determination of M for high molecular weight polyphenyl acrylate[78] in methyl ethyl ketone ($T = 30\,°C$, $\lambda_0 = 436$ nm): upper curves – full angular scan, lower curves – 90° scattering and estimation of intrinsic dissymmetry

dissymmetry. For a value of $[Z_d] = 1.62$, these tables give $\langle h^2 \rangle_d / \lambda = 0.30$. Interpolating to a curve for $\tilde{\epsilon} = 0.15$ in Fig. 21, yields $M/M_d = 1.01$ for the value of $\langle h^2 \rangle_d / \lambda$ in question. Hence the final molecular weight corrected for the excluded volume effect is 1.01 times greater than the molecular weight corrected for dissymmetry, that is, $M = 1.10 \times 10^6$. Consequently, the dissymmetry correction is quite sufficient in this example, but for very large molecules in thermodynamically good solvents ($\langle h^2 \rangle_d / \lambda$ and $\tilde{\epsilon}$ both large), the implementation of the additional excluded volume correction can be important.

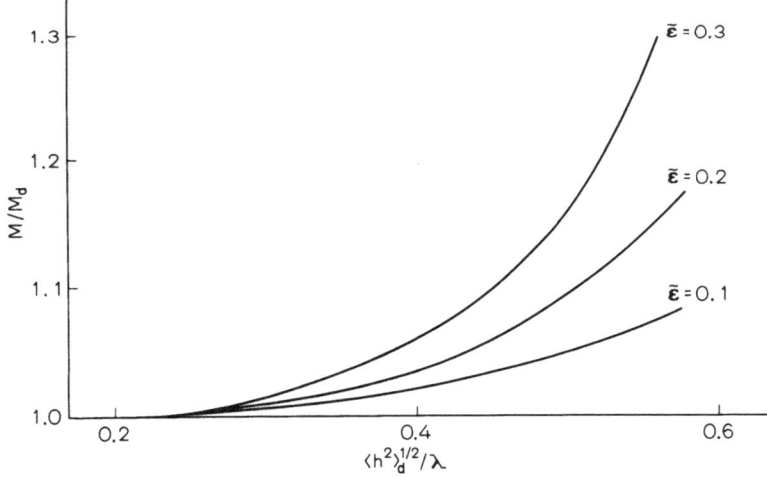

Fig. 21. Ratio between the correct molecular weight (M) and that calculated from dissymmetry (M_d), as a function of the root mean square end-to-end distance calculated from dissymmetry ($\langle h^2 \rangle_d^{1/2}$) for different values of $\tilde{\epsilon}$ [30]

Much of the tedium of deriving M from LS data has been removed by computer programming. This consists in the first instance of converting raw experimental data (*e.g.* galvanometer readings, neutral filter factor, angles, concentrations etc.) into values of Kc/R_θ or $[\sin^2/\theta/2) + k_2 c]$, which can be plotted. Additionally, these can then be treated by computer least squares analysis to yield M directly, without the necessity to plot the results. A programme[82] geared to the Brice-Phoenix instrument was among the first to be used successfully; this was adapted later[83] for use with the Sofica instrument. Most important programmes have been listed and discussed[79]. More recent ones include those of Bryce[84] and of Miller and Stepto[85].

IV. Applications to Different Types of Solute

General simplified outlines have been covered in the preceding sections. In practice there are several complicating features not readily assimilated into a general treatment. Among these may be included charge effects and multicomponent systems. Here we shall illustrate and discuss some applications of LS to the determination of molecular weights and introduce any additional theory where relevant to the system under consideration. In the first instance it will prove convenient to classify the applications according to the approximate order of molecular weight of the solute.

1. Low Molecular Weight Liquids

LS measurements on binary liquid mixtures have been directed primarily as a means of obtaining fundamental thermodynamic information such as chemical potentials and the excess mixing functions. Although molecular weights could in fact be derived from some published data, this has largely not been done by the authors, since such an exercise on substances of known molecular weight would have been subsidiary to the main purpose of their studies.

One of the most elegant LS studies on liquid mixtures is that of Sicotte and Rinfret[17] and it will be instructive to summarise their approach solely with regard to that aspect which is concerned with molecular weight determination. Liquid 1 will be considered as solvent and liquid 2 (of ostensibly unknown molecular weight M_2) as solute. The Rayleigh ratios implied are the isotropic ones, which are obtained for liquid 1 as well as for solutions (subscript 12) *via* the measured Cabannes factors [Eqs. (44) and (45)].

For pure liquid 1 the Rayleigh ratio R_d arises from density fluctuations in accord with Einstein theory:

$$R_d = (\pi^2 k T \beta_1 / 2\lambda_0^4)[(1/\beta_1)(\partial \widetilde{n}_1^2/\partial p)_T]^2 \tag{71}$$

Here, β_1 (instead of the more cumbersome notation β_{T1}) is used for the coefficient of isothermal compressibility of liquid 1. The presence of the second liquid gives rise to an additional scattering due to concentration fluctuations, R_c. As seen earlier [cf. Eq. (36)], this is related to M_2 as follows:

$$R_c = Kc_2/[(1/M_2) + 2 A_2c_2] \qquad (72)$$

The 90° Rayleigh ratio of the mixture, R_{12}, is normally considered as the sum of the two constituent sources of scattering, viz:

$$R_{12} = R_d + R_c \qquad (73)$$

If the density fluctuation of a dilute solution of liquid 2 in solvent 1 is identified with the Rayleigh ratio of pure solvent, then $R_d \equiv R_1$, from which $R_c = R_{12} - R_1$. Hence, using Eq. (72) in the customary manner one obtains $1/M_2$ as the intercept at $c_2 = 0$ in a plot of Kc_2/R_c versus c_2. An example of such a plot[17] appears in Fig. 22(a) for carbon disulphide in carbon tetrachloride at 25 °C and $\lambda_0 = 546$ nm. The intercept yields $M_2 = 125$, compared with the true value of 76. An absolute error of 49 in the molecular weight is, of course, quite negligible for macromolecules in solution. In this connection, it has been estimated that the normal Eq. (72) is unlikely to introduce an error of more than 100 in M_2 provided there is a sufficiently large optical constant K. In the present instance, the error is extremely large and arises from the assumption that the density fluctuation is a constant characteristic of the solvent and is independent of concentration, that is, of the presence of liquid 2.

Bullough[86] has proposed that:

$$R_{12} = R_d(1 + Y) + R_c \qquad (74)$$

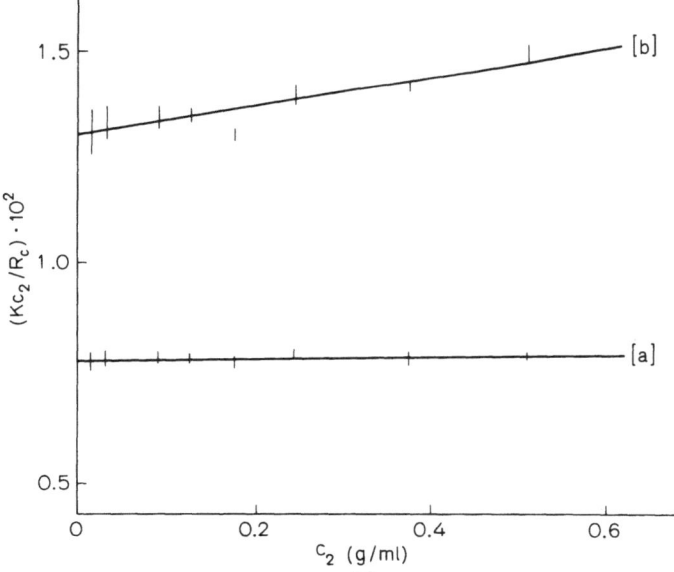

Fig. 22. LS plots for CS_2 in benzene[17] according to (a) – Eq. (72) and (b) – Eq. (76)

where

$$Y \equiv 2(\partial \widetilde{n}_{12}^2/\partial c_2)_{p,T}/(1/\beta_{12})(\partial \widetilde{n}_{12}^2/\partial p)_{c,T} \tag{75}$$

Hence the amended form of Eq. (72), which allows for the effect of concentration on the density fluctuation, is:

$$Kc_2/[R_{12} - R_d(1 + Y)] = 1/M_2 + 2 A_2 c_2 \tag{76}$$

Use of Eq. (76) requires determinations of the isotropic Rayleigh ratio for mixtures of various concentrations c_2. The density fluctuation R_d differs slightly for each solution and is determined from the following expression [Eq. (77)], which is obtained by comparing the forms of Eq. (71) for pure liquid 1 and a mixture, and in which R_1 denotes the isotropic Rayleigh ratio for pure liquid 1:

$$R_d = \left(\frac{\beta_1}{\beta_{12}}\right)\left[\left(\frac{\partial \widetilde{n}_{12}^2}{\partial p}\right)_{c,T}\right]^2 \bigg/ \left[\left(\frac{\partial \widetilde{n}_{12}^2}{\partial p}\right)_T\right]^2 \tag{77}$$

Evaluations of R_d and Y necessitate a knowledge of certain physical properties of the two liquids and the mixtures. The variation of refractive index with concentration is measured readily by refractometry, if $|\widetilde{n}_1 - \widetilde{n}_2|$ is large. The coefficient of isothermal compressibility of a mixture β_{12} requires specialised equipment. Alternatively, it can be determined from the heat capacity and the coefficient of isentropic compressibility[87, 88], the latter being yielded from velocity of sound data[88]. However, provided β_1 and β_2 for the pure compounds are known, β_{12} is evaluated most conveniently on the basis of additivity, thus:

$$\beta_{12} = (1/V_{12})(X_1 V_1 \beta_1 + X_2 V_2 \beta_2) \tag{78}$$

where V and X denote molar volume and mole fraction respectively. The variation of refractive index with pressure has been measured by some workers, but this also requires specialised apparatus. This requirement can be by-passed by an indirect evaluation *via* the following relationships which are quoted in general terms without any specific identifying subscripts:

$$\rho\left(\frac{\partial \epsilon}{\partial p}\right)_T = \frac{1}{\beta}\left(\frac{\partial \epsilon}{\partial p}\right)_T = \frac{2\,\widetilde{n}^2}{\beta}\left(\frac{\partial \widetilde{n}^2}{\partial p}\right) \text{ (since } \epsilon = \widetilde{n}^2\text{)} \tag{79}$$

Among the semi-empirical relationships available for the quantity $\rho(\partial \epsilon/\partial p)_T$ (where ρ denotes density), that due to Eykmann appears[7] to be among the most accurate, viz

$$\rho\left(\frac{\partial \epsilon}{\partial p}\right)_T = 2\,\widetilde{n}^2\,(\widetilde{n}^2 - 1)(\widetilde{n} + 0.4)/(\widetilde{n}^2 + 0.8\,\widetilde{n} + 1) \tag{80}$$

Using this approach the left-hand-side of Eq. (76) may be evaluated and plotted versus c_2 in the normal way to obtain M_2 [see Fig. 22(b)]. The resultant molecular weight of 77 for carbon disulphide is only ca. 1% greater than the true value.

2. Simple Compounds and Oligomers

For solutes which are not macromolecules the requirements for a successful determination of M are (a) higher concentrations than for polymer solutions, (b) selection of solvent yielding a large value of ν and (c) essential implementation of corrective depolarisation factor. Since ν changes with M for low molecular weight polymers, it must consequently be measured separately for each sample it the molecular weight is suspected to be low. Similarly the depolarisation factor ρ_u (or ρ_v for vertically polarised light) is dependent on both concentration and molecular weight, as was illustrated in Fig. 3. For propylene glycol in acetone at 25 °C the Cabannes factor is 1.35 at infinite dilution for vertically polarised incident light, and the molecular weight obtained from LS neglecting this factor would therefore be too high by 35% in relation to the true corrected M. Highly purified samples of simple compounds of known formula have been employed to test the sensitivity of LS photometers. Table 6 demonstrates that, in most studies, remarkably good agreement is possible between the measured and known molecular weights.

Table 6. Molecular weights by LS for pure simple low molecular weight compounds

Substance	M by LS	Theoretical	Ref.
Propylene glycol	67	74	35)
Dipropylene glycol	125	134	35)
Diphenyl	160	154	35)
Dibenzyl	184, 186	182	91)
Dibenzyl	184, 181, 186, 195	182	92)
Methylated poly-oligophenylene	174	182	89)
Glucose	189	190	90)
Methylated poly-oligophenylene	250	273	89)
Sucrose	244	342	90)
Sucrose	332, 344	342	36)
Methylated poly-oligophenylene	360	363	89)
Methylated poly-oligophenylene	458	453	89)
Methylated poly-oligophenylene	543	543	89)
Raffinose	571	594	90)
Sucrose octa-acetate	660, 677	678	92)
Sucrose octa-acetate	690	678	93)
Sucrose octa-acetate	674, 740	678	94)
Methylated poly-oligophenylene	720	723	89)
Tristearin	860	891	35)
Pentaerythritol tetrastearate	1280	1202	35)

The same essential requirements are needed also for measurements on oligomers, which are low molecular species of polymers ($M <$ ca. 10^4). LS has been used to measure M for oligomers of the following: methylated polyphenylenes[89], nylon 66[95], polyethylene glycol[96], poly-α-methyl styrene[97], polystyrene[91], polyheptamethylene urea[98], polyoctadecyl vinyl ether[99], polypropylene glycol[35], poly(trimethyl)hexamethylene urea[98], polyvinyl pyrrolidone[100] and polydimethyl siloxane[101]. Data for the first of these are included in Table 6, since their theoretical molecular weights are known from the mode of synthesis.

In general the true value of M for an oligomer as a basis for comparison is rarely known. However, demonstration of internal consistency and use of a method additional to LS can provide supporting evidence. Thus, for a sample of polyvinyl pyrrolidone, LS gives M = 7150, 7400, 7500 and 7250 from measurements in water at $\lambda_0 = 436$ nm, in water at $\lambda_0 = 546$ nm, in chloroform at $\lambda_0 = 436$ nm and in chloroform at λ_0 546 nm respectively[102]. For a sample of poly-α-methyl styrene[97], LS and sedimentation equilibrium yield M = 1450 ± 28 and 1450 ± 10 respectively, as opposed to a value of $\overline{M}_n$ = 1310 ± 50 afforded by VPO. If the value of $\overline{M}_n$ is assumed to be subject to very low error for this low molecular weight polymer prepared by anionic polymerisation, then one would anticipate a low degree of polydispersity, giving a value of M which is ca. 5% greater than that of $\overline{M}_n$ (it is actually 11% greater, probably due to the fact that techniques for producing very monodisperse samples by anionic polymerisation had not quite attained their present level of sophistication).

3. Polymers

Molecular weights of both synthetic and naturally occurring polymers have been obtained by LS almost as a routine measurement. For reasons which are not wholly evident but which probably have historical and medical connotations, sedimentation in the ultracentrifuge seems to be somewhat preferred by workers in the field of biopolymers, although this technique offers no advantages over LS. Indeed LS can frequently provide additional information and is less time consuming.

Polymer samples of known M and $\overline{M}_n$, usually highly monodisperse in nature, are becoming increasingly available commercially as standards. They include polystyrene, poly-α-styrene, polyethylene, polymethyl methacrylate, polyisoprene (95% cis) and polytetrahydrofuran. The last of these offers a considerable improvement on polystyrene with regard to monodispersity in the low molecular weight region. Polymer fractions of known polydispersity ($M/\overline{M}_n$) are also available, for example polyvinyl chloride, polyethylene and polyphenylene oxide. The quoted value of M for standards represents the mean of a large number of experimental determinations, usually by LS. Some LS plots for a N.B.S. polystyrene standard (No 705) are shown in Fig. 23; they relate to two recent techniques, viz Brillouin spectroscopy and fixed low angle LS on the Chromatix instrument. The resultant values of M together with other values obtained for this sample are assembled in Table 7.

Several anomalies in the LS determinations of molecular weight have been attributed to the presence of a stable supramolecular structure, which can persist even in thermodynamically good solvents[104]. Examples have been reported for solutions of

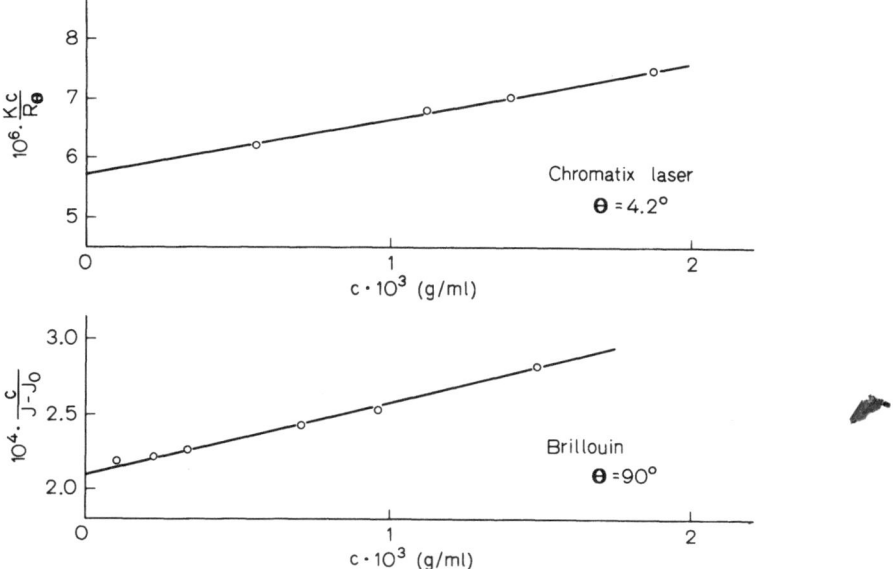

Fig. 23. Determination of M for polystyrene (N.B.S. sample No. 705) by Chromatix low angle scattering[66] at $\theta = 4.2°$ and by Brillouin scattering[41] at $\theta = 90°$

Table 7. Weight average molecular weight of a standard polystyrene sample (N.B.S. 705) by different methods

Solvent	Method	$M \times 10^{-5}$	Ref.
Cyclohexane	LS	1.79	Datum supplied with sample by N.B.S.
Cyclohexane	Sedimentation	1.90	Datum supplied with the sample by N.B.S.
Benzene	LS	1.76	103]
Benzene	Brillouin scattering	1.73	41]
Toluene	Brillouin scattering	1.73	41]
Toluene	Low angle (Chromatix) LS	1.78	66]

polyacrylonitrile, polyethylene, polyvinyl alcohol, methyl cellulose, natural rubber and, most frequently, polyvinyl chloride. The microgel, often present in unfraction-ated species as well as in high molecular weight fractions, manifests itself by abnormally large LS at low angles. The resultant curvature in the plot of $(Kc/R_\theta)_{c=0}$ versus $\sin^2(\theta/2)$ renders an extrapolation to zero angle either impossible or one fraught with uncertainty. Examples in Fig. 24 relate to a copolymer prepared by grafting with γ-irradiation to different doses[105]. The lightly grafted copolymer is normal in behaviour displaying a fairly constant value of $(Kc/R_\theta)_{c=0}$ and small dimensions, whereas more highly grafted species of incompatible constituents induces microgel formation. The downward curvature becomes very pronounced and extra-

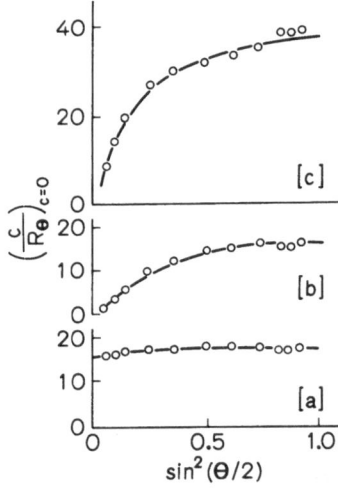

Fig. 24. LS at zero concentration for solutions of nylon-6/acrylic acid copolymers prepared by radiation grafting[105]. The degrees of grafting are (a) 0.038, (b) 0.056, (c) 0.123

polation, if possible, would indicate astronomically high molecular weights. A treatment based on a model system has been put forward by Altgelt and Schulz[106] for rubber microgel and for polyvinyl chloride Kratochvíl[107] proposed a model having a scattering pattern identical with that of the system studied. This model is a two-component one comprising a large weight fraction of small (cf. λ) particles and a small proportion by weight of very large particles. The overall scattering function was expressed in terms of the constituent ones and the approximation was made that $P(\theta) \approx 1$ for the small molecularly dissolved particles displaying normal behaviour. The curvature of experimental points at low angles accords well with the angular dependence for the suggested model. Moreover, it is often just feasible[30, 108] to obtain the value of M by extrapolation from readings relating to moderate-high angles (Fig. 25). Aggregation at normal temperatures may be evidenced by LS and disruption of this by heating[109] (Fig. 26) frequently restores the plot to its normal form. In Fig. 25 the polymer is a rather low molecular weight sample of polyvinyl chloride for which the value of (Kc/R_θ) would be expected to remain rather constant. It is seen that this only obtains after heating.

In certain systems the radiation envelope demonstrates that the polymer is not perfectly and molecularly dissolved, and the downward curvature can only be eliminated by very extended dissolution times and/or heating. As noted, this can be adequate, but the scattering envelope may then develop its normal form at the expense of uniformly decreased scattering, which indicates the result of molecular breakdown by hydrolysis, for example. These effects for solutions of nylon-6[110] are illustrated in Fig. 27. In fact it has been shown that molecular dissolution without aggregation, hydrolysis or polyelectrolyte behaviour (see Section V.2) can only be accomplished by the use of a mixed solvent comprising lithium chloride, water and 2,2',3,3' tetrafluoropropanol[111]. In such a system the value of M for nylons is yielded readily and without ambiguity by LS. A very recent, and hence less commonly appreciated, effect is that of possible cryogenic degradation of polymers in solution[112, 113]. For example, polystyrene ($M = 7.3 \times 10^6$) in p-xylene is reduced in molecular weight to 2.3×10^6

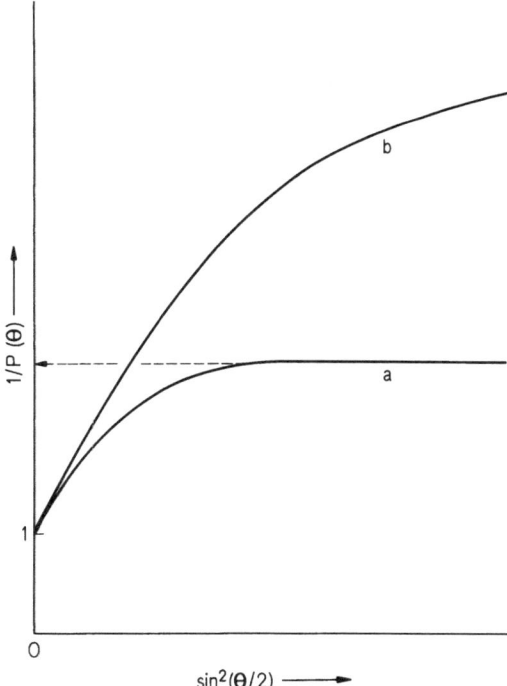

Fig. 25. Typical forms of reciprocal particle scattering functions for systems containing supermolecular structure[30, 108]: (a) small individual macromolecules and a small amount of supermolecular structure sometimes allowing extrapolation to unity (and hence I/M) from moderate – high θ, as indicated, (b) large individual macromolecules or small individual macromolecules but high content of supermolecular structure

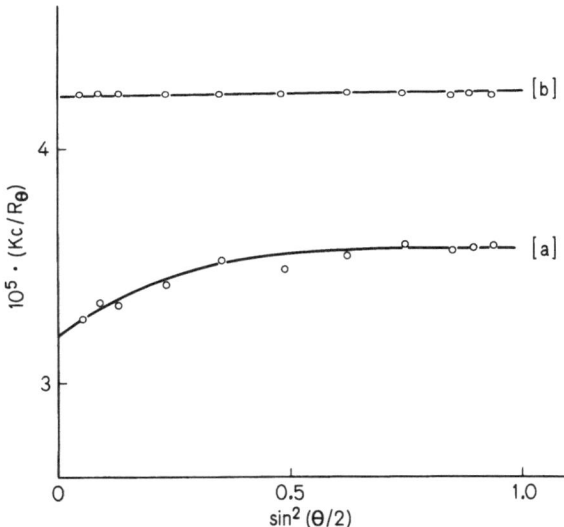

Fig. 26. LS plots for polyvinyl chloride ($M = 28 \times 10^3$) at a concentration $c = 5 \times 10^{-3}$ g/ml in cyclohexanone, (a) before heating, (b) after heating. The sample loses dissymmetry on heating[109]

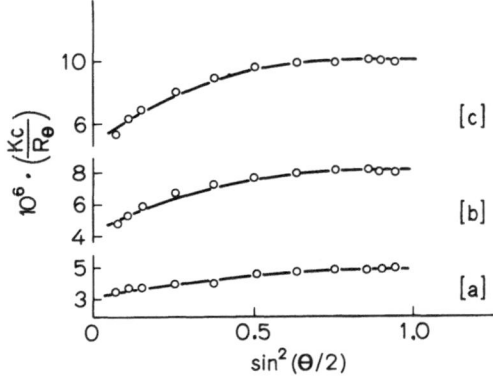

Fig. 27. LS plot for nylon-6 at a concentration c = 4 x 10^{-3} g/ml in 90% formic acid containing 1.0 M KCl: (a) without heating, (b) 7 hr. at 70 °C, (c) 12 hr. at 70 °C. Intensity of LS decreases on heating[110]

after a large number of freezing cycles. This effect is not noted if the initial molecular weight of the polymer is smaller ($< 2 \times 10^6$).

Upward curvature at low angles (that is, decreased scattering) has been observed for solutions of polyvinyl carbanilate[114] (Fig. 28). This was interpreted as being due to interparticle scattering emanating from ordering of the solution, which is broken down at high temperature. According to measurements at high temperature and in other solvents, the abnormal data at room temperature can yield the correct molecular weight, if extrapolation is conducted on data for high angles only.

Multimerisation as an intermolecular process is accompanied by a change in particle mass and the process can thus be studied *via* the concentration dependence

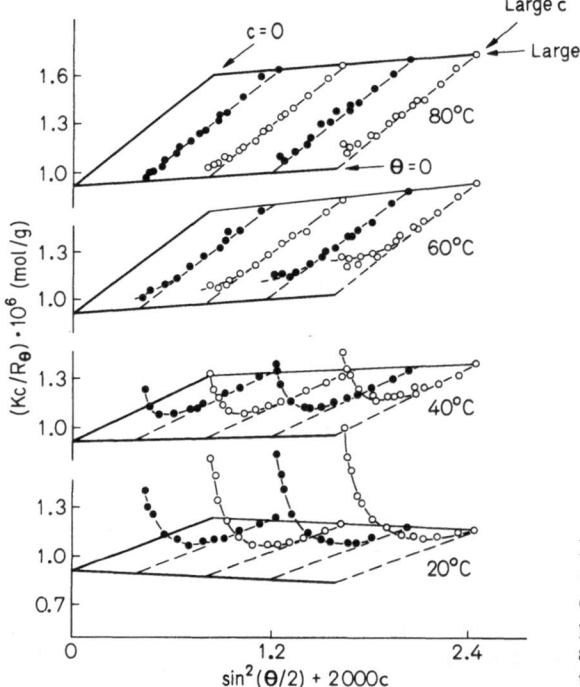

Fig. 28. Zimm plots for polyvinyl-carabanilate (M = 1.1 x 10^6) in diethyl ketone at different temperatures. The upward curvature at low angles is absent only at high temperature[114]

of the apparent molecular weight[115]. An apparent molecular weight in this context is the molecular weight calculated from experimental data at final concentrations using an equation valid for infinite dilution only. The helical aggregation in dioxan of poly-γ-benzyl-L-glutamate has thus been elucidated by LS[116]. The various types of aggregation which may be studied by LS have been classified and reveiwed in great detail by Elias[115]. The work of Sund and Markau[117] places a greater emphasis on biopolymers.

In certain instances the value of M may be required for a polymer which is not accessible in the sense that isolation and dissolution prior to LS measurements would either not be possible or would alter the nature of the polymer. Two examples may be quoted:

Polyimides are thermally stable, heterocyclic aromatic materials of desirable engineering properties. They are, however, insoluble. A typical mode of preparation[118, 119] is given in Fig. 29 where reactants (a) as well as the polyamic acid or pyrrone prepolymers (b) are maintained in solution.

Ring closure without change in chain length occurs in the stage (b) → (c). LS is conducted directly on solution (b) as formed and the value of M may be identified with the desired molecular weight of (c) making a small corrective allowance for the elements of water involved in the cyclisation stage. Due to fluorescence and absorption it is often necessary to operate at λ_0 = 633 nm (red laser) for such solutions.

Anionic polymerisation of hydrocarbon monomers is initiated by lithium butyl to produce a living polymer the association number of which in solution is required to elucidate the kinetics. When the living polymer (for example polystyryl lithium) is terminated, the polystyrene can be isolated and a solution then made to determine its molecular weight, M. If the living polymer is associated in solution, the ratio of its

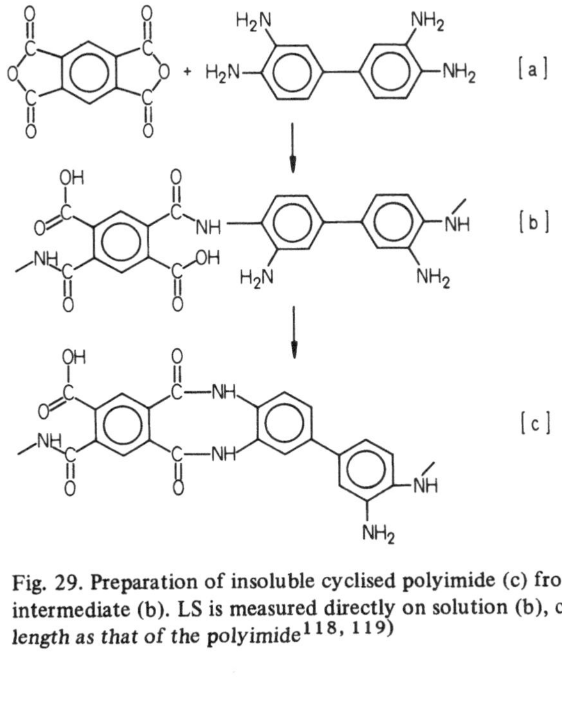

Fig. 29. Preparation of insoluble cyclised polyimide (c) from soluble reactants (a) and soluble intermediate (b). LS is measured directly on solution (b), containing polymer of the same chain length as that of the polyimide[118, 119]

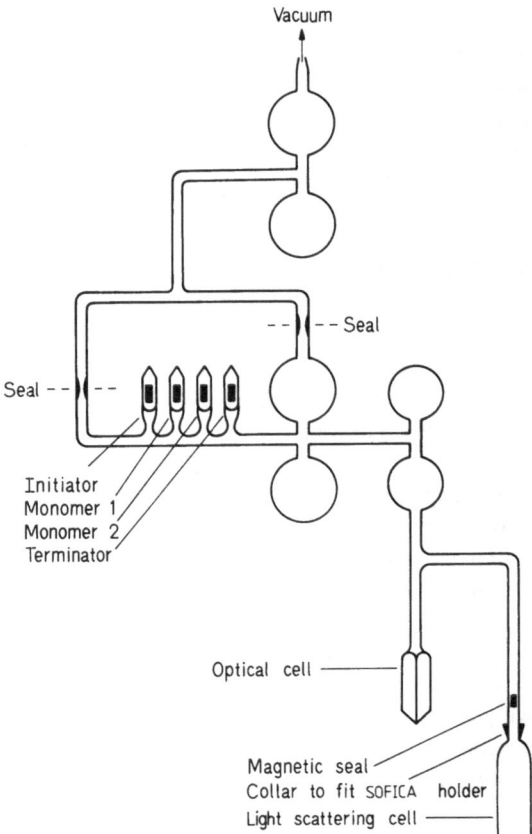

Vacuum

Seal

Seal

Initiator
Monomer 1
Monomer 2
Terminator

Optical cell

Magnetic seal
Collar to fit SOFICA holder
Light scattering cell

Fig. 30. Apparatus for purifying
and manipulating reactants under
high vacuum in order to prepare
solutions of block copolymers via
anionic polymerisation. Concentrations are determined spectrophotometrically in the optical cell
and LS is measured on the solution
in the Sofica cell[120]

molecular weight to M gives the association number. The preparation, LS measurements and estimation of concentrations necessitate high vacuum conditions as indicated in the apparatus of Worsfold and Bywater[120] (Fig. 30). In addition to the
difficult experimental technique involved there is also a problem regarding the determination of molecular weights for some of these systems by LS. For polystyryl
lithium (PS-Li) the use of cyclohexane as solvent at 35 °C constitutes thermodynamically ideal conditions so that $A_2 = 0$ and the molecular weight is obtained readily
by extrapolation according to Eq. (36). With the same solvent and temperature for
solutions of polyisoprenyl lithium (PI-Li) and polybutadienyl lithium (PBD-Li), however, extrapolation is uncertain due to the presence of finite virial coefficients since
the conditions are no longer thermodynamically ideal. In fact there are no thermodynamically ideal solvents for these polymers, which are suitable media for anionic
polymerisation also. The problem was overcome by adding a little isoprene or buta-
diene to the living polymer PS-Li. From the standpoint of associative capacity the
products are characteristic of PI and PBD. Moreover since they comprise predomi-
nantly PS, they behave thermodynamically as this polymer, that is, $A_2 = 0$ in cyclohexane at 35 °C and their molecular weights are obtained readily. By this means an
association number of 4 has been found for PI-Li and for PBD-Li; the value for PS-Li
is 2.

Table 8. Survey of existing complex solvents for cellulose[122]

Solvent	Formula	Designation	Characteristics
Cuprammonium hydroxide	$[Cu(NH_3)_4](OH)_2$	Cuoxam	Good solvating properties, extensive oxidative degradation[1], rather unstable, clear, coloured (blue)
Cupriethylene diamine hydroxide	$[Cu(en)_2](OH)_2$	Cuen and CED	Good solvating properties, extensive oxidative degradation, rather unstable on storage, clear, coloured (blue)
Tri-ethylene diamine Cobalt hydroxide	$[Co(en)_3](OH)_2$	Cooxen	Good solvating properties, extensive oxidative degradation, coloured (claret)
Tri-ethylene diamine nickel hydroxide	$[Ni(en)_3](OH)_2$	Nioxen	Good solvating properties, extensive oxidative degradation, coloured (violet)
Iron – tartaric acid – sodium complex solution	$[(C_4H_3O_6)_3Fe]Na_6$	EWNN	Good solvating properties, extremely high salt concentration, slight oxidative degradation, coloured (green)
Tri-ethylene diamine zinc hydroxide	$[Zn(en)_3](OH)_2$	Zincoxen	Questionable solvating power, stable only at low temperatures, slight oxidative degradation, colourless
Tri-ethylene diamine cadmium hydroxide	$[Cd(en)_3](OH)_2$	Cadoxen and Cden[3]	Good solvating properties, slight oxidative degradation, clear, stable and colourless

[1] Degradation characteristics of the various solvents taken from Ref.[123].
[2] (en) denotes ethylenediamine.
[3] Proposed as a further abbreviation of cadoxen, analogous to Cuen.

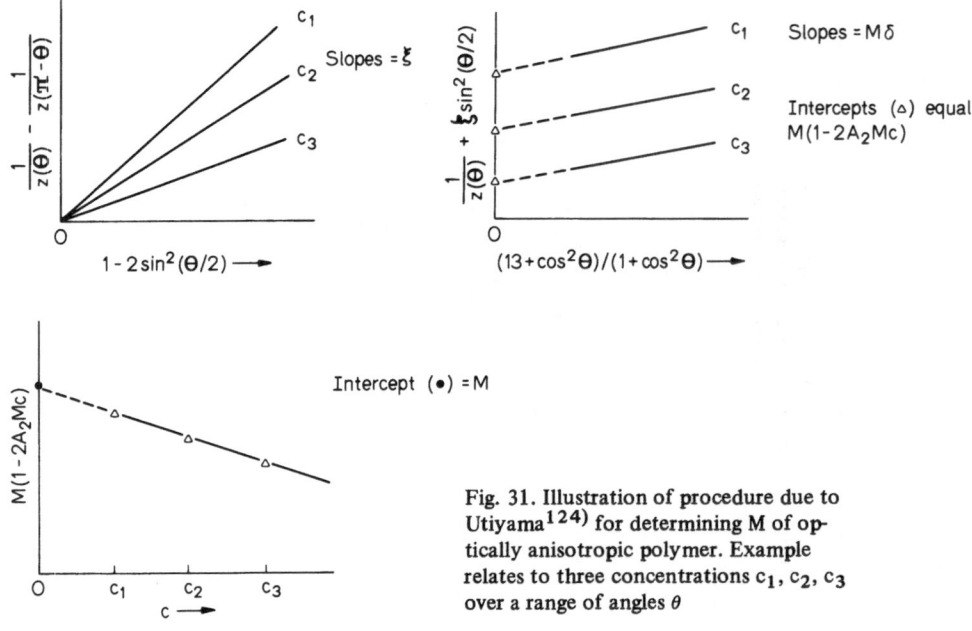

Fig. 31. Illustration of procedure due to Utiyama[124] for determining M of optically anisotropic polymer. Example relates to three concentrations c_1, c_2, c_3 over a range of angles θ

Table 9. LS results for fractions of isotactic polystyrene in chlorobenzene. M* denotes apparent molecular weight obtained directly, δ is the anisotropy parameter and M is the molecular weight obtained after correcting for anisotropy *via* the method of Utiyama[124]

Fraction No	$M^* \times 10^{-5}$	$\delta \times 10^3$	$M \times 10^{-5}$
F–2	4.63	1.78	4.03
F–3	5.99	3.13	5.75
F–4	4.13	53.3	3.41
F–5	2.29	8.34	2.18

The inherently limited solubility of certain (but fortunately, few) polymers poses problems with regard to measurements of both LS and ν. Thus, poly-L-lactic acid is soluble at room temperature in only a few liquids such as chloroform, m-cresol and dichloracetic acid, but the values of ν are very small. Limited solubility in bromobenzene is exhibited at temperatures above 50 °C and LS has been conducted at 85 °C to determine molecular weights[121]. However, even under these conditions, the solubility does not exceed 0.8 g/dl and the moderately acceptable magnitude of ν (= −0.06 ml/g) is subject to a quoted error of ±0.002 ml/g resulting in a possible error of ca. ±7% in M even before any LS measurements. There are no simple solvents for cellulose and the complex ones usually employed are coloured and/or unstable. The most satisfactory one for LS has been found[122] to be Cadoxen (see Table 8). Mixed solvents (Section V.1) frequently offer the only satisfactory means of achieving solubility and the phenomenon of co-solvency is beginning to be investigated more fully. Thus a polymer may dissolve in a binary mixture both components of which are individually non-solvents.

Anisotropy in polymer solutions is rather rare, but, where it occurs, the effect on the derived molecular weight can be large enough to warrant appropriate corrections. The procedures have been developed by Utiyama[124] and by Utiyama and Kurata[125]. Without a polariser or analyser the normal reciprocal scattering function Kc/R_θ can be measured. It is denoted by $Z(\theta)$ the form of which contains an anisotropy parameter δ which is a function of the number of optically anisotropic elements and the principal polarisabilities in the chain:

$$1/Z(\theta) = M(1 - 2 A_2 Mc) - \xi \sin^2(\theta/2) + M\delta (13 + \cos^2\theta)/(1 + \cos^2\theta) \tag{81}$$

where[a]

$$\xi \equiv M\widetilde{\mu}^2 \langle s^2 \rangle (1 - 4 A_2 Mc)/3 \sin^2(\theta/2) \tag{82}$$

Formulation of the corresponding expression in terms of the angle $(180 - \theta)$ and subtraction from Eq. (81) yields

$$1/Z(\theta) - 1/Z(180 - \theta) = \xi[1 - \sin^2(\theta/2)] \tag{83}$$

Hence for each concentration, separate plots of the left-hand-side of Eq. (83) give lines of slope equal to ξ. From Eq. (82) it is evident that the slope decreases with increase in c. Insertion of the values of ξ thereby derived into Eq. (81) and re-arrangement yield

$$1/Z(\theta) + \xi \sin^2(\theta/2) = M(1 - 2 A_2 Mc) + M\delta (13 + \cos^2\theta)/(1 + \cos^2\theta) \tag{84}$$

Since the left-hand-side of Eq. (84) is now known for each concentration, it may be plotted versus $(13 + \cos^2\theta)/(1 + \cos^2\theta)$ to yield lines of intercept equal to $M(1 - 2 A_2 Mc)$ and slope equal to $M\delta$. When these intercepts and now plotted against concentration, the resultant intercept at infinite dilution is M. Because M has been evaluated, the magnitude of δ is also known. These procedures are outlined schematically in Fig. 31. In Table 9 are assembled some data of Utiyama[124] for solutions of isotactic polystyrene in chlorobenzene. For samples having finite values of δ, the discrepancy between the true corrected molecular weight and the uncorrected one (termed "apparent" in this context) can be quite considerable, amounting to 24% for one sample. Another procedure due to Utiyama[124] may be referred to. This consists of using polarised incident light to evaluate δ.

The feasibility of obtaining not only M but also $\overline{M}_n$ from a single set of LS measurements is an inherently attractive proposition. The procedure is based on the

---

[a] The symbols ξ, ξ_1, ξ' and ξ'' appear in this article as abbreviations for complex expressions. The exact forms of these expressions are mostly omitted, but may be found in the appropriate quoted references. It should be stressed, however, that these abbreviations have totally different connotations in different parts of the text.

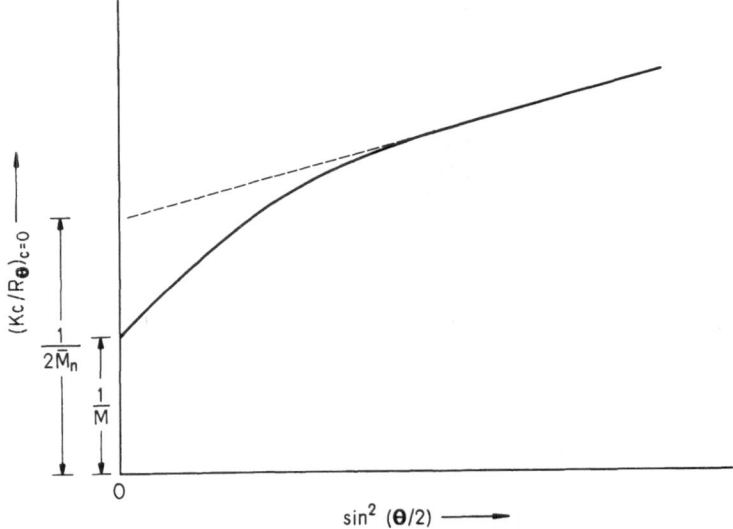

Fig. 32. Molecular weights obtainable in principle from the angular dependence of LS by linear Gaussian coils

asymptotic (rather than initial) behaviour of the particle scattering function $P(\theta)$ at infinite dilution, which yields finally:

$$\lim_{c \to 0} (Kc/R_\theta) = (1/2\,\overline{M}_n)(1 + \langle s^2 \rangle_n \widetilde{\mu}^2) \qquad (85)$$

For a fixed wavelength the variable $\widetilde{\mu}^2$ is equivalent to $\sin^2(\theta/2)$ and hence at high angles the left-hand-side of Eq. (85) should be a linear function of $\sin^2(\theta/2)$, yielding an intercept of $1/2\,\overline{M}_n$ at zero angle. The non-asymptotic region, that is, the initial linear part at lower angles yields an intercept of $1/M$ as previously noted. Figure 32 illustrates the mode of obtaining M and $\overline{M}_n$ in principle. This analysis may be applied only if two prerequisites are fulfilled: (a) all the dissolved polymer molecules are linear Gaussian coils and (b) the dimensions of the molecules are such that for all of them it is possible to measure the initial and the asymptotic regions. Benoit et al. [126] showed for molecules of small-moderate dimensions the asymptotic region is not attainable within the normal angular range of $\theta = 30°-150°$. Moreover, if the dimensions are exceedingly high, the entire experimentally accessible set of data lies in the asymptotic region and it is the initial region which is beyond reach. The procedures and precautions necessary when interpreting the asymptotic behaviour of $1/P(\theta)$ have been discussed by Kratochvíl [30].

For solutes comprising thin rod-like molecules the analysis of Holtzer [127] for the asymptotic reciprocal scattering function yields:

$$\lim_{c \to 0} (Kc/R_\theta) = (2/\pi^2 \overline{M}_n)\,[1 + (\pi \overline{L}_n/2)\widetilde{\mu}] \qquad (86)$$

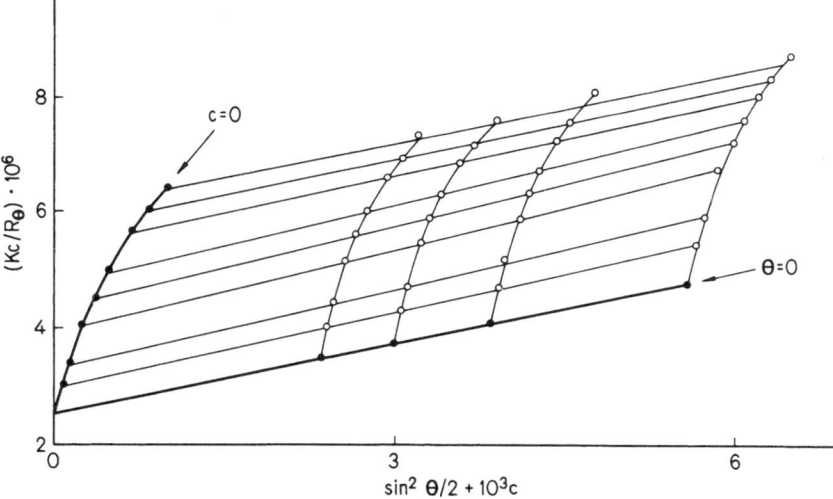

Fig. 33. Zimm plot[128] for a sample of poly-γ-benzyl-L-glutamate in dimethyl formamide at 25 °C

where $\overline{L}_n$ is the number average length of the rod. We note here that the variable is $\tilde{\mu}$ and hence the left-hand-side of Eq. (86) should be a linear function of $\sin(\theta/2)$ and not of $\sin^2(\theta/2)$. The intercept at zero angle should yield $2/\pi^2\overline{M}_n$. The synthetic polypeptide poly-γ-benzyl-L-glutamate (PBLG) adopts a random coil configuration in dichloracetic acid, but is helical or rod-like in many solvents such as dimethyl formamide and chloroform. In Fig. 33 the conventional Zimm diagram is shown for

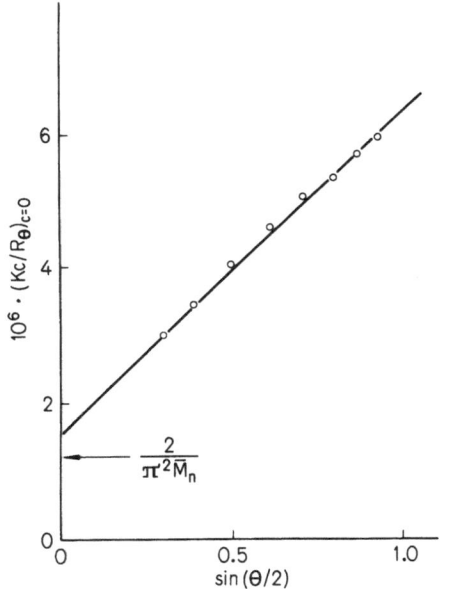

Fig. 34. Plot[128] according to Eq. (86) for sample in Fig. 33

a PBLG sample in dimethyl formamide[128]. The initial region of the angular dependence is linear and its intercept is identical with that of the extrapolated line for $\theta = 0$. The resultant value of M is 4.1×10^5. However, the curvature of $(Kc/R_\theta)_{c=0}$ versus $\sin^2(\theta/2)$ is evident at high angles. The plot[128] according to Eq. (86) is shown in Fig. 34 and from the intercept a value of 1.40×10^5 is derived for $\overline{M}_n$. This agrees reasonably with the value of 1.62×10^5 obtained by direct osmotic pressure measurements on the sample.

On the other hand, Cassassa and Berry[32, 129] conclude from their LS measurements on rod-like fibrinogen in solution that Eq. (86) is followed very well, but the intercept is extremely small or effectively zero. A rather trivial change in the model can be shown to produce a theoretical intercept of zero. Accordingly, they consider that the asymptotic scattering data reveal nothing about $\overline{M}_n$ but do give information from the slope on the ratio $\overline{L}_n/\overline{M}_n$.

In this brief survey on molecular weights of polymers emphasis has been laid on abnormal behaviour and difficulties, which do not of course constitute the norm.

4. Ultra-high Molecular Weight Macromolecules

Provided the specific refractive index increment is large, solutions of ultra-high molecular weight polymers ($M > $ ca. 3×10^6) do not necessitate that the highest

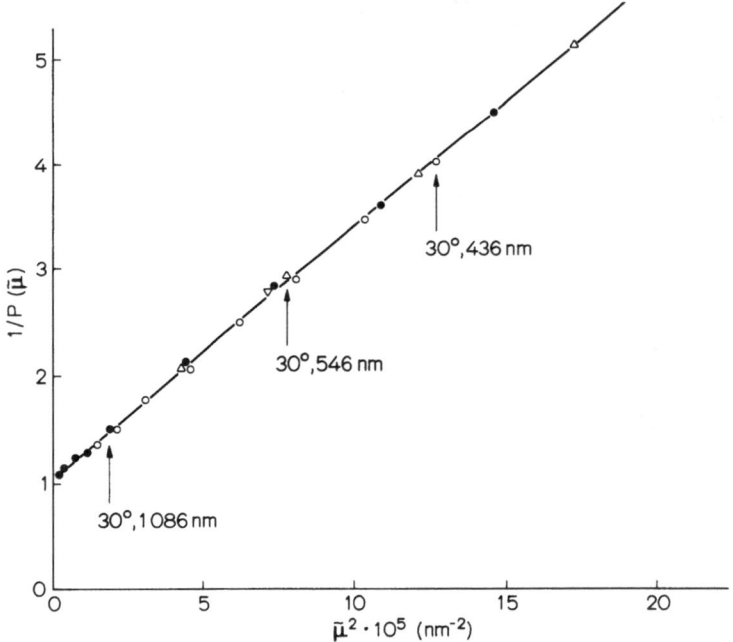

Fig. 35. Reciprocal particle scattering function[60] as a function of $\widetilde{\mu}^2$ for polystyrene of $M = 13.8 \times 10^6$. Different symbols distinguish different LS photometers (Fica − 50 at 436 nm, low angle photometer at 436 nm and near IR photometer of Burmeister and Meyerhoff at 1086 nm)

concentrations be greater than ca. 1×10^{-3} g/ml for LS measurements. Meyerhoff and Burmeister[60] have made a comparative study of the implementation of visible and near IR wavelengths in the determination of such molecular weights. It will be recalled [cf. Eq. (39)] that at infinite dilution the value of Kc/R_θ differs from $1/M$ by a factor $1/P(\widetilde{\mu})$ or, as more usually expressed, $1/P(\theta)$. This factor in turn is greater than unity by an amount $(\widetilde{\mu}^2/3) \langle s^2 \rangle$ and hence will be of considerable magnitude for polymers of large dimensions. Nonetheless, even if $\langle s^2 \rangle$ is large, extrapolation of $(\widetilde{\mu}^2/3) \langle s^2 \rangle$ to zero yields $1/M$. The facility for obtaining very low angle data on the

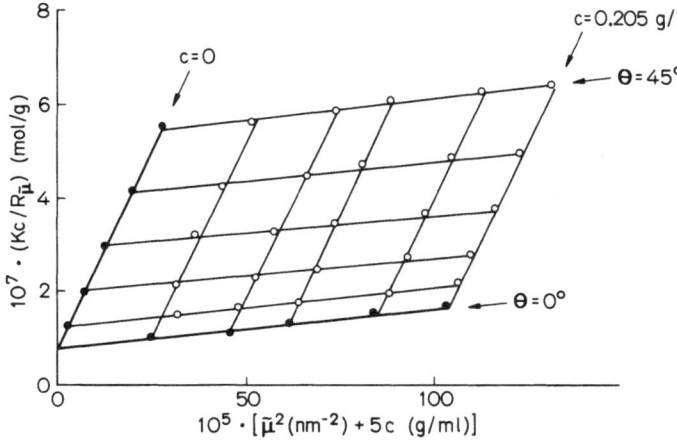

Fig. 36. LS plot[60] in terms of $\widetilde{\mu}$ for sample indicated in Fig. 35. (λ_0 = 436 nm)

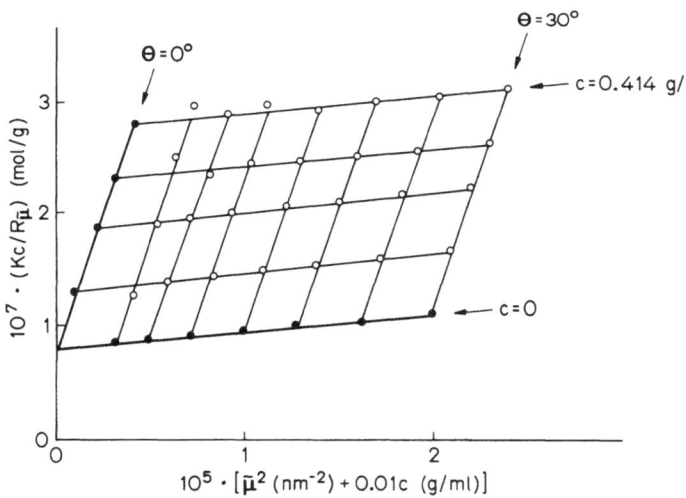

Fig. 37. LS plot[60] in terms of $\widetilde{\mu}$ for sample indicated in Fig. 35. (λ_0 = 1086 nm)

instrument is crucial in this respect and so also is the alternative expedient of rendering $\widetilde{\mu}^2$ small with high wavelength incident radiation. Figure 35 illustrates[60] how $1/P(\widetilde{\mu})$ approaches a value of unity more rapidly for near IR light than for visible wavelengths. The data relate to a solution of a very high molecular weight polystyrene sample having $\langle s^2 \rangle = 7.02 \times 10^4$ nm^2. The actual Zimm plots for determining M are shown[60] in Figs. 36 and 37 for $\lambda_0 = 436$ nm and 1086 nm respectively. A more accurate extrapolation is afforded in the latter situation, since for it a minimal value of only 3×10^{-6} nm^{-2} is reached experimentally, for $\widetilde{\mu}^2$ (at c = 0) compared with a much larger value of 30×10^{-6} nm^{-2} for the same quantity when $\lambda_0 = 436$ nm. Using near IR incident light, molecular weights have been obtained with minimal uncertainty in extrapolation for PS (M = 7×10^6, 13.6×10^6 and 39.3×10^6) as well as for cellulose trinitrate[60] from the alga Valonia (M = 9×10^6).

An alternative mode[31] of extrapolating $1/P(\widetilde{\mu})$ to unity is to conduct experiments with light of different wavelength and to extrapolate $\widetilde{\mu}$ to zero at $\lambda_0 = \infty$. In this procedure the value of $\langle s^2 \rangle$ for the sample is constant and θ is fixed (e.g. at 90°) for all experiments. An example[31] is given in Fig. 38. The resultant value of M = 2.8×10^6 compares well with values lying between 2.7×10^6 and 2.9×10^6 obtained from six separate determinations via Zimm plots in which θ is varied and λ_0 maintained constant at 365, 436, 488, 546, 578, and 1078 nm.

In contrast to these systems, bacterial suspensions demand the use of near IR incident light as an absolute necessity rather than for preference. Such suspensions require maximum concentrations of the order of only 3×10^{-6} g/ml. They scatter light enormously but the relative size parameter exceeds the limits demanded by

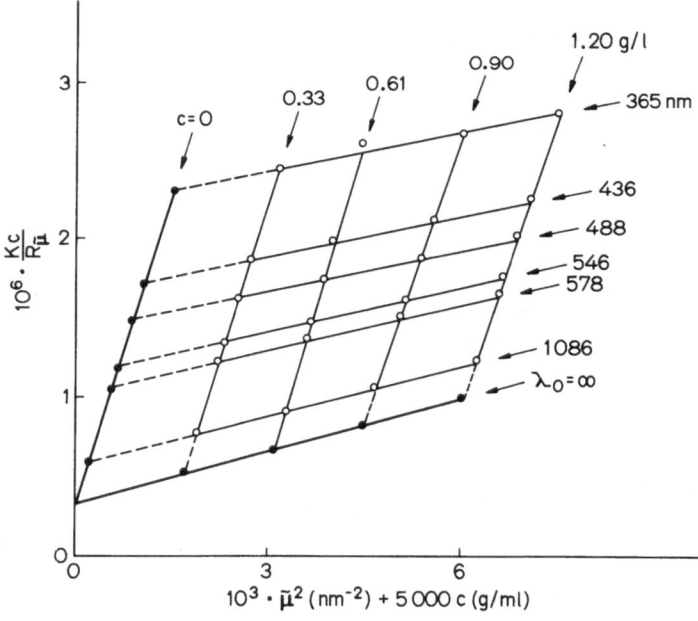

Fig. 38. LS plots[31] for high molecular weight polystyrene (M = 2.8×10^6) at a fixed angle (90°) and variable wavelength

Rayleigh-Debye scattering when visible light is used. The angular scattering profile (cf. Fig. 1) exhibits oscillations and precludes the possibility of obtaining a Zimm diagram. Since bacteria are generally neither spherical nor monodisperse, analysis *via* Mie theory is not possible. With the use of near IR incident light[18, 130] the value of λ_0 is increased so as to approximately halve the relative size parameter with respect to its value with visible light.

V. Multicomponent Systems

Many biopolymeric systems such as nucleic acid solutions (except for rare exceptions) always contain additional low molecular weight salts or buffer components. Consequently, they constitute multicomponent or mixed solvent systems. Similar considerations apply to solutions of polyelectrolytes and neutral polymers, since there are often compelling reasons for adding another component. For example, the ionisation of poly-N-n-butyl-4-vinyl pyridinium bromide is suppressed by the addition of KBr to the aqueous solution. Moreover, some polymers do not dissolve molecularly in single solvents, but true molecular solutions can be achieved by using solvent mixtures. The requirement of a certain quality of the solvent medium (*e.g.* thermodynamic interaction with dissolved polymer or a particular refractive index) is attainable much more readily with mixed than with single solvents. The complications in interpreting LS obtained with such multicomponent systems have been reported and reviewed[4, 14, 15, 72].

1. Mixed Solvents

Essentially, the complex behaviour emanates from the fact that the scattering entity no longer consists of the dissolved molecule alone but of this molecule with some molecules of the selectively adsorbed solvent and of the selectively desorbed other solvent. This selective adsorption is quantified by a coefficient γ_1 for selective adsorption of solvent-1 (the other constituent of the mixed solvent being solvent-3). This is defined in rigorous thermodynamic terms as the change in molality dm_1 of component-1 caused by changing the polymer (subscript-2) concentration by an amount dm_2 with the chemical potential μ of all low molecular weight components remaining constant, viz

$$\gamma_1 = \left(\frac{\partial m_1}{\partial m_2} \right)_\mu$$

If γ_1 is positive, component-1 is selectively adsorbed on the polymer and if polymer is added to the system, the concentration of component-1 must also be increased, if the activity of component-1 is to remain constant. A helpful schematic illustration has been provided by Kratochvíl[131] and is shown in Fig. 39. The example relates to a binary solvent composition of 1 : 1 for the bulk solvent 1 — solvent 3 mixture.

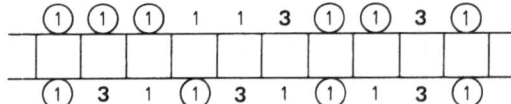

Fig. 39. Selective adsorption to a polymer chain from a bulk solvent medium comprising an equimolar mixture of liquid 1 and liquid 3. Only molecules of liquid 1 (circled) in excess of the equimolar ratio in the solvated shell are part of the selective adsorption complex[131]

Every segment in the polymer chain is solvated by two solvent molecules. Those molecules of solvents 1 and 3 in the solvated sheath which correspond to the bulk composition of $1:1$ are disregarded. Only those molecules of solvent 1 in the solvated sheath in excess of this $1:1$ ratio constitute what may be regarded as the complex with the polymer. In this particular example the value of γ_1 is 1 mol of solvent-1 per mol of segment. If the molar volumes of liquid-1 and polymer segment were equal, the coefficient of selective adsorption could also be expressed here as 1 ml of liquid-1 per ml of monomer units.

The coefficient γ_1 (or γ_3) is easy to visualise physically. It is derived readily from an equivalent coefficient λ_1' which is expressed as ml of liquid-1 preferentially adsorbed per g of polymer[72, 132]:

$$\lambda_1' = \gamma_1 \overline{V}_1 / M_2 \tag{87}$$

where the partial molar volume $\overline{V}_1$ of liquid-1 may be approximated to the molar volume V_1. The true molecular weight of the polymer is M_2. A comparison of the relevant scattering equations for a polymer in a single and in a binary solvent reveals[72, 132] that the molecular weight afforded in the latter instance is an apparent value only, M_2^* viz

$$M_2^* = M_2 \left[1 + \lambda_1' \left(\frac{d\widetilde{n}_0}{d\phi_1} \right) \Big/ \left(\frac{d\widetilde{n}}{dc} \right) \right]^2 \tag{88}$$

In Eq. (88), $d\widetilde{n}_0/d\phi_1$ expresses how the refractive index $\widetilde{n}_0$ of the binary solvent alone varies with its composition expressed as volume fraction ϕ_1 of liquid-1. Clearly, if liquids 1 and 3 are iso-refractive or nearly so, then $M_2^* = M_2$, that is, a LS experiment will yield the true molecular weight irrespective of the composition of the mixed solvent. This situation is exemplified[133] by the system polystyrene (2)-ethylacetate (1)-ethanol (3) for which the molecular weight in mixed solvents of different ϕ_1 is the same as that obtained in pure ethylacetate (Fig. 40). The values of $d\widetilde{n}_0/d\phi_1$ for the mixed solvents are only of the very small order of ca. 0.01, whilst the values of $d\widetilde{n}/dc$ for the polymer solutions are large (ca. 0.22 ml/g).

It should be noted that $d\widetilde{n}_0/d\phi_1$ is generally not zero and moreover its finite value may be positive or negative. Accordingly, the apparent molecular weight obtained experimentally can be equal to M_2, less than M_2 or greater than M_2. Also the binding or selective adsorption coefficient, previously considered to be constant, is now known[132] to vary with molecular weight, the departure from constancy being manifested by a sharp increase at low M. Although the usual object of these studies is to quantify the selective binding of a particular component, other methods are

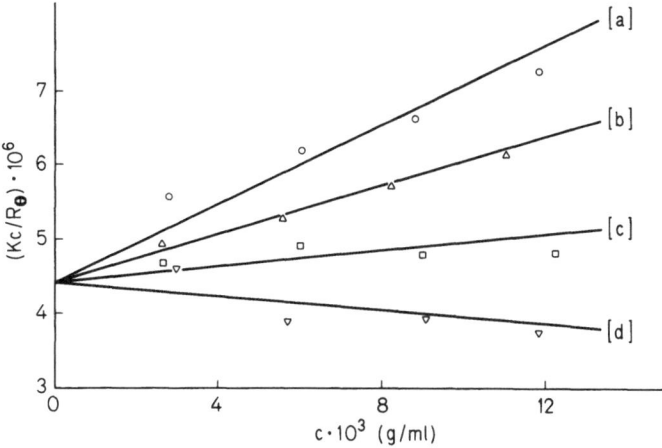

Fig. 40. LS plots for polystyrene in mixed solvents[133] comprising the following % (vol./vol.) of ethanol in admixture with ethyl acetate: (a) 0, (b) 5.19, (c) 8.24, (d) 10.66

Table 10. Typical values of the selective adsorption coefficient γ, as expressed as the ratio of the apparent (M*) to true (M) molecular weight for polymer − mixed solvent systems[131]

Type of polymer	Specification of system	Typical M*/M
Non-ionic	Medium γ	1.1−1.3
Non-ionic	Large γ	3−10
Polyelectrolyte	Equal counterions	1.1−1.3
polyelectrolyte	Different counterions	2−3

available to determine λ'_1 such as differential refractometry and densimetry on dialysed and undialysed solutions. Hence the resultant independent value of λ'_1 can be combined with M_2^* in Eq. (87) to yield the true molecular weight M_2. Full details of these two independent procedures have been reported[134]. Table 10 due to Kratochvíl[131] gives a rough indication of the magnitude of selective adsorption effects for different types of systems.

Equation (88) is the expression used commonly for solutions of synthetic polymers, but, where the nature of adsorption and binding is of critical interest, alternative forms exist. These differ mainly in the modes of expressing concentration (e.g. activity, molality, molarity, mass/unit volume). Interrelations among the units and expressions have been presented very clearly by Timasheff and Townend[15].

One formulation is

$$\lim_{\theta \to 0}\left[K'c_2 \left(\frac{\partial \tilde{n}}{\partial c_2}\right)^2_{T,p,m_3}/R_\theta\right] = \left[1/M_2\left(1 + \xi\right)^2\right]\left[1 + \left(\frac{c_2}{\mathscr{R}T}\right)\left(\frac{\partial \mu_2^e}{\partial c_2}\right)_{T,p,m_3} + \right.$$

$$\left.\left(\frac{\partial \mu_3^e}{\partial c_2}\right)_{T,p,m_3}\left(\frac{\partial m_3}{\partial m_2}\right)_{T,p,\mu_3} + 2\mathscr{R}T\left(\bar{v}_2\right)_{T,p,m_2}\right] \quad (89)$$

199

where

$$\xi = \frac{M_3(1 - c_3\bar{v}_3)_{m_2}}{M_2(1 - c_2\bar{v}_2)_{m_3}} \cdot \left[\left(\frac{\partial \tilde{n}}{\partial c_3}\right)_{T,p,m_2} \middle/ \left(\frac{\partial \tilde{n}}{\partial c_2}\right)_{T,p,m_3}\right]\left(\frac{\partial m_3}{\partial m_2}\right)_{T,p,m_3} \quad (90)$$

$$\left(\frac{\partial m_3}{\partial m_2}\right)_{T,p,\mu_3} = \frac{M_2}{M_3}\left(\frac{\partial g_3}{\partial g_2}\right)_{T,p,\mu_3} = \frac{M_2}{M_3}\frac{(1 - c_2\bar{v}_2)_{\mu_3}}{(1 - c_3\bar{v}_3)_{\mu_2}}\left(\frac{\partial c_3}{\partial c_2}\right)_{T,p,\mu_3}$$

and μ_i, μ_i^e, $\bar{v}_i$, m_i, M_i denote respectively chemical potential, excess chemical potential, partial specific volume, molality and molecular weight of species i. In Eq. (90) one measures the specific refractive index increment with respect to the concentration of component-2, keeping the molality of component-3 constant. An aqueous solution of a protein as component-2 could have urea, a salt or an added organic liquid as component-3. The expression, just as that leading to Eq. (88), has an important consequence, namely; when light scattering data at ($\theta = 0$) are plotted against concentration of protein in the normal manner, the intercept at $c_2 = 0$ is not $1/M_2$, but is the product of this quantity and $1/(1 + \xi^2)$, where ξ is a complicated expression, but essentially measures the interaction of component-3 with the protein. Rather surprisingly, the magnitude and sign of the interaction term turn out to be functions of the concentration units employed. Thus the same system may appear to exhibit preferential adsorption of salt in protein with the molal or g/g scale, but selective desorption when measured on the molar or mol/ml scale.

Figure 41 illustrates the situation for a solution comprising water, bovine serum albumin and 6 M guanidine hydrochloride (GHCl) as components 1, 2 and 3 respectively[15]. The molecular weight M_2^* appears as 38 000 if the molarity of GHCl is kept identical in solution and solvent, whereas it is 103 000 when the molality of GHCl

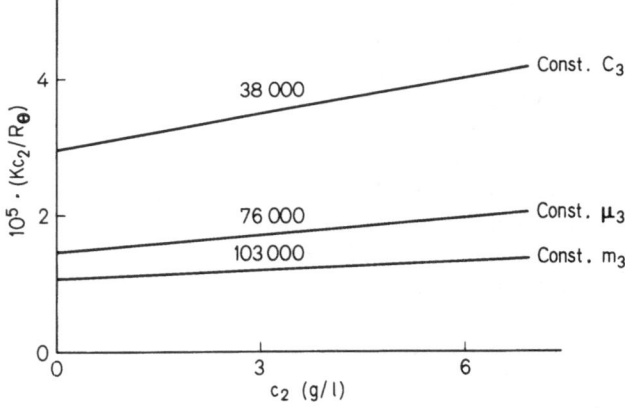

Fig. 41. LS plots for solutions of bovine serum albumin (concentration c_2) in aq. 6 M guanidine hydrochloride, obtained at constant molarity C_3, constant chemical potential μ_3 and constant molality m_3 of guanidine hydrochloride. The resultant molecular weights yielded for this sample are given above each plot[15]

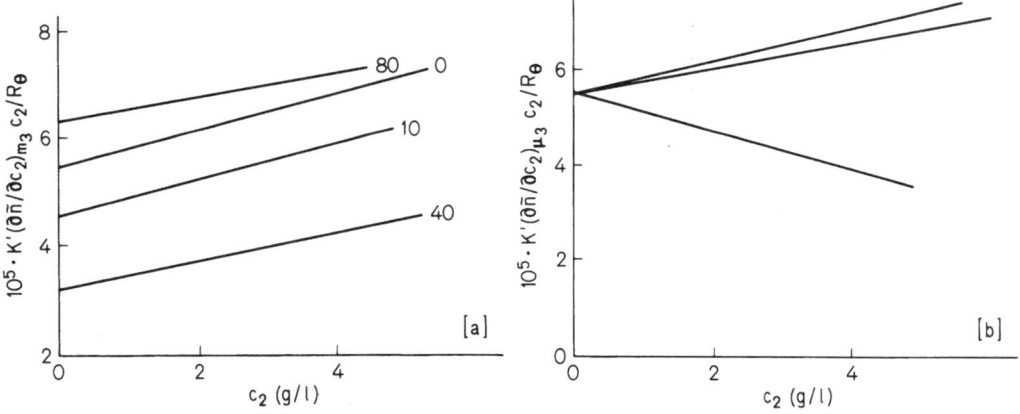

Fig. 42. LS plots for β-lactoglobulin A (concentration c_2) dissolved in water/2-chloroethanol mixtures of compositions (% vol./vol. of 2-chloroethanol) indicated in (a). Plots (a) and (b) correspond to constant molality m_3 and constant chemical potential μ_3 respectively of the organic solvent[135]. The primary data points have been omitted for clarity

is kept constant. Since these molecular weights are respectively less than and greater than the value of 76000 it is clear from Eq. (89) without detailed analysis that the selective adsorptions will differ in sign and magnitude. The correct molecular weight, $M_2 = 76000$, is obtained at constant chemical potential of GHC1.

Figures 42 illustrate forcibly how a variety of apparent molecular weights (M_2^*) of β-lactoglobulin A is obtained from LS plots in aqueous 2-choroethanol mixtures of various composition when the molality of the alcohol is kept constant[135]. If the chemical potential of the alcohol is maintained constant, a common intercept yielding the correct molecular weight, M_2, is obtained for all the solvent mixtures[135].

The mode of reduction of such systems in pseudo two-component ones is embodied in the following modified expression:

$$\underset{\theta \to 0}{\text{Lim}} \left[K'c_2 \left(\frac{\partial \tilde{n}}{\partial c_2} \right)^2_{T,\mu_1\mu_3} \Big/ R_\theta \right] = \left(\frac{1}{M_2} \right) \left\{ 1 + \frac{c_2}{\mathscr{R}T} \left[\left(\frac{\partial \mu_2^e}{\partial c_2} \right)_{T,p,\mu_3} + 2\mathscr{R}T\ \bar{v}_2 \right. \right.$$

$$\left. \left. + 2\mathscr{R}T\bar{v}_3 \frac{M_3}{M_2} \left(\frac{\partial m_3}{\partial m_2} \right)_{T,p,\mu_3} \right] \right\}. \tag{91}$$

It is fortunate that theory has been extended to take into account selective interactions in multicomponent systems, and it is seen from Eq. (91) (which is the expression used for the plots in Fig. 42b) that the intercept at infinite dilution of protein or other solute does give the reciprocal of its correct molecular weight M_2. This procedure is a straightforward one whereby one specifies within the constant K [Eq. (24)] a specific refractive index increment $(\partial \tilde{n}/\partial c_2)_{T,\mu}$. The subscript μ (a shorter way of writing subscripts μ_1 and μ_3) signifies that the increments are to be taken at constant chemical potential of all diffusible solutes, that is, the components other than the polymer. This constitutes the osmotic pressure condition whereby only the macromolecule (component-2) is non-diffusible through a semi-permeable membrane. The quantity

$(\partial \widetilde{n}/\partial c_2)_{T,\mu}$ is measured as outlined in Section III.2 on solutions which have been allowed to attain dialysis equilibrium. The types of dialyser have been reviewed[40] and it has been emphasised recently[14] that it is essential to agitate or mix the polymer solution inside the dialysis bag as well as the mixed solvent outside the bag during the dialysis in order to obtain good results. Reference is especially recommended to authoritative articles on all aspects of the technique[4, 14, 15]. The procedure for obtaining the correct value of M_2 is not restricted to protein solution of course, and has been employed with synthetic polymers and copolymers.

In certain instances, such as in comparing results involving Na DNA and Cs DNA, it is found preferable[4, 14] to use equivalent concentrations C_u (moles of nucleotides or phosphate per litre) instead of weight of polymer per cubic centimetre, c_2. Equivalent concentration units are invariant to the nature of the counterion of the DNA and, instead of the molecular weight M_2, it is the number of nucleotides Z_2 per macromolecule which is yielded by LS experiments. If M_u is the average molecular weight of a paired nucleotide unit in the double helix structure (for example 331 g/mol and 441 g/mol in the Na form and Cs form respectively), then $M_2 = Z_2 M_u$. Similarly, the interconversion between the concentration units is $c_2 = C_u M_u/10^3$. The specific refractive index increment is obtained from measurements on solutions of known C_u and, in the limits of zero angle and zero C_u, the value of Z_2 is given by:

$$\lim_{\substack{\theta \to 0 \\ C_u \to 0}} [K'C_u \cdot 10^3 (\partial \widetilde{n}/\partial C_u)^2_{T,\mu}/R_\theta] = 1/Z_2 \qquad (92)$$

The effect of dialysis on the refractive index increment may be illustrated by the following data[14] for DNA solutions at 25 °C and $\lambda_0 = 546$ nm: $d\widetilde{n}/dc_2$ in water for NaDNA is 0.179 ml/g; in the presence of 0.2 mol/l NaCl the value of $(\partial \widetilde{n}/\partial c_2)_{T,\mu}$ in 0.168 ml/g.

Sonicated DNA of molecular weight of the order of 0.5×10^6 does not present especial difficulties with regard to its molecular weight determination by LS with conventional apparatus. The value of the left hand side of Eq. (92) is only about 10% lower than the value of the quantity under conditions of $C_u = 0$ and $\theta = 30°$. Consequently the necessary extrapolation yielding $1/Z_2$ is quite short.

However, for high molecular weight samples there is a wide disparity among experimental results; the reasons have been discussed[50]. Essentially, the difficulties are not unique to this macromolecule and arise from taking readings at insufficiently low angles. (We have referred to this point earlier in Section III.1.) For $M_2 >$ ca. 3×10^6 use of conventional wide angle LS ($\theta = 30° - 150°$) shows there is downward curvature in the scattering envelope below $\theta = 30°$ resulting in M_2 values which are lower by a factor of at least 2 than those obtained by other methods for high molecular weight DNA samples. Results in the low angle LS range ($\theta = 10° - 30°$) have confirmed the presence of curvature and the magnitude of the error. However, results computed from measurements at $\theta \geqslant 10°$ for a monodisperse bacteriophase T7 DNA of $M_2 = 25 \times 10^6$ were shown to be about 20% too high because of upward curvature in the very low angle region. It is now clear that only very low angle LS data ($\theta \geqslant 9°$) can be extrapolated linearly to $\theta = 0$ so as to yield accurate values of M_2. Table 11 lists[50] molecular parameters of calf thymus DNA obtained recently

Table 11. Influence of accessible angular range on
derived molecular weight and dimensions of Calf-
Thymus DNA[50])

Range of θ (deg)	$M \times 10^{-6}$ (g/mol)	$\langle s^2 \rangle^{1/2}$ (nm)
6–9	20.0	361
6–24	20.6	486
10–35	23.7	522
30–75	6.38	281

over different angular ranges. The Chromatix instrument offers the almost ultimate
relief from extrapolation problems. The plots using this instrument at a fixed angle
of 4.9° yield $M_2 = 25.9 \times 10^6$ for bacteriophage T7 DNA as compared to
$25.5 \ (\pm 1.0) \times 10^6$ obtained by lengthy sedimentation techniques[136]. The molecular
weights of DNA complexes with a non histone chromasomal protein are found to be
of the order of 100×10^6.

2. Polyelectrolytes

Polyelectrolytes are long chain molecules bearing ionisable sites. It is not always
possible to predict with confidence the extent to which polyelectrolytes behaviour
is exhibited. Thus, polyacrylic acid in water is only weakly ionised and in dioxan it
behaves as a typical non-electrolyte. It is usual to overcome the complications imposed by ionic interactions by the inclusion of simple salts and LS studies in salt-free
solutions are rather rare. The problems have been discussed recently by Kratochvíl[137],
whilst the review of Nagasawa and Takahashi[138] constitutes one of the few devoted
exclusively to LS from polyelectrolyte solutions. LS from many biopolymers such as
proteins is, of course, extremely relevant in this context.

It has been observed by Alexandrowicz[139, 140] that in salt-free systems the
practical osmotic coefficient φ_p rather the molecular weight is actually determined.
φ_p is the ratio of the osmotic pressure to the ideal value which would obtain in the
absence of electrostatic interactions. The magnitude of φ_p decreases rapidly with increasing charge density. The expression for Π consequently involves φ_p and, since the
LS depends on the concentration dependence of Π [cf. Eqs. (34) and (35)], the
reduced scattering in terms of C_u, for example, becomes

$$\lim_{\theta \to 0} [K' \cdot 10^3 \, (\partial \widetilde{n}/\partial C_u)_{p,T}/R_\theta] = (1/Z_2) \, (1 + z\varphi_p) \qquad (93)$$

where each polymer chain bears z charges. For highly ionised polyelectrolyte
$z\varphi_p \gg 1$ and the right hand side of Eq. (93) is not $1/Z_2$ but a quantity much greater
than this and the scattering, R_θ, is much reduced. This may be seen by comparing
the large limiting value of Kc/R_θ ($\approx 10^{-5}$ mol/g) for a polyelectrolyte in a single
solvent with the much smaller value ($\approx 10^{-7}$ mol/g) for the same sample in the

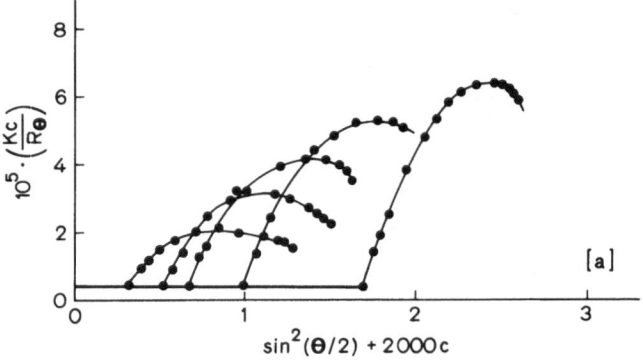

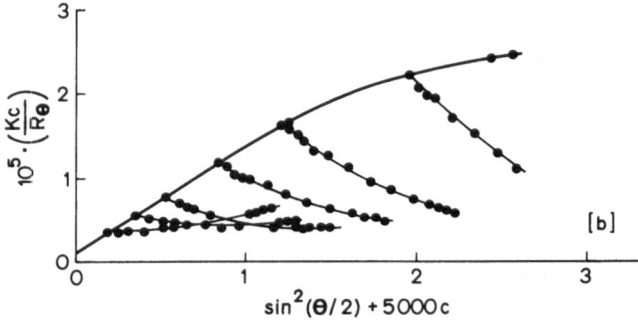

Fig. 43. Zimm diagram[137, 141] for solutions of poly [1 (2-hydroxyethyl pyridinium toluene sul-phonate methacrylate] of molecular weight 1.37×10^6 in (a) water and (b) methanol. (com-pare with Fig. 45)

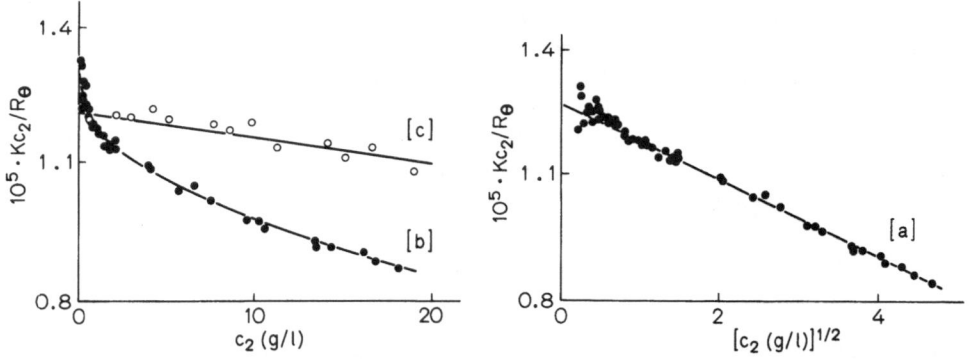

Fig. 44. LS plots[142] for bovine serum albumin in iso-ionic salt free aqueous solution [plot (b)]. Also shown are data (open circles) for solution in 0.001 M NaCl [plot (c)]. Plot (a) is for same data as in (b) plotted against the square root of the bovine serum albumin concentration, c_2, according to Eq. (94)

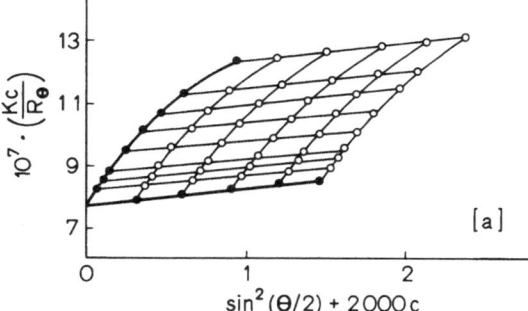

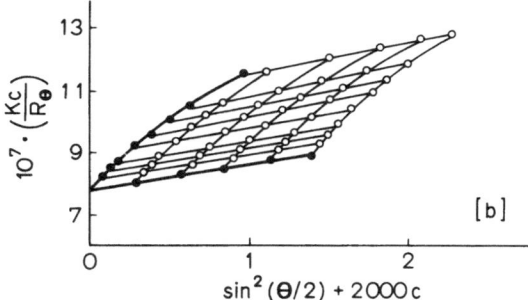

Fig. 45. Zimm diagram[137, 141] for the polymer in Fig. 43 in (a) 0.5 M aq. KCl and (b) 0.5 M methanolic LiCl. (compare with Fig. 43)

presence of added salt (Fig. 43). The polyelectrolyte effect in a solvent other than the usual one, water, is also evident[137, 141] from this figure.

In the special case of polymers capable of carrying both positive and negative charges but having a net charge of zero, there will still be species bearing one or two more charges of both signs as a result of charge fluctuations (analogous with density, concentration and polarisability fluctuations). Hence the mean square net charge will not be zero. Application of Debye-Hückel theory leads to

$$\lim_{\theta \to 0} (Kc_2/R_\theta) = (1/M_2) (1 - \xi c_2^{1/2}),$$ (94)

where ξ is a complex expression involving, *inter alia,* the mean square net charge. Hence the reciprocal reduced scattering extrapolates at infinite dilution of polyelectrolyte to a value of $1/M_2$ and the slope is negative. Figure 44a relating to isoionic salt free aqueous solutions of bovine serum albumen offers a verification[142] of this equation. The plot in terms of c_2 instead of $c_2^{1/2}$ is also indicated and the difficulty in obtaining the molecular weight is apparent. The same molecular weight as that yielded[142] by Eq. (94) is obtained with c_2 as variable, if the system is no longer salt free (Fig. 44b).

Figure 45 should be compared with the distorted plots shown previously[137, 141] (Fig. 43) for a strongly cationic polyelectrolyte. Normal plots yielding the correct molecular weight are obtained when salt is included; KCl for aqueous solutions and LiCl for methanolic ones in view of the insufficiently great solubility of KCl in

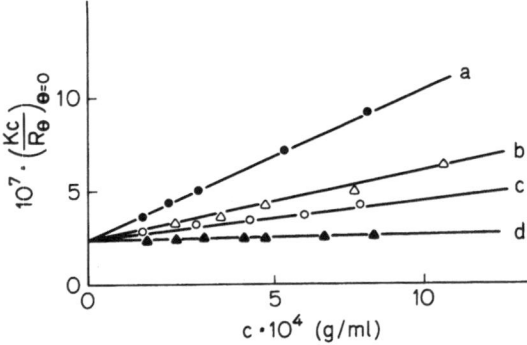

Fig. 46. Determination of molecular weight of poly [1 (2-hydroxyethylpyridinium benzene sulphonate methacrylate] by extrapolation of $(Kc/R_\theta)_{\theta=0}$ to zero polymer concentration in aq. KCl of different molarity[141]: (a) 0.01 M, (b) 0.02 M, (c) 0.1 M, (d) 3.0 M. The resultant molecular weight is 4.3×10^6

methanol. It is emphasised again that, in the presence of salt, the system becomes a three component one. (When the counterion of a polyelectrolyte is not identical with any of the ions of the added salt, the system must be regarded as a four component one.) Molecular weights are found to be independent of added electrolyte concentration, with the important proviso that the specific refractive index increment at constant chemical potential of diffusible components be measured first and included within the optical constant K. Figure 46 illustrates[141] that, although the second virial coefficient [to be denoted strictly as $(A_2)_\mu$ rather than A_2] differs in each case, the plots for different concentration of KCl all extrapolate to the same value for the reciprocal molecular weight[141]. Comparison of these results with those obtained in the presence of KBr and NaF indicates that the moleuclar weight obtained is also independent of the nature of the added salt. Use of $(d\tilde{n}/dc_2)_\mu$ for sodium alginate dialysed against aq. NaCl and magnesium alginate dialysed against aq. $MgCl_2$ yields the same molecular weight[143] for these two alginates, although the large difference in charge densities has a large affect on $(A_2)_\mu$. These recent observations[143] in conjunction with meticulous attention to sample preparation and clarification by centrifugation resolves the problem associated with earlier work whereby the molecular weight of magnesium alginate in aqueous solutions was reported[144, 145] to be very much higher than that of the sodium salt of the same material.

Mention should be made of an expedient for evaluating correct molecular weights without resort to equilibrium dialysis or using $(d\tilde{n}/dc_2)_{T,\mu}$. It has been used by Vrij and Overbeeck[146] for half neutralised polymethacrylic acid in 0.1 molar solutions of sodium halides. The relevant equation is

$$M_2^* = M_2 [1 - \xi M_3 (\partial\tilde{n}/\partial c_3)_{c_2}]^2 \qquad (95)$$

where subscripts 2 and 3 refer to polymer and salt respectively. The constant ξ is a function of the charge, M_2 and the specific refractive index increment of the polymer solution at constant concentration (not constant chemical potential) of salt. The

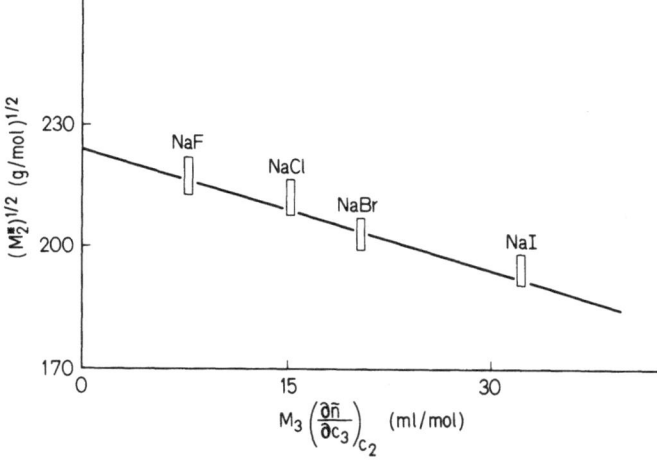

Fig. 47. Square root of apparent molecular weight by LS as a function of molar refractive index increment of salt for sodium polymethacrylic acid in aqueous solution of different sodium halides[146]

value of ξ need not be known. Since $(\partial\tilde{n}/\partial c_3)_{c_2}$ is in units of millilitres per gram, the product $M_3(\partial\tilde{n}/\partial c_3)_{c_2}$ is the molar specific refractive index (ml/mol) increment of salt at constant concentration of polymer. The square root plot[146] corresponding to Eq. (95) is given in Fig. 47 for four different sodium halides. M_2^* is the apparent molecular weight of polymer yielded by experiment and the square root of the true molecular weight (M_2) is given by the intercept.

Micellar properties of bile salts such as sodium taurodeoxycholate and sodium glycodeoxycholate have been studied by several techniques including LS[147]. In very dilute solution (ca. 10^{-3} g/ml) the critical micelle concentration, c_{cmc}, is located. In dilute solution ($<$ ca. 10^{-2} g/ml) measurements of LS yield the weight average micellar molecular weight $\tilde{M}$. For this purpose the concentration is taken to be the difference between the actual concentration of bile salt used and the value of c_{cmc}; similarly the Rayleigh ratio at an angle of $90°$ is taken as the difference

Table 12. Micellar molecular weights $\tilde{M}$ of bile salts by LS and other methods[147] (SE sedimentation equilibrium; SD, sedimentation − diffusion)

Bile Salt	Temp (°C)	Method	$\tilde{M} \times 10^{-3}$
Sodium taurodeoxycholate	–	LS	11.0
Sodium taurodeoxycholate	20	SD, SE	11.9
Sodium taurodeoxycholate	20	SE	11.4
Sodium taurodeoxycholate	36	SE	9.55
Sodium taurodeoxycholate	25	LS	12.0
Sodium glycodeoxycholate	20	SE	8.80
Sodium glycodeoxycholate	36	SE	7.54
Sodium glycodeoxycholate	25	LS	11.5

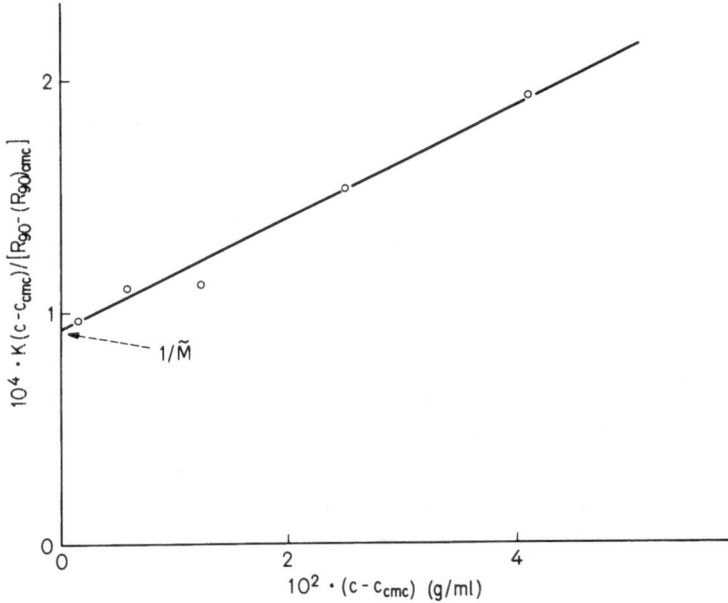

Fig. 48. Plot[147] according to Eq. (96) for solutions of sodium glycodeoxycholate in 0.15 M aq. NaCl. (T = 25 °C, λ_0 = 436 nm)

between the values of R_{90} for the solution and the solution at a concentration c_{cmc}. The data have been treated successfully according to the theory of charged micelles, which leads to

$$K (c - c_{cmc})/R_{90} - (R_{90})_{cmc} = (\xi'/\widetilde{M}) (1 + \xi'' c) \qquad (96)$$

Reference should be made to original papers for the significance and use of the involved functions ξ' and ξ''. However, the magnitude of ξ' is usually assumed to be close to unity so that the plot according to Eq. (96) [see Fig. 48] gives $1/\widetilde{M}$) as the intercept[147]. Some micellar molecular weights obtained by LS and other methods are compared[147] in Table 12. The estimation *via* LS of the particle weight, size and shape has been reported and reviewed by Tuzar and Kratochvíl[148].

3. Copolymers and Terpolymers

A copolymer is a macromolecule comprising two chemically distinct types of monomer unit, A and B, whilst a terpolymer is composed of units A, B and C. The analytically determined composition of a copolymer is expressed as the weight fractions W_A and W_B of its constituents. For LS studies on a copolymer solution it is necessary to know the value of the specific refractive index increment ν, which can be either measured or calculated from:

$$\nu = W_A \nu_A + W_B \nu_B \qquad\qquad (97)$$

The theory of LS from copolymer solutions was pioneered by Stockmayer et al. and later by Bushuk and Benoit, subsequent developments being due to Benoit, Inagaki et al. Molecular weight determination constitutes only a part of the most recent review of the field by Benoit and Froelich[149].

Under stringent conditions of preparation, e.g. via controlled anionic polymerization, characterisation of the copolymer poses relatively few problems. The product is uniform with respect to composition and molecular weight[150]. The former means that W_A is a true measure of the composition of all the molecules and not an average within the range $(W_A + \delta W_A)$ to $(W_A - \delta W_A)$. Measurement or prediction from reaction conditions affords the value of W_A. Uniformity of molecular weight means that M should equal M_n, and here again the molecular weight can be predicted or measured by osmotic pressure, for example[150, 151]. The resultant value should be very close to the molecular weight measured by LS, provided ν is sufficiently large; moreover, the value of M should be independent of the solvent used. Since $M_n = M$, the molecular weights of the constituent portions of the copolymer (M_A and M_B) should be given respectively as $W_A M$ and $W_B M$. From LS measurements on copolymer solutions the values of M_A and M_B are also yielded. Of the numerous systems studied, most work has been devoted to block copolymers of styrene (A) with either methyl methacrylate (B) or isoprene (B). Some typical results are listed[150] in Table 13. A slight, but discernible dependence of M on the nature of the solvent is evident from the data. This will be discussed later.

An alternative and general mode of determining the molecular weight of one constituent is to optically mask the other[152, 153]. Thus, if $\nu_B = 0$, then any observed

Table 13. Characteristics of di-block copolymers of styrene (A) — isoprene (B) containing various weight fractions W_A of styrene[150]

$W_A \times 100$		Molecular weight $\times 10^{-5}$					
predicted[1])	found[2])	M_{mi}[3])	$\bar{M}_n$[4])	M* by LS in toluene	M* by LS in cyclo-hexane	M* by LS in methyl isobutyl ketone	M
25.0	24.7	1.00	1.02	1.29	1.10	1.09	1.07
50.0	52.2	1.00	1.04	1.30	1.14	1.10	1.07
77.8	78.2	1.00	1.03	1.29	1.14	1.12	1.08
25.0	25.5	2.50	2.52	3.00	2.75	2.71	2.62
50.0	48.5	2.50	2.54	2.80	2.70	2.67	2.62
25.0	25.1	5.00	4.99	5.78	5.46	5.42	5.34
50.0	48.7	5.00	5.01	5.64	5.60	5.53	5.41

[1]) From ratio of the weight of styrene to total monomers consumed.
[2]) From uv analysis.
[3]) From ratio of concentrations of monomer to initiator.
[4]) By osmotic pressure. M* is apparent weight average mol.wt. yielded directly from LS experiment; M is true weight average molecular weight obtained from the three values of M*.

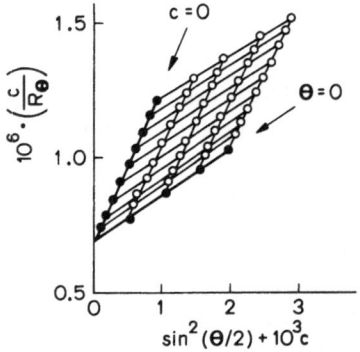

Fig. 49. Zimm plot for a styrene/isoprene block copolymer in methyl isobutyl ketone[153]

Fig. 50. Zimm plot for same sample as in Fig. 49, but
with chlorobenzene as solvent[153]

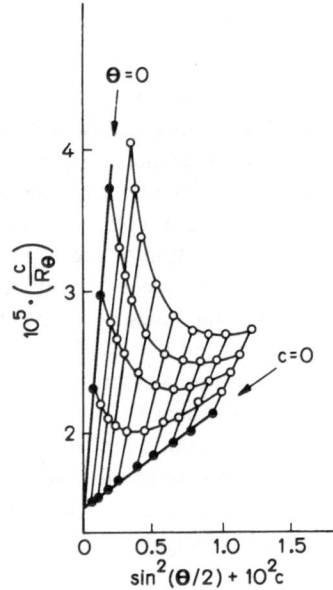

scattering will be due to the presence of A only. Rendering the refractive index increment of one portion zero may be accomplished by altering the temperature and/or
wavelength, but much more usually by selecting a solvent which is iso-refractive with
the portion to be masked. For polystyrene (A) and polymethyl methacrylate (B) in
toluene at 70 °C, ν_A and ν_B are 0.125 and 0.014 ml/g respectively[154]. Hence for
the copolymer, toluene would be almost satisfactory for the purpose of masking
the methyl methacrylate portion and thereby obtaining M_A. At a temperature of
20 °C the system becomes almost perfectly suitable with the same solvent, since
the values of ν_A and ν_B are now 0.110 and 0.001 ($\simeq 0$) ml/g. Several solvents which
are effectively iso-refractive with polymer portions of a copolymer have been listed[153, 154].

In a block copolymer having one portion optically masked the resultant Zimm
plots are sometimes spurious. For example, Figs. 49 and 50 both relate to a styrene
(A) – isoprene (B) block copolymer[153]. In the former figure the solvent is one in
which ν_A and ν_B are both large and the plot is normal. In Fig. 50 the solvent is one
in which ν_A = 0.083 ml/g and ν_B is effectively zero (actually ν_A = −0.004 ml/g); the
plot is abnormal. Here the observed angular LS envelopes are linear at c = 0 but at
low angles they exhibit curvature to an extent which increases with c and also (although no obvious from the data shown for just this single sample) with M. For this
copolymer a similar effect is observed when the other portion is masked instead, that
is, when ν_A = 0 in bromoform. Other examples have been reported. Until comparatively recently this sort of behaviour was known only for some homopolymers, especially charged ones, in single solvents. For block copolymers in single solvents, reasonable explanation has been advanced on the basis of the loss in scattering power caused
by external interference, as known for X-ray scattering from dense groups. Extension
to the present situation leads to the formulation of an apparent particle scattering

function $P^*(\theta)$, which is related to the normal intramolecular scattering function $P(\theta)$ via an intermolecular function $\Phi(\theta)$:

$$P^*(\theta) = P(\theta)/[1 + (c \mathcal{N} V_{exc}/M) \Phi(\theta)], \tag{98}$$

where V_{exc} is the excluded volume and $\Phi(\theta)$ is a function of both angle and the excluded mean square radius of gyration, $\langle s_{exc}^2 \rangle$:

$$\Phi(\theta) = 1 - \tilde{\mu}^2 \langle s_{exc}^2 \rangle/6 + \ldots \tag{99}$$

Appropriate combination of Eqs. (98) and (99) allows the magnitude of V_{exc} and $\langle s_{exc}^2 \rangle$ to be determined. The amended LS equation is expressible now as

$$Kc/R_\theta = 1/MP^*(\theta) = 1/MP(\theta) + 2A_2 c [\Phi(\theta)/P(\theta)] \tag{100}$$

The upward curvature in the scattering envelope due to exceptionally low R_θ has been examined theoretically and experimentally, and correlated with the relevant dimensions[153-155]. Essentially, it emanates from the application of normal (as opposed to apparent) LS laws to a system in which only a part actually scatters light. The intermolecular excluded volume, however, which appears within $P^*(\theta)$ is a function of the size of the whole copolymer molecule, that is, both scattering and non-scattering portions. For normal homopolymers, partical cancellation of the ratio $\Phi(\theta)/P(\theta)$ produces only a very slight increase in the slope of the angular dependence at higher concentrations. For block copolymers in which B, for example, is masked, $P(\theta)$ reflects only the size of A and therefore stays closer to unity than expected for a whole copolymer molecule of a larger total size. On the other hand, $\Phi(\theta)$ is determined by the intermolecular excluded volume of the whole molecule and falls in value below unity more rapidly at high angle and for a large excluded volume cf. Eq. (99). The net effect is an exaggerated decrease in the ratio $\Phi(\theta)/P(\theta)$ at high angles.

From the strictly practical standpoint, it is encouraging to note that $\Phi(\theta)$ and $P/\theta)$ both tend to unity as θ tends to zero, so that in the limit of infinite dilution the intercept remains as $1/M$. Hence, provided ν_A is reasonably large, one can work at low concentration and obtain the molecular weight without too much difficulty. The anomalous behaviour described appears to depend on both types of unit, A and B, being in the same chain and in block form, since no similar effects have been reported for random copolymers or mixtures of two homopolymers. When anomalous behaviour is possible, a larger distortion is expected if the unmasked portion is small relative to the overall dimensions of the whole molecule and also if the deviation of the centre of mass of the unmasked portion from the centre of mass of the whole molecule in large. Thus, if prepared under identical conditions an A−B di-block would give a larger distortion, whilst a B−A−B tri-block a smaller distortion.

Block copolymers of styrene(A)/dimethyl siloxane (B) are even more unusual in the respect that in many solvents ν_A and ν_B are roughly of similar magnitude, but the former is positive and the latter negative[156]. Hence, depending on the composition W_A [cf. Eq. (97)], the value of ν for the copolymer can be positive, negative

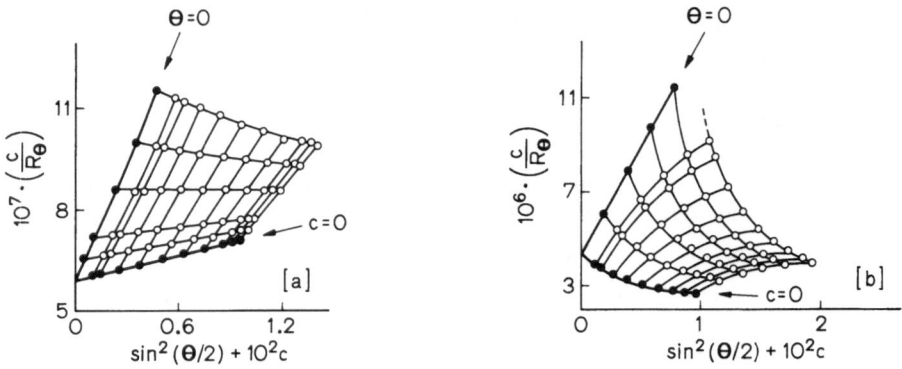

Fig. 51. Zimm plot[156] for a styrene/dimethylsiloxane block copolymer in (a) cyclohexane and (b) toluene

or zero. This dictates the magnitude of a convenient parameter $W_A \nu_A / \nu$, which, as demonstrated theoretically, governs the apparent behaviour of $\langle s^2 \rangle$ or the slope of $1/P(\theta)$ at zero concentration. An enigmatic situation can thus arise when using a solvent in which neither ν_A nor ν_B is zero, whereby the apparent $\langle s^2 \rangle$ is zero or even negative (see Figs. 51). Such behaviour at infinite dilution is in total accord with predictions hitherto unsubstantiated by experimental results. Reference is recommended to the original papers[149, 156], since this behaviour does not introduce any real difficulties in the measurement of the molecular weight.

Most copolymers are heterogeneous in both molecular weight and composition. The latter of these arises from the mechanism of the copolymerisation (particularly at high conversion) and individual copolymer molecules differ slightly in their value of W_A. Solutions of heterogeneous copolymers constitute multicomponent systems

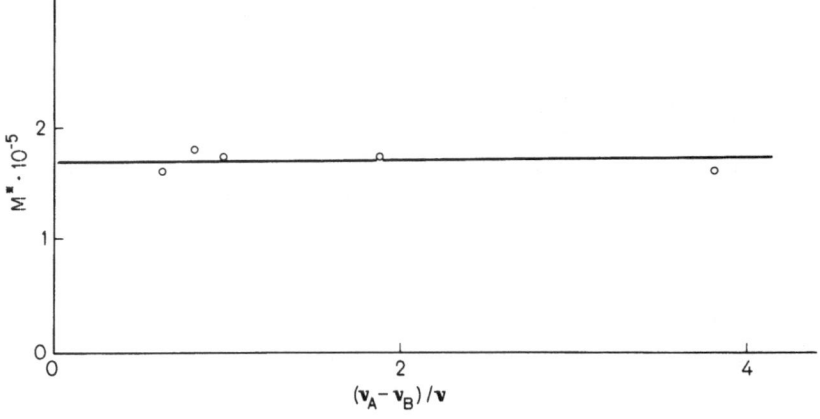

Fig. 52. Demonstration of the constancy of the measured apparent molecular weight M* for a block copolymer, which is monodisperse with respect to composition and molecular weight[157]

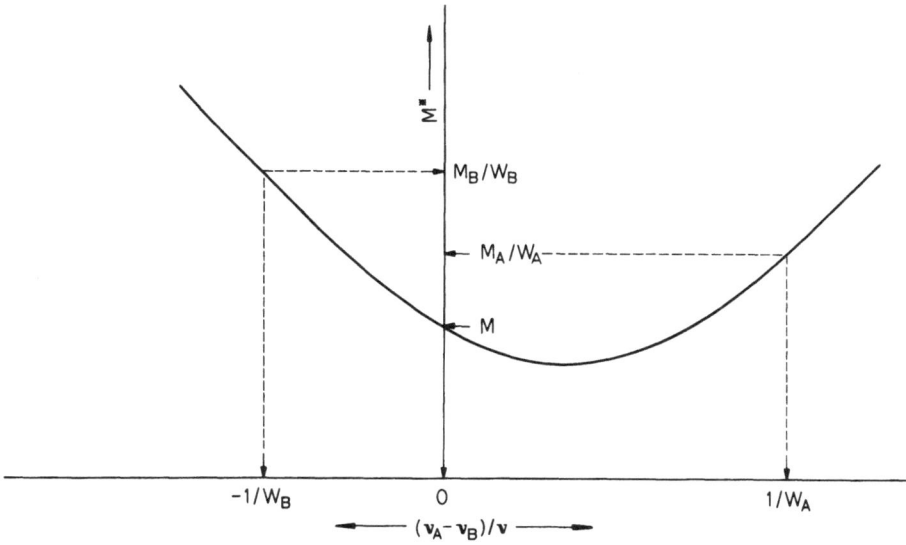

Fig. 53. Parabola according to Eq. (101) showing derivation of molecular weights of copolymer and its constituents[149]

with respect to the polymer. Since W_A differs somewhat among the copolymer molecules, so also does ν, which is an average value (and the only one, in fact, which is accessible experimentally). The molecular weight of the copolymer using the measured ν is only an apparent value M* which differs from the true one M in accord with the following equivalent expressions[149]:

$$M^* = (\nu_A \nu_B/\nu)M + [\nu_A(\nu_A - \nu_B)/\nu^2]W_A M_A - [\nu_B(\nu_A - \nu_B)/\nu^2]W_B M_B \qquad (101)$$

$$M^* = M + 2P[(\nu_A - \nu_B/\nu] + Q[\nu_A - \nu_B)/\nu]^2 \qquad (102)$$

Changing the solvent to one of different refractive index induces a change in $(\nu_A - \nu_B)/\nu$. The form of M* as a function of this parameter is parabolic of curvature dictated by the quantities P and Q. If these, or more conveniently, P/M and Q/M are zero, M*/M will always have a value of unity, that is, the true molecular weight M will always be yielded by LS irrespective of the solvent used (Fig. 52)[157]. The general form of the parabola is shown schematically[149] in Fig. 53 and the required molecular weights are afforded by specific co-ordinates of it, viz for abscissae of $1/W_A$, 0 and $- 1/W_B$ the corresponding ordinates are M_A/W_A, M and M_B/W_B respectively. Several approaches have been made to evaluate the heterogeneity parameters P and Q. These include solution of simultaneous equations[149], recasting data to yield a linear plot[42] and computer least squares fit of data[158].

Since P and Q are characteristics of the copolymer sample, they are independent of the nature of the solvent and hence, in principle, a knowledge of their values can yield the correct molecular weight M from Eq. (102). Many of the reported

Table 14. Refractometric data and apparent LS molecular weight of a styrene (A)/di-n-butyl itaconate (B) copolymer[159] of composition $W_A = 0.169$

Solvent	ν_A(ml/g)	ν_B(ml/g)	ν(ml/g)	$(\nu_A - \nu_B)/\nu$	$M^* \times 10^{-3}$ (g/mol)
Carbon tetrachloride	0.156	0.020	0.043	3.163	144.8
Styrene	0.054	−0.077	−0.055	−2.382	109.3
n-Amyl acetate	0.205	0.074	0.096	1.365	82.3

values of P and Q are now considered to be considerable over-estimates of the truth and the rather stringent conditions necessary for accurate determination of P and Q by LS have been analysed and discussed[131]. On the whole, determination of M *via* the parameters P and Q is not a reliable or recommendable procedure.

The use of simultaneous equations with at least three sets of experimental data can, however, be applied to Eq. (101) or (102) to yield M, M_A and M_B. This is especially useful, if the specific co-ordinates of the whole parabola cannot be realised experimentally as is often the case. The data[159] in Table 14 relate to a random copolymer of styrene (A)/di-n-butyl itaconate (B) in which $W_A = 0.169$. Solution of simultaneous equations [Eq. (101)] yields $M = 67300$, $M_A = 56800$ and $M_B = 64700$. The compositional heterogeneity is such that the data conform to the required theoretical relationship[149]:

$$M > W_A M_A + W_B M_B \tag{103}$$

(that is, $67\,300 > 63\,300$)

The values of $(\nu_A - \nu_B)/\nu$ in Table 14 lie between 3.16 and −2.38. Hence, with these data or with the inclusion of additional ones within the range delineated by these extrema, it would be possible to use the parabola also to determine M and M_B, since the abscissae yielding them are 0 and $-1/(1-0.169)$, *i.e.* −1.20 respectively. However, the range of $(\nu_A - \nu_B)/\nu$ does not extend to a large enough positive value ($1/0.169 = 5.92$) to enable M_A to be derived in a similar fashion.

On the assumption of some finite compositional heterogeneity, Eq. (101) reveals the following criteria necessary for the most sensitive determination of the relevant molecular weights[152]:

$$\nu_A = \nu_B \text{ and both large: to obtain } M \tag{104}$$

$$\nu_B = 0 \text{ and } \nu_A \text{ large: to obtain } M_A \tag{105}$$

$$\nu_A = 0 \text{ and } \nu_B \text{ large: to obtain } M_B \tag{106}$$

Clearly, if one is using only a single solvent, these three requisites are mutually exclusive; the criteria may well be satisfied with different solvent in each case. Expression (104) means that poly-A and poly-B are iso-refractive and that they have a very different refractive index from that of the solvent. Such systems are comparatively

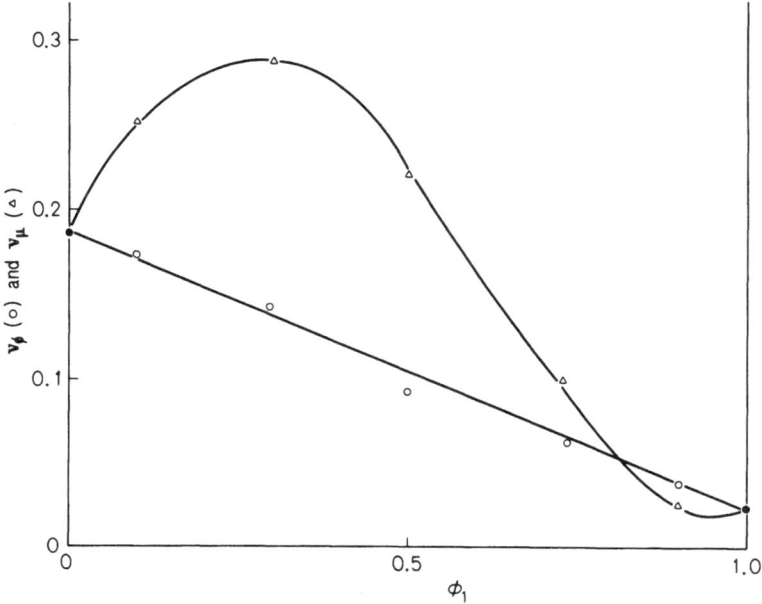

Fig. 54. Specific refractive index increments at constant composition (○) and constant chemical potential (△) for solutions of nylon-6 in 2,2,3,3-tetrafluoropropanol/1-chlorophenol binary mixtures, ϕ_1 is the volume fraction of 1-chlorophenol and filled circles refer to the two pure single solvents[161]

rare, although polystyrene and cellulose tricarbanilate approximate to this situation[160]. When appropriate solvents cannot be found, recourse is made to the far more versatile expedient of using binary solvent mixtures comprising liquids 1 and 3 of variable composition as expressed by the volume fraction ϕ_1 of liquid −1. The refractive index of the mixed solvent $\tilde{n}_0$ depends on ϕ_1 and the value of ν for the copolymer will therefore be governed also by ϕ_1. The consequent complications due to selective adsorption in mixed solvents have already been discussed and the specific refractive index increments employed must be those at constant chemical potential of mixed solvents. Fortunately, Kratochvíl and co-workers[152] have demonstrated that Eqs. (101) and (102) as well as criteria (104), (105), and (106) apply equally well with ν_A, ν_B and ν (or, as they should be more precisely expressed, $\nu_{\phi A}$, $\nu_{\phi B}$ and ν_ϕ) replaced by $\nu_{\mu A}$, $\nu_{\mu B}$, and ν_μ respectively. The difference between the specific increments at constant composition and constant chemical potential can often be quite dramatic, as indicated[161] in Fig. 54.

Another way of determining the composition of the most appropriate mixed solvent for the particular requirement is to actually conduct LS on a polymer in binary solvents of different composition. Since the intensity of scattered light is proportional to the square of the specific refractive index increment at constant chemical potential, it follows that the square root of the light scattering intensity at a given concentration and angle (say 90°) will be proportional to $\nu_{\mu A}$ for poly-A and to $\nu_{\mu B}$ for poly-B in solution. The sign (positive or negative) of the root

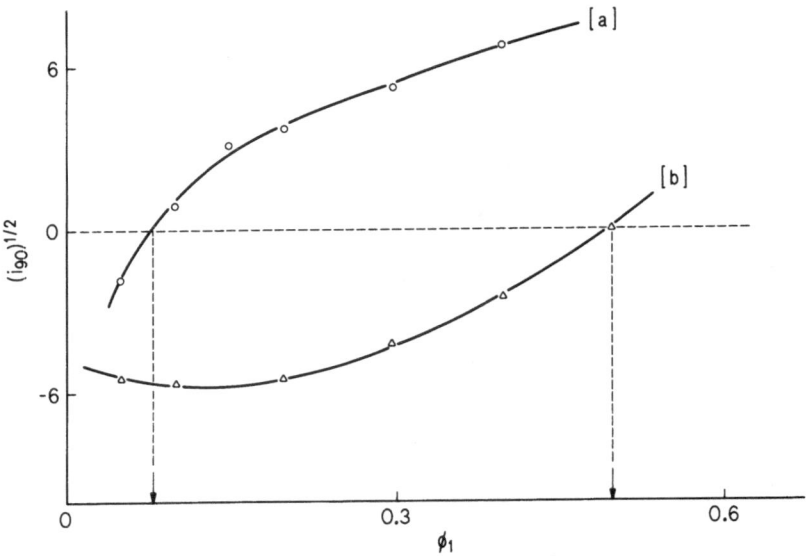

Fig. 55. Dependence of square root of excess LS for solutions of (a) polyethyleneglycol methacrylate and (b) polymethyl methacrylate in mixed solvents comprising 2,2,3,3-tetrafluoropropanol (volume fraction ϕ_1) and benzyl alcohol. (T = 25 °C, λ_0 = 546 nm). Broken lines indicate composition of mixed solvent yielding no excess scattering from the polymer in each case[152]

of the scattered intensity must be inferred by reasoning[152]. Examples are given in Fig. 55 from which it is seen that for a copolymer comprising the two species in question, a binary solvent comprising a low volume fraction of tetrafluoropropanol will optically mask one portion and yield the molecular weight of the other (polymethyl methacrylate); similarly, if the volume fraction of the tetrafluoropropanol is large (≈ 0.5), the other portion of the copolymer would be masked and the molecular weight of poly-2-hydroxyethyl methacrylate in the copolymer yielded[152].

Characterisation of A/B/C terpolymers is still in its infancy, but there are already clear indications that, in a compositionally heterogeneous terpolymer, the determination of molecular weights M_A, M_B, M_C and M may prove an intractable problem in the general case. Apart from some semi-quantitative studies, most effort in this direction has been made by Kambe, Kambe and collaborators[162, 163], who extended the original approach of Bushuk and Benoit[164].

The concepts and definitions involved are more numerous and complex than those involved in the derivations of Eqs. (101) and (102). The full treatment of Y. Kambe has not been published but has been kindly made available to the author. Mention will be made here of only very small selected aspects.

    (i)   Instead of one parameter P, there are three P_A, P_B and P_C,

    (ii)   similarly, parameters Q_A, Q_B and Q_C are invoked,

    (iii) additional heterogeneity parameters R_{AB}, R_{BC} and R_{CA} are introduced,

    (iv) deviations in composition of A, B and C from their mean values are required,

    (v)   molecular weights of portions (M_{AB}, M_{BC} and M_{CA}) are invoked,

    (vi) the molecular weight M of the terpolymer is defined as

$$M = \Sigma W_i M_i$$
$$= W_A M_A + W_B M_B + W_C M_C + 2(W_{AB}M_{AB} + W_{BC}M_{BC} + W_{CA}M_{CA}) \qquad (107)$$

Unfortunately there is no expression corresponding to Eq. (101) for copolymers in which the measured apparent molecular weight M^* appears solely as a function of M_A, M_B, M_C and the true molecular weight M. The only available and tractable expression in the following one[162] [Eq. (108)]; in principle it is capable of yielding M and the values of the five heterogeneity parameters. It does not yield M_A, M_B or M_C.

$$M^* = M + 2P_A(\nu_A - \nu_C)/\nu + 2P_B(\nu_B - \nu_C)/\nu + Q_A(\nu_A - \nu_C)^2/\nu^2 \qquad (108)$$
$$+ Q_B(\nu_B - \nu_C)^2/\nu^2 + 2R_{AB}(\nu_A - \nu_C)(\nu_B - \nu_C)/\nu^2$$

By analogy with Eq. (97), the specific refractive index of a terpolymer is given by the following expression, which has been verified experimentally;

$$\nu = W_A \nu_A + W_B \nu_B + W_C \nu_C \qquad (109)$$

The existence of compositional heterogeneity may be evidenced by the dependence of the measured ν on the extent of conversion of monomers to terpolymer for a random terpolymer[163] (Fig. 56). Other data on the same diagram demonstrate that composition and ν remain sensibly constant with conversion for a partial azeotrope. These observations are corroborated by the LS results in Table 15, which show that for the random terpolymer, M^* varies between 2.63×10^5 and 4.05×10^5 (that is, by about 50%) according to the solvent used. In contrast, for the partial

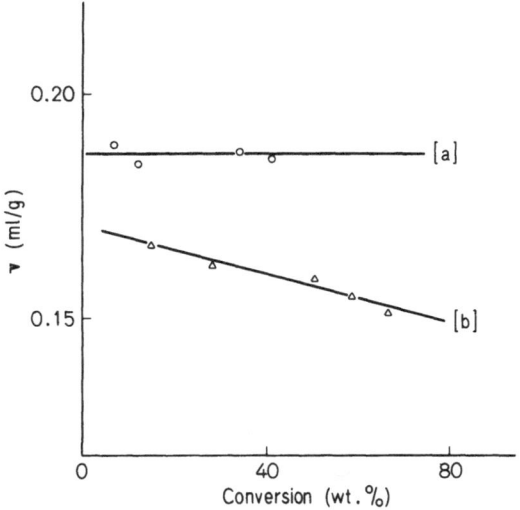

Fig. 56. Dependence of specific refractive index increment on conversion of monomers to polymer for a styrene/acrylonitrile/methyl methacrylate terpolymer in methyl ethyl ketone at 20 °C and 436 nm.: (a) – partial azeotrope, (b) terpolymer with composition distribution[163]

Table 15. Apparent molecular weights (M*) by LS in different solvents for terpolymers comprising acrylonitrile, styrene and methyl methacrylate

Solvent	$M^* \times 10^{-4}$ Partially azeotropic terpolymer	Compositionally heterogeneous terpolymer
Dioxan	45.5	26.3
Acetonitrile	insoluble	29.0
Tetrahydrofuran	47.6	29.1
Methyl ethyl ketone	44.5	30.6
Dimethyl formamide	50.0	40.5
Chloroform	58.8	50.0
Toluene	52.6	insoluble

azeotrope wherein the compositional heterogeneity is low, one anticipates that Eq. (108) should reduce approximately to the form $M^* = M$ irrespective of solvent (that is, irrespective of the values of ν, ν_A, ν_B and ν_C). The data in the table suggest that this is so, since M^* ranges only from 4.65×10^5 to 5.21×10^5 (that is, by ca. 12%).

The azeotropic terpolymer, being atypical in general, one is faced with the prospect of solving at least five simultaneous sets of Eq. (108) in order to obtain M. This manipulation is exact, but the inherent inaccuracies in five different measurements of M^* are coupled with those involved in measuring the refractive index increments. It is as true for a terpolymer as it is for a copolymer that the difference between specific increments (for example, $\nu_A - \nu_B$ or $\nu_B - \nu_C$) must be measured for high

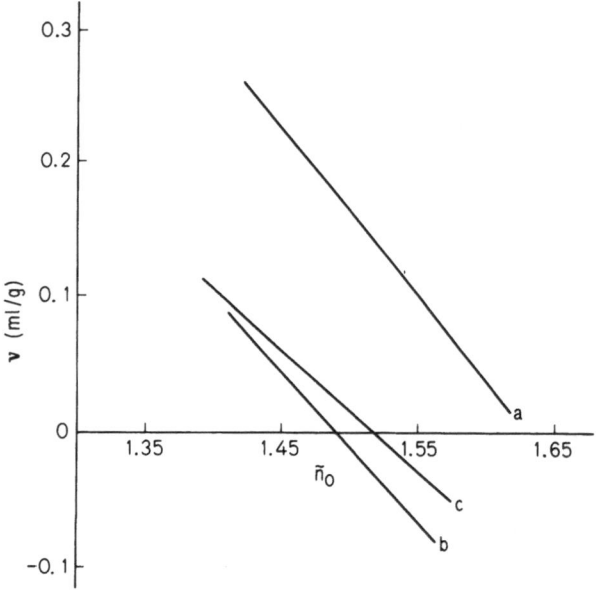

Fig. 57. Gladstone – Dale plots [Eq. (110)] for solutions of (a) polystyrene, (b) poly-n-butyl methacrylate, (c) polymethyl methacrylate[165] ($\lambda_0 = 436$ nm)

accuracy, since this difference varies slightly with the nature of the solvent. According to the Gladstone-Dale equation.

$$\nu = \bar{v}_2 \, (\tilde{n}_2 - \tilde{n}_0) \tag{110}$$

the specific increment of a homopolymer is a linear function of the refractive index $\tilde{n}_0$ of the solvent, the slope being $-\bar{v}_2$ where $\bar{v}_2$ is the partial specific volume of the polymer in solution. Actually, $\bar{v}_2$ itself can vary according to the solvent so that in any selected solvent of refractive index $\tilde{n}_0$ the difference in specific refractive index increments for two polymers, $\nu_A - \nu_B$, may be independent of $\tilde{n}_0$ (as in Fig. 57(a) and (b) relating to solutions of homopolymers A and B) or may change markedly with $\tilde{n}_0$ [as in Fig. 57(c)][165].

The individual values of ν_A, ν_B and ν_C have been excluded from Table 15 and indeed for this system the value of ν_B can only be measured in three solvents due to the insolubility of polyacrylonitrile in the others. For the purpose of calculating $(\nu_A - \nu_B)/\nu$, the value would have to be interpolated rather less accurately via Eq. (110). In the same order as the solvents listed in Table 15, the values of $(\nu_A - \nu_C)$ are 0.114, ——, 0.117, 0.115, 0.109, ——, 0.110, 0.110 and 0.116 ml/g, that is, a variation of only 8% between the largest and smallest values. The only circumstance in which the procedure could be grossly simplified would be that of a terpolymer in which two of the components, say B and C, happen to be iso-refractive. Insertion of $B \equiv C$ into Eq. (108) thereby reduces the expression to the one for an A/B copolymer [Eq. (101)].

4. Mixtures

We have noted that LS yields the weight average molecular weight M. If the solution comprises as solute a mixture of two polymers A and B each of identical molecular weight, LS distinguishes between these species only by virtue of their different specific refractive index increments ν_A and ν_B. Under these conditions a mixture of isorefractive polymers behaves as a single species solute of concentration equal to the sum of the two individual concentrations.

In general not only will ν_A and ν_B differ, but also the individual molecular weights M_A and M_B will not be the same. If these four quantities are obtained first from experiments on the individual polymers then the molecular weight M_{AB} of the mixture of known composition is calculated directly from

$$M_{AB} = W_A M_A + W_B M_B \tag{111}$$

The equation differs from the corresponding one for a copolymer. It is in fact simply an application of the definition of weight average molecular weight [Eq. (10)]. Provided the magnitude of M_{AB} can be established for a mixture of known composition and the molecular weight of one of the polymers is known, Eq. (111) can provide the molecular weight of the other polymer if this is unknown.

Table 16. Molecular weights (M_A) by LS of polymer A alone and in the presence of polymer B when the latter is optically masked by conditions wherein $\nu_B = 0$ or $\nu_{\mu B} = 0$

Polymer(s)	W_A	$M_A \times 10^{-3}$	Ref.
Polystyrene(A)	1.0	120	
Polystyrene(A) + polymethyl methacrylate	0.5	118	166)
Polystyrene(A) + polybutadiene (B)	0.5	125	
Polymethyl methacrylate(A)	1.0	200	166)
Polymethyl methacrylate(A) + polystyrene(B)	0.5	199	
Polybutadiene(A)	1.0	148	166)
Polybutadiene(A) + Polystyrene(B)	0.5	145	
Nylon-6(A)	1.0	49.7	161)
Nylon-6(A) + Polymethyl acrylate (B)	0.573	47.4	

For meaningful implementation of Eq. (111) it is necessary to use the true molecular weight M_{AB} and not the apparent value M_{AB}^* always yielded by a LS experiment, since a mixture constitutes a system of maximum compositional heterogeneity. Equations (101) and (102) for a copolymer apply also to a mixture[152] provided that for the latter one uses M_{AB}^* (instead of M^*), M_{AB} (instead of M), and the specific refractive index of the mixture in solution ν_{AB} (instead of ν). If mixed solvents are used, the specific refractive index increments at constant chemical potential of diffusible solvents are involved. The analogy with copolymers is furthered by optical masking[152, 161] (usually, but not necessarily, by using mixed solvents) one polymer to obtain the molecular weight of the other. Thus, if the mixed solvent is one in which $\nu_{\mu B} = 0$, the molecular weight of polymer A is yielded by

$$M_{AB}^{*2} = (\nu_{\mu A}^2/\nu_{\mu AB}^2)\, W_A M_A \tag{112}$$

Examples have been reported for several systems[161, 166]; some results are listed in Table 16. In summary, polymer mixtures can be treated as copolymers of extreme heterogeneity in chemical composition. Under suitable conditions the techniques used in analysing data of LS from copolymer solutions can be applied to yield with good accuracy the molecular weights of components of the binary polymer mixtures.

The increasing interest in polymer blends has acted as a stimulus not only to the aspects just outlined but also to a study of interactions between polymer A and polymer B in solution as a route to quantifying their thermodynamic compatibility. Hyde[167] has reviewed the relevant theory whilst Kratochvíl and co-workers[168, 169] have been responsible for much of the latest experimental studies.

Essentially, separate experiments on each polymer in the same solvent yield ν_A, M_A and the second virial coefficient $(A_2)_A$ as well as the corresponding quantities for polymer B. When a mixture of the two polymers in which the composition of the polymers are W_A and W_B is dissolved in the same solvent, there are two approaches.

In the first, the ratio c_A/c_B is kept constant and LS is conducted on solutions of varying total concentration c ($= c_A + c_B$). As for experiments on single polymers, dilution is effected with pure solvent. Theoretical analysis shows that:

$$\text{Lim}_{\theta \to 0} (K'c/R_\theta) = 1/(\nu_A^2 W_A M_A + \nu_B W_B M_B) + \xi'c \tag{113}$$

where ξ' is a function of $\nu_A, \nu_B, W_A, W_B, M_A, M_B$ as well as certain interaction parameters. Although they are of prime importance in a different context, we shall ignore these parameters in view of the emphasis purely on molecular weights. The intercept at $c = 0$ in the linear plot according to Eq. (113) is seen to be a composite quantity involving both M_A and M_B. If, for example, polymer B is isorefractive with the solvent, than $\nu_B = 0$ and the system is optically masked in part. The intercept becomes $1/\nu_A^2 W_A M_A$ thus yielding the molecular weights of polymer A as expected and as indicated previously. Similarly, if $\nu_A = 0$, the intercept yields the value of M_B.

The second approach is radically different in the respect that the concentration c_B is kept constant. Hence the whole solution may be regarded as comprising polymer A as solute and polymer B at a fixed concentration in the liquid as the complex solvent. The Rayleigh ratio R_θ' is obtaining by measuring the total LS (from polymer A + polymer B + liquid) and subtracting from it the LS due to solvent, that is, the scattering due to polymer B at a concentration c_B in the liquid. The variable is thus c_A and the relevant equation[168] is

$$\text{Lim}_{\theta \to 0} (K'\nu_A^2 \, c_A/R_\theta') = (1/\xi M_A) + (\xi_1/\xi^2 M_A^2)c_A \tag{114}$$

where ξ and ξ_1 are both functions of interaction parameters as well as of ν_A, ν_B, M_A, M_B, and c_B. Under the special circumstance that $\nu_B = 0$, the form of ξ dictates that in the linear plot according to Eq. (114), the intercept at $c_A = 0$ in simply $1/M_A$.

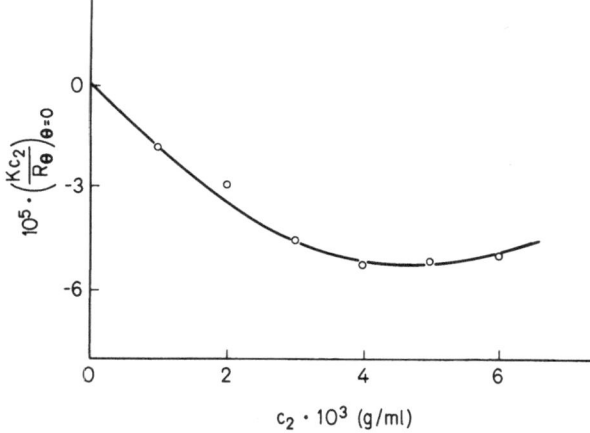

Fig. 58. Concentration dependence of negative LS for solutions of polymethyl methacrylate (concentration c_2) dissolved in a solution of polystyrene in dioxan[169]

However, in general, the intercept affords no simply derivable information on molecular weights because of the presence of interaction parameters within ξ. This system in which one component of the solvent is a polymer and the more common system (Section V.1) in which both components of the binary solvent are low molecular weight compounds form the basis of some novel and interesting comparisons by Kratochvíl and Vorlíček[169]. In both instances the experimental data yield only an apparent molecular weight of the polymer solute, as seen in Eqs. (88) and (114). The ratio of the apparent molecular weight to its true value is denoted in either case by ξ. The significance of ξ for a polymer dissolved in a binary liquid mixture has been referred to, whilst if the solute itself contains a polymer, ξ depends on interaction parameters. Hence in the latter situation the interaction of unlike (and also identical) macromolecules in solution may be regarded as a form of selection adsorption. More precisely, it is selective desorption or exclusion of one polymer from the domains of the other. Selective exclusion of bovine serum albumin[170] by hyaluronic acid, for example, has been demonstrated by LS.

For a mixture of polymers A and B in one liquid the parameter characterising polymer A/polymer B interaction is positive and finite. Without reproducing the complex expression for ξ (which contains this parameter), it suffices to note the important consequence that addition of polymer A to a solution comprising polymer B at a concentration c_B in the liquid need not change the intensity of scattered light; in fact it might even decrease it. Recalling that scattered light intensity refers to excess of scattering between solution and solvent, it is apparent that zero or negative scattering is feasible in principle. Figure 58 provides one of the few published examples[169] to date of negative LS and shows the effect of the concentration c_A of polymethyl methacrylate ($M_A = 1.66 \times 10^5$) on a solvent comprising polystyrene ($M_B = 1.75 \times 10^5$) at a concentration $c_B = 2 \times 10^{-3}$ g/ml in dioxan.

VI. Concluding Remarks

The voluminous literature on LS from solutions indicates that the technique is already one worthy of serious consideration for all concerned with macromolecular characterisation. In this review an attempt has been made to emphasise its versatility with respect to the possible number of variables in the system. It is only fair to re-iterate that the manipulations involved should be approached with caution in view of the stringent requirements of absolute cleanliness of equipment and solutions.

Until comparatively recently one of the most crucial limiting factors in the practice of LS was the quantity of monochromatic incident radiation which it is possible to collimate into the cell using conventional light sources. The resultant scattered light intensity is, of course, dependent on this. Hence there is a practical lower limit on the solute concentration which can be successfully studied. The advent of continuous laser light sources goes a long way to freeing this restriction and the minimum concentrations of solutes detectable with the Chromatix instrument have already been quoted in Fig. 9. As a useful rule of thumb, one aims to obtain a scattered light intensity from a solution which is at least twice that of the solvent.

1. Comparison of LS with Other Methods

In his recent book Billingham[16] has discussed the relative merits and disadvantages of some common modes of measuring the molecular weights of polymers. Many of these are equally applicable to ordinary small solutes and oligomers. Indeed, the conventional colligative properties such as lowering of freezing point, elevation of boiling point and vapour pressure osmometry become more accurate the lower the molecular weight of the solute. Membrane osmometry, though admirable in principle at low molecular weight, cannot be used for $\overline{M}_n <$ ca. 15×10^3 because of possible diffusion of low molecular species present in a polydisperse sample or comprising all of a monodisperse one. Moreover, despite the use of pressure transducers and strain gauges there remains a practical upper limit of ca. 1×10^6 for $\overline{M}_n$ in view of the very small osmotic pressures exerted. Thus, under ideal conditions, a polymer of $\overline{M}_n = 1 \times 10^6$ and a concentration of 0.01 g/ml in water at room temperature would exert an osmotic height of only ca. 0.25 cm; a realistic uncertainty of ±0.05 cm has a drastic impact on the molecular weight derived.

The use of LS as a preferred technique must be assessed in relation to the following factors:

(a) availability, complexity and cost of equipment
(b) time demanded by the experiment
(c) number and type of ancillary experiments necessary for final evaluation of molecular weight
(d) quantity of solute required; this can be a crucial consideration for certain biopolymers
(e) feasibility of recovering solute after measurements
(f) range of molecular weight amenables to determination
(g) type of molecular weight average yielded.

In addition to these, the polymer scientist may be influenced by the possibility of deriving further information such as chain dimensions from the same LS data as those gained primarily to afford M. With respect to the third factor above, (c), a determination of ν or ν_μ is essential in LS, but the remaining factors place LS in a very favourable position in comparison with other techniques. Such a comparison has been reported[171] and comprises part of the larger one assembled here in Table 17.

It is illuminating to quote some early results on aqueous sucrose solutions obtained by Maron and Lou[36]. They demonstrate clearly the complimentarity between osmotic pressure and LS. Reference to Eqs. (34) and (35) shows that

$$Kc/R_{90} = (1/\mathscr{R} T) (\partial \Pi/\partial c)_T \tag{115}$$

and in the limit of infinite dilution both sides of Eq. (115) are equal to the reciprocal of the molecular weight. Measured values of R_{90}, corrected with the Cabannes factor, are plotted versus concentration in Fig. 59. For each value of c the right hand side of Eq. (115) was computed from published osmotic pressure data on these solutions and the corresponding variation of $(1/\mathscr{R} T) (\partial \Pi/\partial c)_T$ with c appears as the full curve[36]. It is seen that there is excellent agreement between the two sets of data

Table 17. Comparison of different techniques for measuring molecular weight

Method	Advantages	Disadvantages	Typical Accuracy %
Chemical end-group analysis	Absolute; single determination	Destructive, restricted to low mol. wt.	1–2
Analysis of radioactively labelled end-groups	Absolute; single determination; wide range of mol. wt.	Restricted to certain types of polymer; requires knowledge of mode of chain termination	1–2
Vapour pressure osmometry	Requires very small sample size	Several concentrations necessary; restricted to mol. wt. < ca. 2×10^4; calibration constant must be established with solute of known mol. wt.	1–2
Membrane osmometry	Absolute; small sample size; reasonable range of mol. wt. $(2 \times 10^4 - 1 \times 10^6)$	Several concentrations necessary	5
Sedimentation equilibrium	Absolute; requires only centrifuge measurements and partial specific volume of solute	Time-consuming measurement and interpretation; somewhat limited mol. wt. range	5
Sedimentation-diffusion	Absolute; fairly fast	Requires 3 separate determinations (diffusion coefficient, sedimentation constant and partial specific volume)	<5
Neutron scattering	Can be used with solid sample	Expensive equipment requires deuteration of sample and calibration with substance of known mol. wt.	2
Low angle X-ray scattering in solution	Non-destructive	Time-consuming	5
Brillouin scattering	Absolute	Requires knowledge of certain physical properties of solvent; several concentrations necessary	2
Light scattering in solution	Fairly fast; wide range of mol. wt.	Must measure $d\tilde{n}/dc$; several concentrations necessary; for most (but not all) apparatuses requires calibration and rigorous exclusion of dust	5

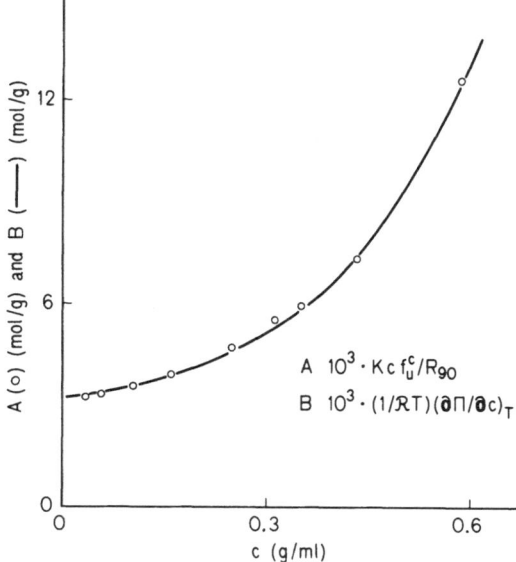

Fig. 59. Agreement between LS data and osmotic pressure data for solutions of sucrose in water[36] at 25 °C

and that the intercept at $c = 0$ yields a molecular weight (338 ± 6) which is in good accord with the theoretical one of 342.

Standards suitable for molecular weight comparison by LS have been listed by Dietz and Green[19]. These comprise mainly pure low molecular weight materials. Periodically, polymer samples have been subjected to "round-robin" characterisation by different techniques (including LS) in different laboratories throughout the world. In the first of these the results were disappointing and the co-ordinators of the second batch of data were also faced with a fairly wide spread of M values all obtained from a single batch of polymer. A critical examination of procedures adopted led to no obvious single source of such a spread of results. Among the reasons advanced was a possible injudicious selection of the range of concentrations used and uncertainty in extrapolation to infinite dilution[172]. The most recent survey of this nature was conducted in several Japanese laboratories[173]. These results were much more self-consistent and also displayed good agreement among the different experimental techniques used. Relevant data are given in Table 18[173].

2. Errors

The determination of M by LS involves several experimental procedures, each of which has some error associated with it. A typical breakdown of errors is the following due to Jolly and Campbell[174] in relation to their value of $M = 3.1 \times 10^6 \pm 6\%$ for superhelical DNA from bacteriophage ϕ X174: −

(a) Ludox calibration constant, ±2%
(b) Concentration measurements, ±2%
(c) Instrument readings, ±1%
(d) Value of $d\tilde{n}/dc$, ±1%

Table 18. Values[173] of (molecular weight $\times 10^{-4}$) for two polystyrene samples: PS – PX prepared by Asahi Dow Chem. Co. Ltd. and PS – IU distributed by Macromolecular Division of I.U.P.A.C. see footnote for further details and explanation of abbreviations

Mol. wt. average.	Method	PS – IU Observed	Reported	Calculated	PS – PX Observed	Calculated
$\bar{M}_n$	O.P.	8.0	7.92	–	38.6	38.6
$\bar{M}_n$	D. – 4th	6.5	–	6.5	41.5	–
$\bar{M}_{Dm}$	D. – 2nd	9.7	–	11.7	50.1	50.3
$\bar{M}_{DA}$	D. – Area	16.1	–	18.4	67.5	66.3
$\bar{M}_v$	V.	19.5	20.0	20.5	71.3	71.4
$\bar{M}_s$	S.V.	20.6	–		70.8	
M	L.S.	25.9, 24.3 24.1	23.5		83.7, 81.5, 81.6	
	S.E.	24.5	21.7	24.7	79.5	81.3
	A.	–	–		80.3	
$\bar{M}_z$	S.E.	44.8	39.7	42.9	124	124
$M/\bar{M}_n$	L.S./O.P.	3.09	3.22	3.8	2.11	2.11

(i)   Osmotic pressure measurements in methyl ethyl ketone at 30.2 °C for PS – PX and in toluene at 25.0 °C for PS – IU. The remainder relate to cyclohexane at 35.0 °C, and for L.S. results in triplicate derive from use of two additional temperatures.
(ii) Osmotic pressure (O.P.) yields number average mol. wt. $\bar{M}_n$.
(iii) Light scattering (L.S.) yielding weight average mol. wt. M. was calibrated with benzene, taking $R_{90} = 46.5 \times 10^{-6}$ cm^{-1} at $\lambda_0 = 436$ nm.
(iv) S.E. denotes sedimentation equilibrium yielding M and Z average mol. wt. $\bar{M}_z$.
(v)   A. denotes Archibald method.
(vi) S.V. denotes sedimentation velocity constant yielding mol. wt. $\bar{M}_s$.
(vii) Diffusion coefficient (D.) method yields mol wt. $\bar{M}_{DA}$ by the area method, mol. wt. $\bar{M}_{DM}$. by the second moment method and number average mol. wt. $\bar{M}_n$ by the fourth moment method.
(viii) Intrinsic viscosity (V.) gives viscosity average mol. wt. $\bar{M}_v$.
(ix) Reported values in table are averages of data obtained in several different laboratories and collated elsewhere in an earlier report.
(x)   Calculated values in table are obtained from established interrelations between mol. wt. and the appropriate experimental quantity at infinite dilution.

The error in (a) is stated to compare favourably with calibration from benzene, since the absolute value of R_{90} is hardly known to this accuracy. In (b) the concentration of DNA was measured spectrophotometrically *via* the molar phosphorous extinction coefficient of 6415 (with a standard deviation of 2%). The low error in (c) arises from low levels of dust achieved as well as the integration over a period of 10 secs of the readings on a digital output. The specific refractive index increment used in (d) was an experimental one from the literature. In point of fact the assessment of an error of ±1% for it is a very realistic one, but since the specific increment appears as a squared term, the overall error in M should actually be assessed as ±7%.

It is likely that ±1% in the instrument reading and ±2% in $(d\widetilde{n}/dc)^2$ cannot be improved on. However, for an absolute photometer, the error in (a) will be effectively zero. Concentration determinations are often specific to the system. For example, alignates in solution[143] have been determined colorimetrically, calibration being effected on solutions prepared from alginate previously dried for 24 hr. at 100 °C. However, successful calculation of the concentrations in this instance depends markedly on a prior knowledge of the fact that 10% of tightly bound water is present in the dried material[175]. In simple systems involving non-volatile solvents, with standard procedures for using volumetric glassware and checks on dried aliquots of solution the error in concentration need not be greater than ± 0.5%. Hence under the most favourable circumstances of instrument and polymer/solvent system cognizance of factors (a) – (d) leads to an overall error in M of ±3.5%.

It has been shown previously (Section V.3) that the true molecular weight of a copolymer M is obtainable in general by calculation involving experimental values of the apparent molecular weight M* in three solvents differing in refractive index. Simultaneous equations of expressions (101) or (102) are involved. Abbreviating the variable $(\nu_A - \nu_B)/\nu$ to $\widetilde{\nu}$ and distinguishing the three sets of data by subscripts 1, 2 and 3 yields the following expression for M after solution of Eq. (102) by determinants

$$M = \frac{M_1^* \widetilde{\nu}_2 \widetilde{\nu}_3 (\widetilde{\nu}_3 - \widetilde{\nu}_2) + M_2^* \widetilde{\nu}_3 \widetilde{\nu}_1 (\widetilde{\nu}_1 - \widetilde{\nu}_3) + M_3^* \widetilde{\nu}_1 \widetilde{\nu}_2 (\widetilde{\nu}_2 - \widetilde{\nu}_1)}{\widetilde{\nu}_2 \widetilde{\nu}_3 (\widetilde{\nu}_3 - \widetilde{\nu}_2) + \widetilde{\nu}_3 \widetilde{\nu}_1 (\widetilde{\nu}_1 - \widetilde{\nu}_3) + \widetilde{\nu}_1 \widetilde{\nu}_2 (\widetilde{\nu}_2 - \widetilde{\nu}_1)} \tag{116}$$

Equation (116) is not readily amenable to assigning a relative error (dM/M) to the derived true molecular weight, even if numerical estimates can be afforded to the six different experimental parameters within the expression for M. It is clear, however, that it is imperative to ensure large (positive or negative) differences between parameters $\widetilde{\nu}$. [If the constituent homopolymers are isorefractive, then $\widetilde{\nu}_1 = \widetilde{\nu}_2 = \widetilde{\nu}_3 = 0$ and Eq. (116) seems to imply that M is indeterminate. Actually the converse is true and M is always obtained. For such a situation there are simply no variables and $M_1^* = M_2^* = M_3^* = 0$]. In other words, as previously indicated, the true molecular weight is yielded directly from one measurement in one solvent. The value obtained for the molecular weight of the copolymer comprising isorefractive homopolymer units is subject to no more or no less error than that of ca. ±7% liable in a single determination of M for a polymer.

In certain multicomponent systems it may be necessary to derive the molecular weight x from two other measured molecular weights y and z together with a knowledge of the composition by weight W of the polymers. Reference to Eq. (111) provides an example of such a system. Here x, y and z correspond to the required molecular weight M_A of polymer A in the mixture, the true molecular weight M_{AB} measured for the mixture and the known or measured molecular weight M_B of polymer B in the mixture respectively. The composition W corresponds to W_A in Eq. (111). Another example of a system to which Eq. (111) and the symbols x, y, z and W may be applied is copolymer formation by radiation grafting to a pre-formed substrate of known measured molecular weight prior to grafting. This molecular weight corresponds to y. A certain weight fraction W participates in grafting and becomes the backbone (of unknown molecular weight x) of the resultant copolymer. That fraction $(1 - W)$ of the original polymer which remains untouched and does not participate in grafting, may be isolated by extraction and its molecular weight (z) measured.

For both examples one derives the required molecular weight x in terms of measured quantities. Thus, Eq. (111) assumes the form

$$y = Wx + (1 - W)z \qquad (117)$$

$$x = (1/W)y - [(1 - W)/W]z \qquad (118)$$

If ϵ_x, ϵ_y and ϵ_z are the respective errors in x, y and z, then it is readily shown that

$$(\epsilon_x/x)^2 = \{1/[y - (1 - W)z]^2\} [\epsilon_y^2 + (1 - W) \epsilon_z^2] \qquad (119)$$

Expressed in terms of relative errors r_x $(= \epsilon_x/x)$, r_y $(= \epsilon_y/y)$ and r_z $(= \epsilon_z/z)$, the relative error in the required molecular weight x is:

$$r_x^2 = \{1/[y - (1 - W)z]^2\} [y^2 r_y^2 + (1 - W)^2 z^2 r_z^2] \qquad (120)$$

A rough estimate of the conditions leading to a small value of r_x may be made from Eq. (120) by assigning $r_y = r_z = 0.07$ and then by considering small, moderate and large values of y/z, each taken in conjunction with $W = 0.1, 0.5$ and 0.9. The result is what might have been predicted intuitively, viz. (i) for any particular composition W, the molecular weight x should be large compared with z and (ii) for a specified molecular weight combination, W should be large.

Hence for a mixture of polymers A and B, the molecular weight M_A can be derived most accurately from the measured quantities M_{AB} and M_B if A comprises a large proportion by weight of the mixture. The corresponding condition for a grafting system is an unfortunate one, because it is rarely realisable in practice[176]. The backbone of the copolymer should comprise a large fraction by weight of the original polymer prior to grafting. However, although statistical considerations do dictate that longer chains participate in grafting (and hence x is large, as required for a small r_x), the weight fraction of polymer which is grafted is usually less than 0.1 (most of the polymer remains ungrafted) and thus W in Eq. (120) does not attain a desirable large value.

VII. References

[1] Serdyuk, I. N.: Dokl. Akad. Nauk SSSR *217*, 231 (1974)
[2] Serdyuk, I. N., Grenader, A. K.: FEBS Letters *59*, 133 (1975)
[3] Serdyuk, I. N., Fedorov, B. A.: J. Polym. Sci-Polym. Letts. Ed. *11*, 645 (1973)
[4] Eisenberg, H.: Biological macromolecules and polyectrolytes in solution. Oxford: Oxford Univ. Press. 1976, Chapt. 4

[5] McIntyre, D., Gornick F. (eds.): Light scattering from dilute polymer solutions. New York: Gordon and Breach 1964

[6] Huglin, M. B.(ed.): Light scattering from polymer solutions. London/New York: Academic Press 1972

[7] Fechner, B. M., Strazielle, C.: Makromol. Chem. *160*, 195 (1972)

[8] Miller, G. A.: J. Phys. Chem. *71*, 2305 (1967)

[9] Scholte, Th. G.: Eur. Polym. J. *6*, 1063 (1970)

[10] Scholte, Th. G.: J. Polym. Sci. Pt. A-2, *9*, 1553 (1971)

[11] Scholte, Th. G.: Report MS 929, Dutch State Mines, Geleen 1971

[12] Schulz, G. V., Lechner, M.: In Ref.[6], Chapt. 12

[13] Delmas, G., Patterson, D.: Polymer *7*, 513 (1966)

[14] Eisenberg, H., in: Procedures in nucleic acid research. Cantoni, G. L., Davies, D. R. (eds.). New York: Harper and Row, 1971, Vol. 2, p. 137 et seq.

[15] Timasheff, S. N., Townend, R., in: Physical principles and techniques of protein chemistry. Leach, S. J. (ed.). New York: Academic Press 1970, Part B, p. 147 et seq.

[16] Billingham, N. C.: Molar mass measurements in polymer science. London: Kogan Page 1977, Chapt. 5

[17] Sicotte, Y., Rinfret, M.: Trans. Faraday Soc. *58*, 1090 (1962)

[18] Coles, H. J., Jennings, B. R., Morris, V. J.: Phys. Med. Biol. *20*, 225 (1975)

[19] Dietz, R., Green, J. H. S.: Pure and Appl. Chem. *48*, 243 (1976)

[20] Oster, G., in: Techniques of chemistry. Weissberger, A. (ed.). New York: Wiley 1972, Vol. 1. Part III A, Chapt. 2

[21] Jennings, B. R.: Brit. Polym. J. *1*, 252 (1969)

[22] Flory, P. J.: Principles of polymer chemistry. Ithaca N. Y.: Cornell Univ. Press 1953, Chapt. 7

[23] Morawetz, H.: Macromolecules in solution. New York: Wiley 1965

[24] Tanford, C.: Physical chemistry of macromolecules. New York: Wiley 1961

[25] Tsvetkov, V. N., Eskin, V. E., Frenkel, S. Ya.: Struktura Makromolekul V Rastvorakh. Moscow: Nauka 1964

[26] Kerker, M.: Scattering of light and other electromagnetic radiation. New York: Academic Press 1969

[27] Stacey, K. A.: Light scattering in physical chemistry. London: Butterworths 1956

[28] Deželić, Gj.: Croat. Chem. Acta *33*, 99 (1961)

[29] Utiyama, H.: In Ref.[6], Chapt. 4

[30] Kratochvíl, P.: In Ref.[6], Chapt. 7

[31] Meyerhoff, G., Burmeister, A.: Makromol. Chem. *175*, 3029 (1974)

[32] Casassa, E. F., Berry, G. C., in: Polymer molecular weights. Vol. 4 of techniques and methods of polymer evaluation. Slade, P. E. Jr. and Jenkins, L. T. (eds.); Slade, P. E. Jr. (ed.). New York: Dekker 1975, Chapt. 5

[33] Rempp, P.: J. Chim. Phys. *54*, 421 (1957)

[34] Elias, H.-G., Lys, H.: Makromol. Chem. *92*, 1 (1966)

[35] Meyerhoff, G., Moritz, U.: Makromol. Chem. *109*, 143 (1967)

[36] Maron, S. M., Lou, R. L. H.: J. Phys. Chem. *59*, 231 (1955)

[37] Brice, B. A., Nutting, G. L., Halwer, M. A.: J. Am. Chem. Soc. *75*, 824 (1953)

[38] Burchard, W., Cowie, J. M. G.: In Ref.[6], Chapt. 17

[39] Valtasaari, L.: Tappi *48*, 627 (1965)

[40] Huglin, M. B.: In Ref.[6], Chapt. 6

[41] Miller, G. A., San Filippo, F. I., Carpenter, D. K.: Macromolecules *3*, 125 (1970)

[42] Huglin, M. B.: Pure & Appl. Chem. *49*, 929 (1977)

[43] Ultiyama, H.: In Ref.[6], Chapt. 3

[44] Utiyama, H., Tsunashima, Y.: Appl. Optics *9*, 1330 (1970)

[45] Kratohvil, J. P.: J. Coll. Interface Sci. *21*, 498 (1966)

[46] Kratohvil, J. P., Oppenheimer, L. E., Shubsa, F. E.: Kolloid Z. Z. Polymere *222*, 157 (1968)

[47] Roche, R. S., Tanner, A. G.: Angew. Makromol. Chem. *13*, 183 (1970)

48) Aughey, W. H., Baum, F. J.: J. Opt. Soc. Amer. *44*, 883 (1954)

49) Livesey, P. J., Billmeyer, F. W., Jr.: J. Coll. Interface Sci. *30*, 447 (1969)

50) Levine, H. I., Fiel, R. J., Billmeyer, F. W., Jr.: Biopolymers *15*, 1267 (1976)

51) Utiyama, H., Sugi, N., Kurata, M., Tamura, M.: Bull. Inst. Chem. Res. Kyoto Univ. *46*, 198 (1968)

52) Doty, P.: J. Cell. Comp. Physiol *49*, 22 (Suppl. 1.) (1957)

53) Sharp, P., Bloomfield, V.: Biopolymers *6*, 1201 (1968)

54) Fleischman, J. B.: J. Mol. Biol. *2*, 226 (1960)

55) Schmid, C. W., Rinehart, F. P., Hearst, J. E.: Biopolymers *10*, 883 (1971)

56) Butler, J. A. V., Laurence, D. J. R., Robins, A. B., Shooter, K. V.: Proc. Roy. Soc. Ser. A, *250*, 1 (1959)

57) Hays, J. B., Magar, M. E., Zimm, B. H.: Biopolymers *8*, 531 (1969)

58) Small angle light scattering photometer bulletin: Nederlandsche Optiek en Instrumenten-fabriek, Dr. C. E. Bleeker nv., Thorbeckelaan 3, Zeist, Holland

59) Chromatix GmbH, D-6903 Neckargemünd – Dilsberg, Untere Str. 45a, W. Germany

60) Burmeister, A., Meyerhoff, G.: Ber. Bunsen, Gesellschaft für physik. Chem. *78*, 1366 (1974)

61) Tabor, B. E.: In Ref.[6], Chapt. 1

62) Bernardi, G.: Makromol. Chem. *72*, 205 (1964)

63) Huglin, M. B., in: The polymer handbook. Brandrup, J., Immergut, E. H. (eds.). New York: Wiley 1975, 2nd Edit., Chapt IV – 10

64) Coles, H. J., Jennings, B. R., Morris, V. J.: Phys. Med. Biol. *20*, 310 (1975)

65) Schoenes, F. J.: Z. Angew. Physik *28*, 363 (1970)

66) Ref.[59]: Application Note LS – 1

67) Washburn, E. W. (ed.): Internat. Critical Tables. New York: McGraw Hill 1930

68) Mächtle, W., Fischer, H.: Angew. Makromol. Chem. *7*, 147 (1969)

69) Schultz, A. R.: J. Am. Chem. Soc. *76*, 3422 (1954)

70) Range of other published values quoted with references in Ref.[66]

71) Kaye, W., Havlik, A. J.: Appl. Optics *12*, 541 (1973)

72) Strazielle, C.: In Ref.[6] Chapt. 15

73) Krachtohvil, J. P., in: Characterisation of macromolecular structure. McIntyre, D. (ed.). Publication 1573, Nat. Acad. Sci., Washington D. C., 59 (1968)

74) Jennings, B. R., Plummer, H.: Brit. J. Appl. Physics (J. Phys. -D), Ser. 2, *1*, 1201 (1968)

75) Scornaux, J., Van Leemput, R.: Makromol. Chem. *177*, 2721 (1976)

76) Hermans, J. J., Levinson, S.: J. Opt. Soc. Amer. *41*, 460 (1951)

77) Miller, G. A., Lee, C. S.: J. Phys. Chem. *72*, 4644 (1968)

78) Hosseinalizadeh-Khorassani, M.-K., Huglin, M. B.: Univ. of Salford, to be published

79) Evans, J. M.: In Ref.[6], Chapt. 5

80) Wijk van R., Staverman, A. J.: J. Polm. Sci. Pt. A, *2*, 1011 (1966)

81) Berry, G. C.: J. Chem. Phys. *44*, 4550 (1966)

82) Evans, J. M., Huglin, M. B., Lindley, J.: J. Appl. Polym. Sci. *11*, 2159 (1967)

83) Urwin, J. R., Girolamo, M.: Makromol. Chem. *142*, 161 (1971)

84) Bryce, W. A. J.: Polymer *10*, 804 (1969)

85) Miller, W., Stepto, R. F. T.: Eur. Polym J. *7*, 65 (1971)

86) Bullough, R. K.: J. Polym. Sci. *46*, 517 (1960)

87) Hölle, H. J., Kirste, R. G., Lehnen, B. R., Steinbach, M.: Progr. Coll. and Polym. Sci. *58*, 30 (1975)

88) Isdale, J. D., Brunton, W. C., Spence, C. M.: NEL Rept. No. 591, Nat. Eng. Labs., Dept. of Industry, East Kilbride, Glasgow 1975

89) Ziegler, I., Freund, L., Benoit, H., Kern, W.: Makromol. Chem. *37*, 217 (1960)

90) Rousset, A., Lochet, R.: J. Polym. Sci. *10*, 319 (1953)

91) Outer, P., Carr, C. I., Zimm, B. H.: J. Chem. Phys. *18*, 830 (1950)

92) Carr, C. I., Zimm, B. H.: J. Chem. Phys. *18*, 1616 (1950)

93) Trap, H. J. L., Hermans, J. J.: Rec. Trav. Chim. Pays-Bas. *73*, 167 (1954)

94) Brice, B. A., Halwer, M., Speiser, R.: J. Opt. Soc. Amer. *40*, 768 (1950)

95) Elias, H.-G., Schumacher, R.: Makromol. Chem. *76*, 23 (1964)

96) Elias, H.-G.: Chemie-Ing.-Techn. *33*, 359 (1961)

97) Elias, H.-G., Schlumpf, H.: Makromol. Chem. *85*, 118 (1965)

98) Feist, J., Elias, H.-G.: Makromol. Chem. *82*, 78 (1964)

99) Fee, J. G., Port, W. S., Whitnauer, L. P.: J. Polym. Sci. *33* 95 (1958)

100) Levy, G. B., Frank, H. P.: J. Polym. Sci. *10*, 371 (1953)

101) Adank, G., Elias, H.-G.: Makromol. Chem. *102*, 151 (1967)

102) Levy, G. B., Frank, H. P.: J. Polym. Sci. *17*, 247 (1955)

103) Smith, T. E.: Ph. D. Thesis, Georgia Inst. Technol., Atlanta, U.S.A., quoted in Ref.[41]

104) Kratochvíl, P., Bohdanecký, M., Šolc, K., Kolinsky, M., Ryska, M., Lím, D.: J. Polym. Sci. Pt. C *23*, 9 (1968)

105) Dziedziela, W. M., Huglin, M. B., Richards, R. W.: Eur. Polym. J. *11*, 53 (1975)

106) Altgelt, K., Schulz, G. V.: Kunststoffe-Plastics *5*, 325 (1958)

107) Kratochvíl, P.: J. Polym. Sci. Pt. C *23*, 143 (1968)

108) Kratochvíl, P.: Coll. Czech. Chem. Commun. *30*, 1119 (1965)

109) Kratochvíl, P.: Coll. Czech. Chem. Commun. *29*, 2767 (1964)

110) Tuzar, Z., Kratochvíl, P.: Coll. Czech. Chem. Commun. *32*, 2255 (1967)

111) Tuzar, Z., Kratochvíl, P., Bohdanecký, M.: J. Polym. Sci. Pt. C *16*, 633 (1967)

112) Abbås, K. B., Porter, R. S.: J. Polym. Sci. − Polym. Chem. Ed. *14*, 553 (1976)

113) Abbås, K. B., Kirschner, T., Porter, R. S.: Eur. Polym. J. *14*, 361 (1978)

114) Burchard, W.: Polymer *10*, 29 (1969)

115) Elias, H.-G.: In Ref.[6], Chapt. 9

116) Gupta, A. K., Strazielle, C., Marchal, E., Benoit, H.: Biopolymers *16*, 1159 (1977)

117) Sund, H., Markau, K.: Intern. J. Polym. Mater. *4*, 251 (1976)

118) Tsimpris, C. W., Mayhan, K. G.: J. Polym. Sci. − Polym. Phys. Ed. *11*, 1151 (1973)

119) Berry, G. C., Ruman, K. A.: J. Polym. Sci. A-2 *8*, 2089 (1970)

120) Worsfold, D. J., Bywater, S.: Macromolecules *5*, 393 (1972)

121) Tonelli, A. E., Flory, P. J.: Macromolecules *2*, 225 (1969)

122) Henley, D.: Arkiv Kemi *18*, 327 (1961)

123) Vink, H.: Arkiv Kemi *14*, 195 (1959)

124) Utiyama, H.: J. Phys. Chem. *69*, 4138 (1965)

125) Utiyama, H., Kurata, M.: Bull. Inst. Chem. Res. Kyoto Univ. *42*, 128 (1964)

126) Benoit, H., Holtzer, A. M., Doty, P.: J. Phys. Chem. *58*, 635 (1954)

127) Holtzer, A.: J. Polym. Sci. *17*, 432 (1955)

128) Fujita, H., Teramoto, A., Okita, K., Yamashita, T., Ikeda, S.: Biopolymers *4*, 769 (1966)

129) Casassa, E. F.: J. Am. Chem. Soc. *78*, 3980 (1956)

130) Morris, V. J., Coles, H. J., Jennings, B. R.: Nature (Lond.) *249*, 240 (1974)

131) Kratochvíl, P.: J. Polym. Sci., Pt. C. *50*, 487 (1975)

132) Dondos, A., Benoit, H.: Intern. J. Polym. Mater *4*, 175 (1976)

133) Johnsen, R. M.: Chemica Scripta *1*, 81 (1971)

134) Hert, M., Strazielle, C.: Makromol. Chem. *175*, 2149 (1974)

135) Inoue, H., Timasheff, S. N.: J. Am. Chem. Soc. *90*, 1890 (1968)

136) Freifelder, D.: J. Mol. Biol. *54*, 567 (1970)

137) Kratochvíl, P.: Faserforsch. und Textiltechnik *24*, 5 (1973)

138) Nagasawa, M., Takahashi, A.: In Ref.[6], Chapt. 16

139) Alexandrowicz, Z.: J. Polym. Sci. *40*, 91 (1959)

140) Alexandrowicz, Z.: J. Polym. Sci. *43*, 337 (1960)

141) Stejskal, J., Beneš, M. J., Kratochvíl, P., Peška, J.: J. Polym. Sci., − Polym. Phys. Ed. *12*, 1941 (1974)

142) Timasheff, S. N., Dintzis, H. M., Kirkwood, J. G., Coleman, B. D.: Proc. Nat. Acad. Sci. U.S.A. *41*, 710 (1955)

143) Dingsøyr, E., Smidsrød, O.: Brit. Polym. J. *9*, 56 (1977)

144) Cooper, R. E., Wassermann, A.: Nature (Lond.). *180*, 1072 (1957)

145) Buchner, P., Cooper, R. E., Wassermann, A.: J. Chem. Soc. *1961*, 3974

146) Vrij, A., Overbeek, J. Th. G.: J. Colloid Sci. *17*, 570 (1962)

147) Kratohvil, J. P., Dellicolli, H. T.: Canad. J. Biochem. *46*, 945 (1968)

[148] Tuzar, Z., Kratochvíl, P.: Adv. Coll. and Interface Sci. *6*, 201 (1976)
[149] Benoit, H., Froelich, D.: In Ref.[6] Chapt. 11
[150] Girolamo, M., Urwin, J. R.: Eur. Polym. J. *8*, 1159 (1972)
[151] Ho-Duc, N., Prud'homme, J.: Intern. J. Polym. Mater *4*, 303 (1976)
[152] Tuzar, Z., Kratochvíl, P., Straková, D.: Eur. Polym. J. *6*, 1113 (1970)
[153] Prud'homme, J., Bywater, S.: Macromolecules *4*, 543 (1971)
[154] Tanaka, T., Kotaka, T., Inagaki, H.: Macromolecules *9*, 561 (1976)
[155] Tanaka, T., Kotaka, T., Inaga, H.: Macromolecules *7*, 311 (1974)
[156] Zilliox, J. G., Roovers, J. E. L., Bywater, S.: Macromolecules *8*, 573 (1975)
[157] Freyss, D., Rempp, P., Benoit, H.: J. Polym. Sci. Pt. B, 217 (1964)
[158] Spatorico, A. L.: J. Appl. Polym. Sci. *18*, 1793 (1974)
[159] Veličković, J. S., Filipović, J. M.: Bull. Soc. Chim. Beograd. *35*, 459 (1970)
[160] Guthrie, J. T., Huglin, M. B., Phillips, G. O.: Polymer *18*, 521 (1977)
[161] Huglin, M. B., Richards, R. W.: Polymer *17*, 587 (1976)
[162] Kambe, H., Kambe, Y., Honda, C.: Polymer *14*, 460 (1973)
[163] Kambe, Y., Honda, C.: Angew. Makromol. Chem. *25*, 163 (1972)
[164] Bushuk, W., Benoit, H.: Canad. J. Chem. *36*, 1616 (1958)
[165] Kambe, Y.: Angew. Makromol. Chem. *54*, 71 (1976)
[166] Kratochvíl, P., Sedláček, B., Straková, D., Tuzar, Z.: Makromol. Chem. *148*, 271 (1971)
[167] Hyde, A. J.: In Ref.[6], Chapt. 10
[168] Kratochvíl, P., Vorlíček, J., Straková, D., Tuzar, Z.: J. Polym. Sci. – Polym. Phys. Ed. *13*, 2321 (1975)
[169] Kratochvíl, P., Vorlíček, J.: J. Polym. Sci. – Polym. Phys. Ed. *14*, 1561 (1976)
[170] Ogston, A. G., Preston, B. N.: J. Biol. Chem. *241*, 17 (1966)
[171] Cummins, H. Z., Pike, E. R. (eds.): Photon correlation and light beating spectroscopy. Proc. NATO A.S.I., Series B: Physics, Vol. 3. New York: Plenum 1974
[172] Strazielle, C., Benoit, H.: Pure and Appl. Chem. *26*, 451 (1971)
[173] Kotera, A., Yamaguchi, N., Yoshizaki, K., Onda, N., Kano, K., Dai, M., Furuya, A., Fukutume, A., Murakami, S.: Rept. Progr. Polym. Phys. Japan *18*, 1 (1975)
[174] Jolly, D. J., Campbell, A. M.: Biochem. J. *128*, 569 (1972)
[175] Haug, A.: Rept. No. 30, Norwegian Inst. Seaweed Research, Trondheim (1964)
[176] Collins, R., Huglin, M. B., Richards, R. W.: Eur. Polym. J. *11*, 197 (1975)

Received February 15, 1978, February 27, 1978

Author Index Volumes 26–77

The volume numbers are printed in italics

Albini, A., and Kisch, H.: Complexation and Activation of Diazenes and Diazo Compounds by Transition Metals. *65*, 105–145 (1976).

Altona, C., and Faber, D. H.: Empirical Force Field Calculations. A Tool in Structural Organic Chemistry. *45*, 1–38 (1974).

Anderson, D. R., see Koch, T. H.: *75*, 65–95 (1978).

Anderson, J. E.: Chair-Chair Interconversion of Six-Membered Rings. *45*, 139–167 (1974).

Anet, F. A. L.: Dynamics of Eight-Membered Rings in Cyclooctane Class. *45*, 169–220 (1974).

Ariëns, E. J., and Simonis, A.-M.: Design of Bioactive Compounds. *52*, 1–61 (1974).

Aurich, H. G., and Weiss, W.: Formation and Reactions of Aminyloxides. *59*, 65–111 (1975).

Balzani, V., Bolletta, F., Gandolfi, M. T., and Maestri, M.: Bimolecular Electron Transfer Reactions of the Excited States of Transition Metal Complexes. *75*, 1–64 (1978).

Bardos, T. J.: Antimetabolites: Molecular Design and Mode of Action. *52*, 63–98 (1974).

Barnes, D. S., see Pettit, L. D.: *28*, 85–139 (1972).

Bauer, S. H., and Yokozeki, A.: The Geometric and Dynamic Structures of Fluorocarbons and Related Compounds. *53*, 71–119 (1974).

Baumgärtner, F., and Wiles, D. R.: Radiochemical Transformations and Rearrangements in Organometallic Compounds. *32*, 63–108 (1972).

Bayer, G., see Wiedemann, H. G.: *77*, 67–140 (1978).

Bernardi, F., see Epiotis, N. D.: *70*, 1–242 (1977).

Bernauer, K.: Diastereoisomerism and Diastereoselectivity in Metal Complexes. *65*, 1–35 (1976).

Bikerman, J. J.: Surface Energy of Solids. *77*, 1–66 (1978).

Boettcher, R. J., see Mislow, K.: *47*, 1–22 (1974).

Bolletta, F., see Balzani, V.: *75*, 1–64 (1978).

Brandmüller, J., and Schrötter, H. W.: Laser Raman Spectroscopy of the Solid State. *36*, 85–127 (1973).

Bremser, W.: X-Ray Photoelectron Spectroscopy. *36*, 1–37 (1973).

Breuer, H.-D., see Winnewisser, G.: *44*, 1–81 (1974).

Brewster, J. H.: On the Helicity of Variously Twisted Chains of Atoms. *47*, 29–71 (1974).

Brocas, J.: Some Formal Properties of the Kinetics of Pentacoordinate Stereoisomerizations. *32*, 43–61 (1972).

Brunner, H.: Stereochemistry of the Reactions of Optically Active Organometallic Transition Metal Compounds. *56*, 67–90 (1975).

Buchs, A., see Delfino, A. B.: *39*, 109–137 (1973).

Bürger, H., and Eujen, R.: Low-Valent Silicon. *50*, 1–41 (1974).

Burgermeister, W., and Winkler-Oswatitsch, R.: Complexformation of Monovalent Cations with Biofunctional Ligands. *69*, 91–196 (1977).

Burns, J. M., see Koch, T. H.: *75*, 65–95 (1978).

Butler, R. S., and deMaine, A. D.: CRAMS – An Automatic Chemical Reaction Analysis and Modeling System. *58*, 39–72 (1975).

Caesar, F.: Computer-Gas Chromatography. *39*, 139–167 (1973).

Čársky, P., and Zahradník, R.: MO Approach to Electronic Spectra of Radicals. *43*, 1–55 (1973).

Inorganic Chemistry Concepts

Editors:
M. Becke, C. K. Jørgensen, M. F. Lapert,
S. J. Lippard, J. L. Margrave, K. Niedenzu,
R. W. Parry, H. Yamatera

The ensuing advances in inorganic chemistry will be the subject of this new series. The focus will be on fields of endeavor which as yet remain open and are therefore topical. The series is edited by an international panel of experts.

Volume 1
R. Reisfeld, C. K. Jørgensen
Lasers and Excited States of Rare Earths
1977. 9 figures, 26 tables. VIII, 226 pages
ISBN 3-540-08324-3

The authors discuss both numerous spectroscopic questions, including the influence of the neighbour atoms on positions, and intensities of transitions in trivalent Pr, Nd, Sm, Eu, Gd, Tb, Dy, Ho, Er, Tm, and Yb and other effects of chemical bonding in vitreous and crystalline solids, as well as the parameters determining the applications in lasers, such as the radiative and non-radiative transitions probabilities, the strategy for suppressing radiationless processes competing with luminescence, and the energy transfer from trivalent cerium, the uranly ion, vanadate, tungstate, ions of thallium, lead and bismuth isoelectronic with mercury atom (and from many other species) to the rare earth. In a final chapter, general questions of radiation from opaque objects, communications, geodesy and astrophysics are discussed, and the use of neodymium glass lasers for initiating nuclear fusion reviewed. Many new suggestions are presented in this novel interdisciplinary field.

Volume 2
R. L. Carlin, A. J. van Duyneveldt
Magnetic Properties of Transition Metal Compounds
1978. 149 figures, 7 tables. XV, 264 pages
ISBN 3-540-08584-X

This book introduces chemists and physicists to magnetic ordering phenomena, Chemists have long been interested in magnetic interactions in clusters, but have shied away from the cooperative phenomena that have always fascinated physicists. Part of the reason for this is that the most remarkable phenomena occur at low temperatures, a region where many chemists have only recently started making measurements. This work starts with a study of paramagnetisms, and shows the specific heat results are as valuable a magnetic measurement as a susceptibility. Emphasized throughout is the valuable information to be obtained by measuring single crystals, rather than powders: The subject of paramagnetic relaxation is also introduced. Numerous illustrative examples from literature are cited throughout. The book concludes with an extensive literary survey of both the single-ion properties of the iron-series ions, and ten interesting examples of magnetic systems to which the principles introduced in the beginning of the book have been applied.

Volume 3
P. Gütlich, R. Link, A. Trautwein
Mössbauer Spectroscopy and Transitions Metal Chemistry
1978. 160 figures, 19 tables, 1 folding plate.
X, 280 pages
ISBN 3-540-08671-4

The book is directed to chemists, physicists and other scientists and is suitable for beginners wishing to acquire an understanding of the method of Mössbauer spectroscopy. It should also be particularly apreciated by the advanced scientists who wishes to apply the method to compounds and alloys of transition metals. In describing numerous examples the authors demonstrate the kind of problems which may be solved using the Mössbauer effect technique. The chapters dealing with non-iron Mössbauer active transition elements include complete lists of references to original work on solid-state chemistry and physics. In the chapter on ^{57}Fe-spectroscopy the authors have shown various ways of chemical bond and molecular structure investigations in connection with quantum-chemical methods. The last chapter introduces the reader into a number of special applications of Mössbauer spectroscopy.

Springer-Verlag
Berlin
Heidelberg
New York

Topics
in Current
Chemistry

**Fortschritte
der chemischen Forschung**

Managing Editor:
F. L. Boschke

Volume 70
Structural Theory of Organic Chemistry
By N. D. Epiotis, W. R. Cherry, S. Shaik, R. L. Yates, F. Bernardi
1977. 60 figures, 58 tables.
VIII, 250 pages
ISBN 3-540-08099-6

Contents: Theory. – Non-bonded Interactions. – Geminal Interactions. – Conjugative Interactions. – Bond Ionicity Effects.

Volume 71
**Inorganic Chemistry
Metal Carbonyl Chemistry**
1977. 51 figures, 54 tables.
IV, 190 pages
ISBN 3-540-08290-5

Contents: Tetranuclear Carbonyl Clusters. – Thermochemical Studies of Organo-Transition Metal Carbonyls and Related Compounds. – The Vibrational Spectra of Metal Carbonyls. – Inorganic Applications of X-Ray Photoelectron Spectroscopy.

Volume 72
Medicinal Chemistry
1977. 80 figures, 9 tables.
IV, 157 pages
ISBN 3-540-08366-9

Contents: Modes of Action of Antimicrobial Agents. – Ansamycins: Chemistry, Biosynthesis and Biological Activity. – Syntheses and Activity of Heteroprostanoids. – Hypolipidaemic Aryloxyacetic Acids. – Tilorone Hydrochloride. The Drug Profile. – Author Index Volumes 26–72.

Volume 73
Organic Chemistry
1978. 19 figures, 42 tables.
IV, 271 pages
ISBN 3-540-08480-0

Contents: Stereochemistry of Penta- and Hexacoordinate Phosphorus Derivatives. – Complex Bases and Complex Reducing Agents. – New Tools in Organic Synthesis. – Aromatic and Heteroaromatic Compounds by Electrocyclic Ringclosure with Elimination. – Some Newer Aspects of Mass Spectrometric Ortho Effects. – Author Index Volumes 26–73.

Volume 74
Organic Compounds
Syntheses/Stereochemistry/ Reactivity
1978. 37 figures. IV, 133 pages
ISBN 3-540-08633-1

Contents: Stereochemistry of Multibridged, Multilayerd, and Multistepped Aromatic Compounds: Transanular Steric and Electronic Effects. – Isotope Effects in Hydrogen Atom Transfer Reactions. – N-Methylacetamide as a Solvent. – EROS – A Computer Program for Generating Sequences of Reactions.

Volume 75
Organic Chemistry and Theory
1978. 39 figures. IV, 187 pages
ISBN 3-540-08834-2

Contents: Biomolecular Electron Transfer Reactions of Excited States of Transition Metal Complexes. – Photochemical Reactivity of Keto Imino Ethers. Type I Rearrangement and (2+2)-Photocycloaddition to the Carbon-Nitrogen Double Bound. – Computational Methods of Correlation Energy. – A Quantitative Measure of Chemical Chirality and Its Application to Asymmetric Synthesis.

Volume 76
Aspects of Molybdenum and Related Chemistry
1978. 71 figures. Approx. 180 pages
ISBN 3-540-08986-1

Contens: Inorganic Sulfur Compounds of Molybdenum and Tungsten. – High-Temperature Sulfide Chemistry.

Springer-Verlag
Berlin
Heidelberg
New York